Laboratory Animal Science and Medicine

Volume 2

Series Editors

Aurora Brønstad, Laboratory Animal Science, University of Bergen, Bergen, Norway

José Sanchez Morgado, Bioresearch and Veterinary Services, The University of Edinburgh, Edinburgh, UK

This book series aims at providing an easily accessible and complete toolbox for researchers, Veterinarians and technicians who design animal studies or/and work with research animals. The series equips the readers with the theoretical and practical knowledge to successfully run an animal facility, to monitor and maintain animal health and wellbeing in compliance with international ethical guidelines, to proof the genetic status of laboratory rodents and furthermore it profoundly introduces on how to design reproducible animal experiments. In a unique way, each volume focuses on a distinct topic which is always explored in a comprehensive manner.

This series is endorsed by the European Society for Laboratory Animal Veterinarians (ESLAV). As a leading voice in European Laboratory Animal Medicine, ESLAV's objectives are to promote and disseminate expert veterinary knowledge within the field of laboratory animal science, with a special focus on advancing skills in subjects connected with the breeding, health, welfare and use of laboratory animals.

Fernando Benavides •
Axel Kornerup Hansen
Editors

Rodent Quality Control: Genes and Bugs

Monitoring Health and Genetics of Laboratory Animals

Springer

Editors
Fernando Benavides
Dept of Epigenetics & Mol Carcinogenesis
The University of Texas MD Anderson Cancer Center
Houston, TX, USA

Axel Kornerup Hansen
Dept of Veterinary and Animal Sciences
University of Copenhagen
Frederiksberg C, Denmark

ISSN 2730-7859 ISSN 2730-7867 (electronic)
Laboratory Animal Science and Medicine
ISBN 978-3-031-59102-0 ISBN 978-3-031-59103-7 (eBook)
https://doi.org/10.1007/978-3-031-59103-7

This Springer imprint is published by the registered company Springer Nature Switzerland AG
The registered company address is: Gewerbestrasse 11, 6330 Cham, Switzerland

Preface

As a part of the *Laboratory Animal Science and Medicine* series, endorsed by the European Society for Laboratory Animal Veterinarians (ESLAV), this book aims to offer a comprehensive and easily accessible toolbox for researchers, veterinarians, and technicians working with laboratory rodents, specifically mice and rats. This book thoroughly describes quality assurance programs for laboratory animal facilities, focusing on microbiological and genetic monitoring. It encompasses common protocols aimed at ensuring the desired level of microbiological and genetic quality controls. Additionally, the content includes sections detailing standardized strains and genetically altered rodents, with recommended quality controls, standardized nomenclature, and cryopreservation. Health monitoring programs are addressed with details on how to do viral, bacterial, and parasitological diagnostics, accompanied by considerations for statistical analysis. The quality controls presented in this book play a crucial role as an essential component of the 3Rs (replacement, reduction, and refinement), contributing to the refinement and reduction in the use of laboratory mice and rats. It is our hope that this book will be helpful for laboratory animal staff involved in the quality assurance of laboratory rodents.

Houston, TX, USA — Fernando Benavides
Frederiksberg, Denmark — Axel Kornerup Hansen

Contents

Editors and Contributors

About the Editors

Fernando Benavides Fernando Benavides is a Professor and serves as the Director of Laboratory Animal Genetic Services at MD Anderson Cancer Center, Houston, Texas, USA. He holds a D.V.M and a Ph.D. in genetics from the University of Buenos Aires, Argentina. His Ph.D. training on mouse genetics was completed at *Institut Pasteur*, Paris, France. A diplomate of the American College of Laboratory Animal Medicine (ACLAM) since 2003, he has been integral to MD Anderson's Institutional Animal Care and Use Committee (IACUC) since 2006. He is currently a member of the International Committee on Standardized Genetic Nomenclature for Mice. With a focus on mouse genetics and laboratory animal medicine, he has extensive experience in developing novel mouse models through both spontaneous mutations and targeted genetic modifications. His expertise in positional cloning of spontaneous mutations has led to the discovery and publication of numerous mouse models. In his capacity as the director of a genetic quality control service since 2006, he has overseen numerous background characterizations and speed congenic projects in mice, utilizing genetic markers such as microsatellites and SNPs. His dedication to enhancing genetic quality control in laboratory rodents has led to a variety of publications, encompassing peer-review articles, books, chapters, and reviews, as well as teaching sessions and presentations at international congresses.

Axel Kornerup Hansen Axel Kornerup Hansen graduated in 1985 as DVM from the Royal Veterinary and Agricultural University in Copenhagen, Denmark. After 2 years in a small animal hospital, he became head of laboratories at Møllegaard Breeding Center, today a part of Taconic Ltd. In 1993, he became Associate Director at the Department of Experimental Medicine at the Faculty of Health and Medical Sciences at the University of Copenhagen. In 1996, he was awarded the degree of Dr. Vet. Sci. based upon his thesis on the impact of bacteria in laboratory rats. From 1997, he has been Professor in laboratory animal science and welfare at the Royal Veterinary and Agricultural University, which in 2007 became a part of University of Copenhagen. From 2007 to 2022, he was head of the Section of Experimental Animal Models. In 2000, he together with six others founded the European College of Laboratory Animal Medicine. His research area has primarily been on how infections and the microbiota have an impact on laboratory animal models for human diseases and how this works in conjunction with the diet, but he has also published work on other issues in laboratory animal welfare. He has published 251 peer-reviewed papers in scientific journals and two full textbooks, i.e., *The Laboratory Swine* and *Handbook of Laboratory Animal Bacteriology*. He is chairing the board of the Danish National Center for the 3Rs and Denmark's National Committee for Animal Experimentation and Alternatives.

Contributors

Lars Andresen Department of Veterinary and Animal Sciences, University of Copenhagen, Frederiksberg, Denmark

Fernando Benavides Department of Epigenetics and Molecular Carcinogenesis, The University of Texas MD Anderson Cancer Center, Houston, TX, USA

André Bleich Institute for Laboratory Animal Science and Central Animal Facility, Hannover Medical School, Hannover, Germany

Stephanie Buchheister Institute for Laboratory Animal Science and Central Animal Facility, Hannover Medical School, Hannover, Germany

Henrik Christensen Department of Veterinary and Animal Sciences, University of Copenhagen, Frederiksberg, Denmark

Martina Crispo Laboratory Animal Biotechnology Unit, Institut Pasteur de Montevideo, Montevideo, Uruguay

Fabien Delerue Department of Genetics, The University of Texas MD Anderson Cancer Center, Houston, TX, USA

Axel Kornerup Hansen Department of Veterinary and Animal Sciences, University of Copenhagen, Frederiksberg, Denmark

Jean Jaubert Central Animal Facility and Mouse Genetics Laboratory, Institut Pasteur, Paris, France

Henrik Elvang Jensen Department of Veterinary and Animal Sciences, University of Copenhagen, Frederiksberg, Denmark

Louise Kruse Jensen Department of Veterinary and Animal Sciences, University of Copenhagen, Frederiksberg, Denmark

Páll Skúli Leifsson Department of Veterinary and Animal Sciences, University of Copenhagen, Frederiksberg, Denmark

María Noel Meikle Laboratory Animal Biotechnology Unit, Institut Pasteur de Montevideo, Montevideo, Uruguay

Werner Nicklas Microbiological Diagnostics, German Cancer Research Centre, Heidelberg, Germany

Torben Sølbeck Rasmussen Department of Food Sciences, University of Copenhagen, Frederiksberg, Denmark

Thomas Rülicke Department of Biomedical Sciences, University of Veterinary Medicine Vienna, Vienna, Austria

Andrew R. Williams Department of Veterinary and Animal Sciences, University of Copenhagen, Frederiksberg, Denmark

Abbreviations

AALAS	American Association for Laboratory Animal Science
AAV	Adeno-associated virus
API	Analytical profile index
ARRIVE	Animal Research Reporting of In Vivo Experiments
ART	Assisted reproductive technology
ATP	Adenosine triphosphate
CAR	*Filobacterium rodentium* (cilia-associated respiratory bacillus)
CF	Complement fixation test
CLSI	Clinical and Laboratory Standards Institute
CPA	Cryoprotective agent
Cre	Cre recombinase
CRISPR	Clustered regularly interspaced short palindromic repeats
CTAB	Cetyltrimethylammonium bromide
DBS	Dried blood spots
DNA	Deoxyribonucleic acid
DSB	Double-strand break
EAD	Exhaust air dust
ECTV	Ectromelia virus
EDIM	Mouse rotavirus (epizootic diarrhea of infant mice)
EDTA	Ethylenediaminetetraacetic acid
EIA	Enzyme immunoassay
ELISA	Enzyme-linked immunosorbent assay
ES	Embryonic stem
F1, F2, etc.	Filial generations (first, second, etc.)
FEC	Fecal egg count
FELASA	Federation of European Laboratory Animal Science Associations
GEM	Genetically engineered mouse
GMS	Grocott's methenamine silver reaction
GP	Guinea pig
GV-SOLAS	Gesellschaft für Versuchstierkunde-Society of Laboratory Animal Science
GWAS	Genome-Wide Association Study
H	Hamster
HDR	Homology-directed repair
HE	Hematoxylin and eosin

HET	Heterozygous
HI	Hemagglutination inhibition test
HOM	Homozygous
HR	Homologous recombination
ICLAS	International Council for Laboratory Animal Science
ICSI	Intracytoplasmic sperm injection
IDIR	Infectious diarrhea of infant rats
IFA	Immunofluorescence assay
ITS	Internal transcribed spacer
IVC	Individually ventilated cages
IVF	In vitro fertilization
KI	Knockin
KO	Knockout
LABA	Laboratory Animal Breeders Association
LAMP	Loop-mediated isothermal amplification
LASA	British Laboratory Animal Science Association
LCMV	Lymphocytic choriomeningitis virus
LDV	Gammaarterivirus lacdeh (lactate dehydrogenase elevating virus)
LOF	Loss of function
LOXP	Locus of X-over P1
LPSN	List of Prokaryotic Names with Standing in Nomenclature (https://lpsn.dsmz.de/)
M	Mouse
MALDI-TOF	Matrix-assisted laser desorption/ionization-time of flight
MAP	Mouse antibody production test
MCMV-1	Mouse cytomegalovirus 1
MDA	Multiple displacement amplification
MFIA	Multiplex fluorescence immunoassay
MHV	Mouse hepatitis virus
MIC	Minimal inhibitory concentration
MKPV	Mouse kidney parvovirus
MLST	Multilocus sequence typing
MNV	Mouse norovirus
MPF	Murine pathogen free
MPtV	Murine pneumotropic virus
MPV	Mouse parvovirus
MPyV	Murine polyomavirus
mRNA	Messenger RNA
MuAstV	Murine astrovirus
MVM	Minute virus of mice
NAD	Nicotinamide adenine dinucleotide
NGS	Next-generation sequencing
NHEJ	Non-homologous end joining
OHC	Optimal hygienic conditions
ORF	Open reading frame
PAM	Protospacer adjacent motif
PAS	Periodic acid-Schiff

PBS	Phosphate-buffered saline
PCR	Polymerase chain reaction
PFA	Paraformaldehyde
PTAH	Phosphotungstic acid hematoxylin
PVM	Pneumonia virus of mice
qPCR	Quantitative PCR
QTL	Quantitative trait locus
R	Rat
RAP	Rat antibody production test
RatPyV2	Rat polyomavirus
RFLP	Restriction fragment length polymorphism
RIA	Radioimmune assay
RMV	Rat minute virus
RNA	Ribonucleic acid
RPV	Rat parvovirus
RT-PCR	Real-time PCR
RV	(Kilham) rat virus
SCID	Severe combined immunodeficiency
SFB	Segmented filamentous bacteria
sgRNA	Single-guide RNA
SNP	Single-nucleotide polymorphism
SNV	Single-nucleotide variant
SOPF	Specific opportunistic pathogen free
SPF	Specific pathogen free
TALEN	Transcription activator-like effector nuclease
TG	Transgenic
TMEV	Theiler's murine encephalomyelitis virus
TYGS	Type Strain Genome Server
UTR	Untranslated region
VAF	Virus antibody free
VLP	Viral-like particle
WGS	Whole genomic sequencing
WHO	World Health Organization
WT	Wild type
ZFN	Zinc finger nuclease

Genetically Standardized Laboratory Rodents

1

Fernando Benavides and Jean Jaubert

Abstract

This chapter provides a comprehensive overview of the current and expanding knowledge base concerning standardized laboratory strains of rodents, with a specific focus on the mouse (*Mus musculus*) and the rat (*Rattus norvegicus*), the predominant species utilized in biomedical research. We include fundamental information on various genetically standardized strains, with a primary emphasis on inbred strains and their derivatives. The production and maintenance processes of these strains are explained, accompanied by a historical context and an exploration of their principal applications. Addressed to veterinarians, animal facility technicians, researchers, and students employing mouse and rat models, this chapter provides an up-to-date account of the different genetically defined stains associated with these widely used laboratory rodents.

Keywords

Rodents · Inbreeding · Inbred strains · Outbred stocks · Congenics

1.1 Introduction

At the beginning of the twentieth century, when genetics emerged as an experimental science, laboratory rodents were extremely limited. Three rodent species were the first to be used to test the recently rediscovered Mendel's laws and to prove that it was possible to develop genetically pure strains. These attempts began in the first decade of the 1900s in the USA and involved the use of the guinea pig by Sewall Wright and George Rommel, the rat by Helen Dean King, and the mouse by Clarence C. Little and Abbie Lathrop. In France, biologist Lucien Cuénot conducted pioneering genetic experiments with mice, specifically focusing on coat color inheritance, providing crucial insights into the principles of Mendelian inheritance in mammals.

Wright and Rommel were the first to start inbreeding experiments in 1906 with guinea pigs at the United States Department of Agriculture. Guinea pig strains 2 and 13 originated after these experiments and are still in use today (2/N and 13/N strains). Clarence Little, while at Harvard University, was the first to try to develop pure mouse lines by inbreeding. The first mouse

F. Benavides (✉)
Department of Epigenetics and Molecular Carcinogenesis, The University of Texas MD Anderson Cancer Center, Houston, TX, USA
e-mail: fbenavid@mdanderson.org

J. Jaubert
Central Animal Facility and Mouse Genetics Laboratory, Institut Pasteur, Paris, France
e-mail: jean.jaubert@pasteur.fr

F. Benavides, A. K. Hansen (eds.), *Rodent Quality Control: Genes and Bugs*, Laboratory Animal Science and Medicine 2, https://doi.org/10.1007/978-3-031-59103-7_1

inbred strain, "dba," was started in 1909 by Little through inbreeding mice homozygous for three recessive coat color alleles (named at the time as *a*, *b*, and *d*). Simultaneously, Helen King worked toward developing the first inbred rat lines at the Wistar Institute in Philadelphia, eventually creating the Wistar King Albino (WKA) inbred rat strain. A mutant rat in one of her colonies is also the origin of the Brown Norway inbred strain.

1.2 Inbred Strains

According to the definition by the International Committee on Standardized Nomenclature for mice, "Strains can be termed *inbred* if they have been mated brother x sister for 20 or more consecutive generations, and individuals of the strain can be traced to a single ancestral pair at the 20th or subsequent generation." At this stage, the genomes of the animals will typically exhibit only 1% residual heterozygosity on average. Consequently, they can be considered genetically identical for most practical purposes. The breeding protocol commonly used to produce an inbred strain is called brother x sister (bxs) or sibling x sibling (sxs) and consists of mating a male and a female from the same litter in successive generations [44] (Fig. 1.1).

In practice, most of the mouse and rat strains that are commonly used in research laboratories nowadays have undergone several tens of generations of bxs matings (indicated with an "F," for filial), and some among the most ancient have passed 200 generations (DBA/2, for example, is over F224). It is important to provide some clarifications on the definition of an inbred strain. Generally, animals within the same inbred strain are genetically almost identical, except for sex-linked traits and de novo mutations. To describe this important characteristic, geneticists say that the mice in question are *isogenic*. This is due to the rigorous inbreeding process that results in all individuals within a given strain becoming homozygous for all loci that were segregating in the founder ancestors (i.e., the original or ancestral breeding pair). As a result, they all carry the same alleles on both maternal and paternal chromosomes. This is also known as *autozygosity* because the two alleles are copies of the same ancestral allele.

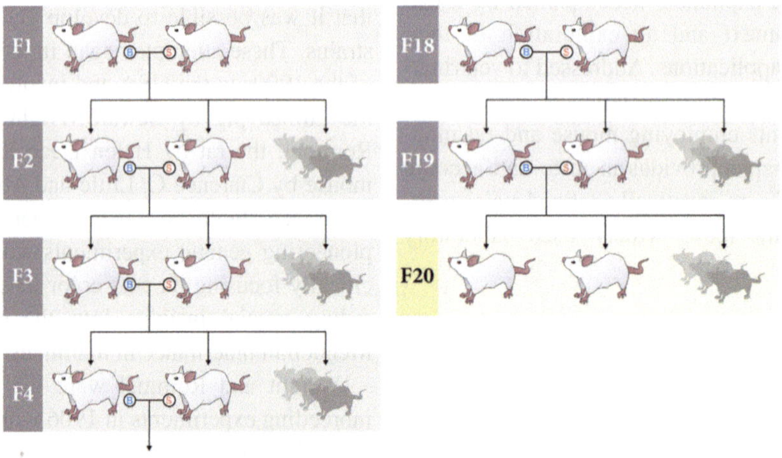

Fig. 1.1 *Inbreeding crosses.* This drawing schematically represents the breeding protocol commonly used to produce an inbred strain: mating a male and a female from the same litter (brother x sister) in successive generations. The uppercase letter F followed by a number represents the number of inbreeding generations. When this number is not known, a question mark is used: F? + 22, for example, would indicate that the number of brother x sister matings was not known when the strain was imported, but 22 generations of unrelaxed inbreeding have been added since this time. Strains can be termed *inbred* if they have been mated brother x sister for 20 or more consecutive generations

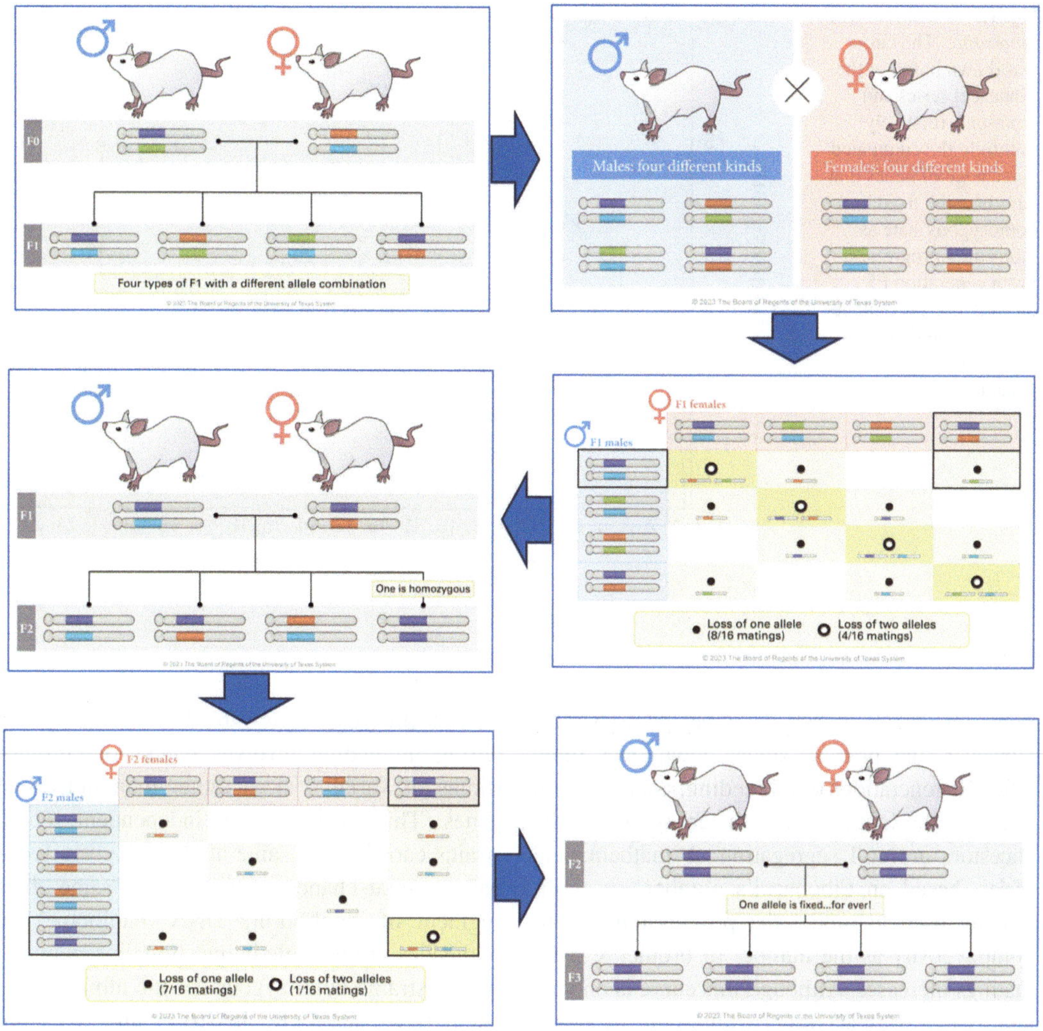

Fig. 1.2 *Alleles are fixed during inbreeding.* The brother x sister mating system results in the fixation or loss of certain alleles in subsequent generations. Using only one pair of chromosomes as an example, and four different colors (blue, green, orange, and violet) representing four alleles, these figures show the probability (percentage of matings) of losing alleles on the initial generations. In the example, the random F1 mice crossed already lost one allele (green) (8/16 probability), later the randomly crossed F2 mice lost two alleles (blue and orange) (1/16 probability). At the F3 generation, one allele (violet) becomes permanently fixed within this inbred colony

The process of homozygosity through progressive allele loss, or fixation, is easy to understand. When an allele present in generation Fn is not transmitted to at least one member of the breeding pair at generation Fn + 1, it is permanently lost (Fig. 1.2). As inbreeding continues, alleles are consistently lost and none are introduced, except for rare de novo mutations. This process leads to both homozygosity and isogenicity. The categorization of alleles that are lost or retained at each generation largely depends on chance. If the inbreeding protocol were reset using the same founder animals, it would result in a strain with a different genetic makeup after undergoing the 20 generations. Thus, an inbred strain represents a unique and random assortment of alleles.

During the process of inbreeding, the progression toward homozygosity is not linear. It occurs relatively quickly during the first few genera-

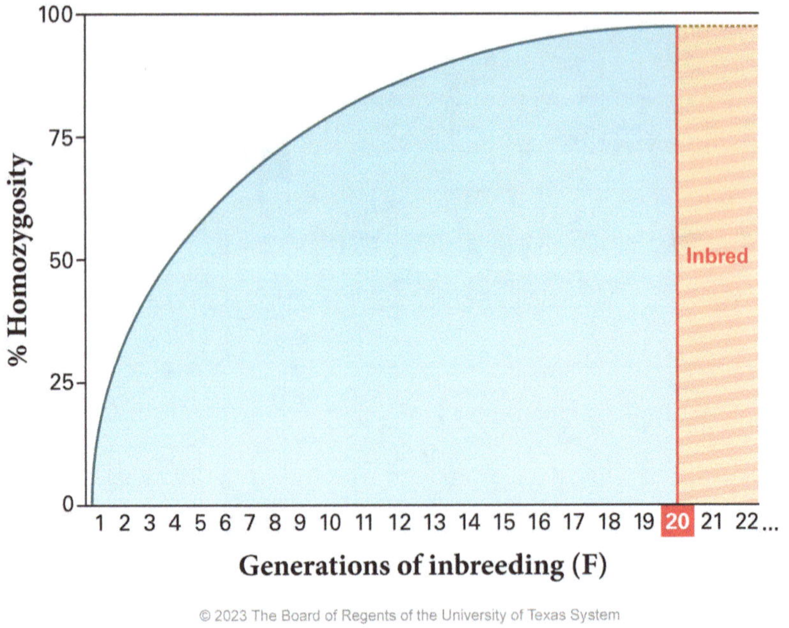

Fig. 1.3 *Progression of inbreeding*. The curve was drawn based of the Fibonacci series and represents relatively faithfully the cumulated percentage of genes that have become fixed in the homozygous state as inbreeding progresses. From generation F5 onward, this percentage is incremented by ~19.6% at each generation

tions, as many genes become homozygous. However, the process slows down over time. After 20 generations of inbreeding, not more than 1–2% of the loci that were heterozygous in the ancestors are still segregating. A mathematical series, based on Fibonacci's numbers, is traditionally used to model the progression toward homozygosity as the number of brother x sister matings increases. Although this curve is only an approximation, it accurately represents the evolution of heterozygosity over time (Fig. 1.3).

When breeding inbred individuals, it is crucial to prevent the formation of independent lines within the colony. These lines may undergo progressive divergence from each other as a result of genetic drift (Fig. 1.4a). Additionally, if a strain stops breeding (a common occurrence in practice), it is permanently lost. The example depicted in Fig. 1.4b, which involves establishing three new pairs of inbred mice from generation N to breed mice of generation N + 1, is considered the best option. However, when the progenies are small (another common occurrence), it may not always be feasible to establish three new pairs of siblings. Therefore, the substitute system depicted in Fig. 1.4c is common practice.

The progression toward full homozygosity during inbreeding involves blocks of chromosomes of variable sizes rather than individual genes. This explains why independent inbred strains carrying the same allele at a given locus have a great chance of sharing the same short segment of neighboring DNA (haplotype) on both sides of the allele in question. For example, mouse strains homozygous for the albino (Tyr^c) allele (e.g., A; AKR; BALB/c; and SJL) are probably homozygous for the same short segment of chromosome 7 flanking the albino mutation (Tyr^c), because the mutation shared by these strains results from the same mutational event that occurred well before the creation of these strains (i.e., identical by descent or IBD). It is important to keep in mind this unique characteristic of inbred strains, as it can have advantages or disadvantages when designing an experimental protocol. We will revisit this point in the section on congenic strains.

In most mammalian species, inbreeding of a natural population often has deleterious effects of variable intensity and phenotypic expression. In some cases, newborns exhibit growth retardation and finally die. In other instances, there is a

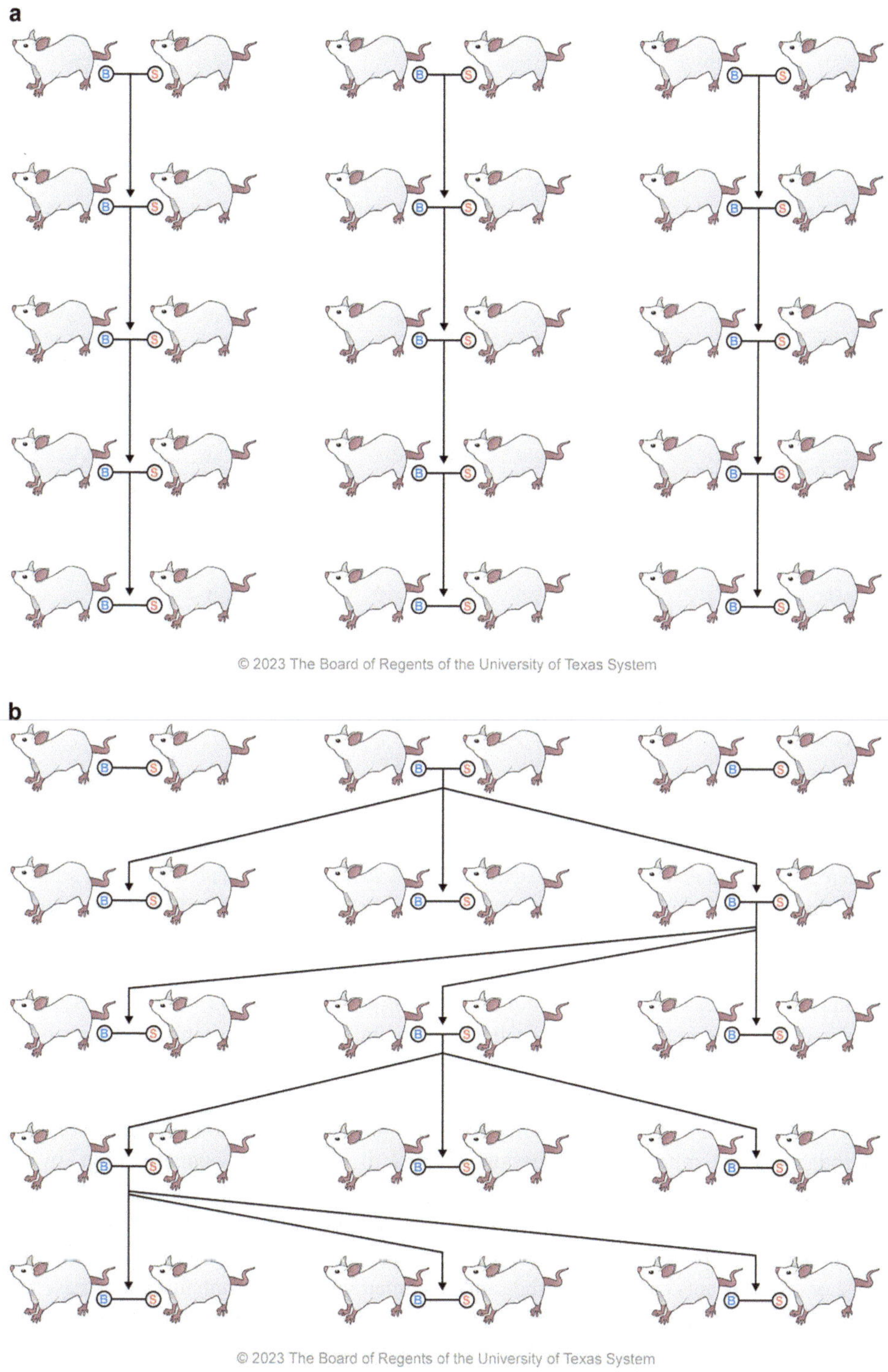

Fig. 1.4 *Breeding inbred mice.* (**a**) The system represented here is not recommended because it will lead to the establishment of three (and not one) independent inbred strains, which are progressively divergent from one another due to genetic drift. In addition, if a strain stops breeding (a common situation in practice), it is then permanently lost. (**b**) The system represented is certainly the best one, since, at each generation, three new pairs are established from a single cage at generation N to breed mice of generation N + 1. However, when the progenies are very small in size (a situation that is also common), it is not always possible to set the three new pairs of brothers and sisters. (**c**) Finally, the modified system represented here is the one that is generally used in practice

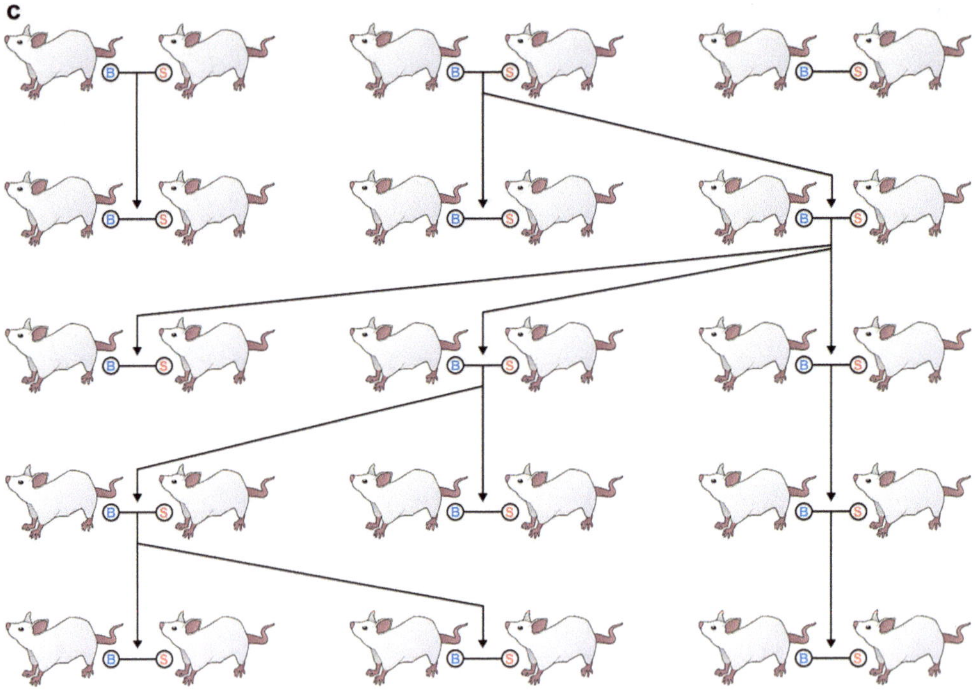

Fig. 1.4 (continued)

decrease in fitness or fertility, which is sometimes severe to the point that it leads to the extinction of the strain. All these adverse manifestations are commonly referred to as *inbreeding depression*. Genetic studies indicate that inbreeding depression is mainly caused by the accumulation of recessive deleterious mutations that are fixed in the homozygous state as inbreeding continues. However, epistatic interactions may also provide alternative explanations. Interestingly, in the case of some rodent species, inbreeding depression is not a significant concern if the breeders come from the same natural population of closely related individuals. This is likely because wild mice, rats, or other rodents caught from the same natural area already have a relatively high level of consanguinity, which may enhance their resilience to inbreeding depression.

Because individuals of the same inbred strain are isogenic, they are also histocompatible, or *syngeneic*, meaning they can permanently accept tissue transplants from any individual of the same strain (and sex). In the mouse, this unique characteristic has been extensively used to study the fate of cells with immunological functions in various contexts involving cellular cooperation. Moreover, this feature has been widely used (and continues to be used) for serial transplantation of tumoral cells in mouse inbred strains.

While inbred strains are generally considered relatively stable in the long term, their genetic profile may change for two main reasons. The first is accidental contamination by another strain, while the second is due to the progressive and insidious accumulation of novel mutations, which we will discuss below. It is possible that numerous genetic contaminations have occurred in the past, and with the current increase in genetically altered (GA)[1] lines, there is concern that the use of models with undefined or mixed back-

[1]We use the term genetically altered (GA) to include genetically engineered as well as spontaneous or chemically induced mutations.

grounds may become even more prevalent. In Chap. 4, we will explore the various options available for genetic quality control. The standard nomenclature of inbred strains will be discussed in Chap. 3, offering comprehensive insights into this critical aspect of research.

1.2.1 Inbred Mouse Strains

As we said in the introduction, Clarence Little was the first to attempt the development of "genetically pure" mouse lines by inbreeding. The first mouse inbred strain, DBA, was started by inbreeding mice homozygous for three recessive coat color alleles (*d*: dilute, now *Myo5a*^{d}; *b*: brown, now *Tyrp1*^{b}; and *a*: non agouti). Little was one of the so-called mousers at the time, who were students of Ernest E. Tyzzer and William E. Castle at Harvard University. Similarly, strain C57BL/6 was established in 1921 also by Little from a cross between two "black" mice obtained from Abbie Lathrop, a mouse supplier and accidental genetics pioneer from Massachusetts [90]. A few other strains were developed at about the same time by other scientists, in particular, Leonell C. Strong in Cold Spring Harbor and Dobrovolskaia-Zavadskaia in Paris [28, 63, 74]. In addition to these North American and European researchers, Japanese scientists also established a number of colonies from fancy mice [62]. Readers who are interested in the history of mouse genetics are invited to read the excellent book edited by H.C. Morse III [63], which contains several chapters written by mouse geneticists of the early days, including Strong himself. In his chapter, Strong reveals the recipes he used for the successful development of his lines, including wild mice trapped and used to introduce some "vigor" in his colonies. Other interesting books on the early years of mouse genetics are *Making Mice* by Karen Rader [74] and *Twentieth Century Mouse Genetics: A Historical and Scientific Review* by Robert Erickson [28].

The fact that all members of the same inbred strain are genetically identical is certainly the major reason why inbred mice become so prevalent in biomedical research. Scientists working with the same inbred strain, but in different laboratories or at different time periods, can perform experiments where the variations in the experimental results, by definition, will not be the consequence of possible differences in the genetic constitution of the animals. Another significant advantage of isogenic strains is that researchers can comprehensively define the phenotypic characteristics of each strain by collecting phenotypic data from multiple laboratories and storing them in a common database. For example, The Jackson Laboratory has developed a program to collect baseline phenotypic data on the most popular inbred strains of mice through a coordinated international effort. Information collected in this program (The Mouse Phenome Database-MPD) is freely available at http://phenome.jax.org/ [8, 9, 65]. The creation and maintenance of this database were feasible solely because inbred mice are isogenic and exhibit long-term genetic stability.

For reference, in 1998, the MGI website listed 426 mouse strains with a brief description (http://www.informatics.jax.org/external/festing/mouse/STRAINS.shtml), but it is probable that many of these strains have since been lost or terminated. However, even within this extensive collection, less than 20 strains are frequently used worldwide in biomedical research. Multipurpose mouse inbred strains can be used in a wide range of research applications, including genetics, reproductive biology, microbiology, and behavior. Examples of multi-purpose mouse inbred strains include C57BL/6, BALB/c, C3H/He, and DBA/2. In contrast, other inbred strains are designed for use in specialized research areas, such as immunology, or cancer research [73, 97]. For many decades, especially before the arrival of genetically engineered mice, inbred strains were a good source of cancer models due to the spontaneous tumors they develop (especially when old). For example, hepatomas and late onset mammary tumors in C3H/He, late onset mammary tumors in BALB/c, testicular teratomas in 129 strains, and lung adenomas in A/J. Other examples of specialized inbred strains include NOD for type 1 diabetes research and various 129 strains for embryonic stem cell research. Overall, the choice of an inbred strain for a research application will depend on the specific experimental objectives and the genetic background and phenotypic characteristics required for the study.

1.2.1.1 Classical Mouse Inbred Strains Are Derived from a Small Number of Ancestors

We know that most mouse inbred strains, particularly those considered "classical," are derived from a small number of ancestors [6]. Already in 1982, observations were published reporting the remarkable structural homogeneity of the mitochondrial DNA (mtDNA) among different strains of laboratory mice [30]. This was quite surprising because the mitochondrial genome, a 16 kb double-stranded circular DNA molecule, was known in most species (including humans) to have a relatively high evolution rate. Considering that the mtDNA is exclusively inherited from the female parent, the most likely explanation to account for these observations is that all the classical inbred strains share the same maternal lineage. A few years after these observations, similar experiments were made concerning the Y chromosome. Using five probes identifying Y chromosome-specific restriction fragments, six distinct Y chromosomes were identified among 39 standard inbred strains, indicating that a minimum of six male mice contributed to the formation of the common inbred strains [100]. DNA sequencing has confirmed that all laboratory mouse strains are derived from a female of the *Mus musculus domesticus* subspecies [40]. All these findings align with historical records, suggesting that the classical inbred strains were derived from a single female, while the interstrain polymorphisms represent the contribution of males, some of them from other *Mus musculus* subspecies [94].

1.2.1.2 Classical Mouse Inbred Strains Are a Mixture of Different *Mus* Subspecies

Inbred strains are often said to be artificial populations because their genetic constitution (isogenicity and homozygosity) has no natural equivalent. In fact, they can also be regarded as an artificial population, as we now understand that they do not originate from a single subspecies of the *Mus* genus but rather from at least three distinct subspecies [43]. The *polyphyletic* origin was confirmed and substantiated further after the complete high-resolution sequencing of the genomes of a large panel of inbred strains [34, 105, 109, 109, 110]. The GRCm39 (Genome Reference Consortium Mouse Reference 39) from C57BL/6J is the most updated (2020) reference. Other inbred strains with complete genomes are 129S1/SvlmJ, A/J, AKR/J, BALB/cJ, C3H/HeJ, C57BL/6NJ, CAST/EiJ, CBA/J, DBA/2J, FVB/NJ, LP/J, NOD/ShiLtJ, NZO/HILtJ, PWK/PhJ, and WSB/EiJ.

In summary, the genomes of laboratory inbred strains are a mosaic of chromosomal regions with distinct subspecific origins. Current estimates indicate that classical inbred strains were predominantly derived from *Mus musculus domesticus* (~94%), with variable contributions from *Mus musculus musculus* (~5%) and *Mus musculus castaneus* (<1%) subspecies [110].

Over the last 40 years, a variety of strains derived from small groups of wild specimens trapped in well-defined geographical regions and belonging to well-characterized taxonomic groups, have been established in various laboratories [43]. In the 1990s, with the widespread use of PCR amplification for the detection of genetic polymorphisms, inbred strains derived from wild mouse populations became valuable for gene mapping [15, 95, 111, 112]. Examples of theses wild-derived strains are PWK/PhJ (*Mus m. musculus*), MOLD/RkJ (*Mus m. molossinus*), and CAST/EiJ (*Mus m. castaneus*). Special mention must be made of those derived from *Mus spretus* (SEG/Pas; SPRET/Ei; and STF/Pas) because this species is one of the most distantly related to the laboratory strains that can still produce fertile hybrids with them [25].

1.2.2 Inbred Rat Strains

The Brown Norway rat (*Rattus norvegicus*) was the first mammalian species domesticated for scientific research [39, 50]. The name "Norway rat" is a misnomer because there is no fossil evidence for any *Rattus* species from Norway [49]. Rats have been used for research purposes since before 1850, however, the first genetic studies in rats were conducted in the 1870s, and the first inbred strain of rats was established in 1909 by Helen King at the Wistar Institute. Today, rats are the second most frequently used rodent species in research. Like mice, rats are relatively easy and inexpensive to care for and have a high reproductive potential. The genetics of the rat is well-characterized, and many inbred strains of rats, with specific phenotypes, are available. Inbred rat strains that exhibit traits modeling specific human conditions are extremely valuable for identifying and studying quantitative trait loci (QTLs) that lead to disease phenotypes.

More than 200 inbred strains of rats have been developed in the last 100 years by careful phenotypic characterization and selective breeding. Examples of widely used multi-purposed inbred rat strains are the albino Lewis (LEW), Fischer (F344), and Wistar (WI) as well as the hooded Long-Evans (LE). These multi-purposed strains can be used for a variaty of research areas including aging, diabetes, neurology, behavior, cardiovascular disease, and cancer. Specialized inbred strains of rats have been also developed for various research purposes, including the Spontaneously Hypertensive Rat (SHR) for hypertension and cardiovascular research, the Zucker Diabetic Fatty (ZDF) for diabetes and metabolic disorders, and the Dark Agouti (DA) for autoimmune disease research [1, 47]. Just like in the mouse, wild-derived rat inbred strains are also available, although to a lesser extent [55].

A high-quality draft, covering more than 90% of the genome of the BN/NHsd/Mcwi inbred strain was published in 2004 [39] and is constantly updated [102]. In 2020, the latest enhanced rat reference genome, mRatBN7.2, was sequenced and assembled [48, 103]. The genome sequences of the DA and F344 inbred strains were published in 2013 [45], followed recently by the stroke prone hypertensive SHRSP strain [51]. Ninety per cent of rat genes have orthologues in the human and the mouse. Orthologues are genes in different species that originated from a common ancestral gene. In the rat, there has been expansion of gene families related to pheromone production, olfaction, and detoxification. In addition to the genomic sequence, there are other valuable rat genetic resources, including a large number of genetic markers for QTL and disease mapping, microsatellite maps, cytogenetic maps, and rat microarray platforms [1, 56, 87, 102, 103].

The Rat Genome Database-RGD (http://rgd.mcw.edu) provides access to a wide variety of curated rat data, including QTL, microsatellites, strains, and pathogenesis information, as well as tools helpful in analyzing this information [27, 52, 103]. The PhenoMiner tool inside the RGD integrates phenotypic data collected from different rat strains under differing experimental conditions. Other online rat resources are also available [102], including the Rat Resource and Research Center at the University of Missouri (https://www.rrrc.us/) [83] and the National BioResource Project-Rat at Kyoto University (https://www.anim.med.kyoto-u.ac.jp/nbr/) [81]. Unlike laboratory mice, all laboratory rat strains have exclusively originated from *Rattus norvegicus*, without any recognized subspecies.

1.2.3 Genetic Drift and the Formation of Substrains

Mutations are a source of genotypic change and are significant to consider for two reasons: first, their occurrence is entirely beyond our control; and second, they are insidious and generally impossible to detect through simple phenotypic observation or routine genetic monitoring. However, while a proportion of these new mutant alleles are effectively eliminated by inbreeding, another proportion may become progressively fixed in the homozygous state, replacing the original allele. This phenomenon is known as *genetic drift,* and it is a very slow process that is completely unavoidable. It contributes inexorably to strain divergence (and to the generation of substrains) when the same strain is propagated independently in different places. Therefore, it is important to use appropriate genetic quality control measures, such as regular monitoring of genetic markers and phenotypic traits, to ensure that the substrains remain true to the original inbred strain and are consistent across different research laboratories (see Chap. 4). The issue lies in the unpredictable consequences of novel mutations, as well as the potential impact of concealed mutations in the genomes of substrains on the outcome of an experiment [53]. Hence, it is advisable to maintain consistency with the same inbred substrain throughout a series of experiments and to refrain from changing vendors midway through a study.

A colony of an inbred strain can be considered a substrain in the following cases: (i) when residual heterozygosity (incomplete inbreeding) is still present at the time of separation; (ii) when undetected spontaneous mutations become fixed in a colony; and (iii) when a strain is separated from its parent colony for a total of 20 or more consecutive generations of inbreeding (e.g., parent colonies have each been bred for ten generations). The latter is the more prevalent scenario, and it is not necessary to identify specific genetic differences; the mere separation for over 20 generations is sufficient to designate it as a substrain, which entails the use of a specific ILAR-approved laboratory code (https://www.nationalacademies.org/ilar/institute-for-laboratory-animal-research) [7, 11, 42, 77, 106].

There are many examples of substrains that, despite originating from the same original inbred strain, have developed new and distinctive phenotypic traits as a result of genetic drift. Below we will discuss some of these examples in detail. In Chap. 4, we will discuss the effects of genetic background (defined as all the genes other than the gene of interest) on the phenotypes of models and the importance of performing background characterizations. It is well known that genetic variations between inbred strains can significantly impact the phenotype of animal models, leading to differences in behavior, physiology, and disease susceptibility [21, 26, 35, 57, 91, 111].

Concerning the mouse, C57BL/6 was historically a common inbred strain, but since the arrival of the gene targeting technologies, its use has exploded because it is the most common choice as background for knock-out (KO) and knock-in (KI) lines. C57BL/6J was chosen by the *International Mouse Sequencing Consortium* as the inbred strain to generate the first draft of the mouse genome (the second mammal to be sequenced), published in 2002 [105]. Interestingly, the number 57 in the name of the strain is originated from a female with ID 57 that was crossed by Clarence Little in 1921 with a male with ID 52. The ancestors of "C57BL" mice were homozygous for the nonagouti allele and hence exhibited a black coat [98]. The ancestral group of C57BL/6 (official symbol: B6) mice is the origin of the substrain known today as C57BL/6J (J is the Lab Code for JAX) (stock # 000664). Over the years, several substrains have been created due to the distribution of JAX C57BL/6 mice to different places. For example, C57BL/6N (B6/N) was separated from C57BL/6J (B6/J) at generation F32 in 1951 and is the substrain that was relocated to the National Institute of Health (NIH). ES cells developed from this C57BL/6N substrain are used by the International Knockout Mouse Consortium (IKMC) to create a KO line for every protein-coding gene (~22,000 genes). During the 1970s, new colonies of C57BL/6N were started in Charles River Laboratories (Charles River) and Harlan/Envigo (now Inotiv). These are the C57BL/6NCrl and C57BL/6NHsd substrains, respectively. The same B6/N substrain was later acquired by Taconic (C57BL/6NTac), Janvier (C57BL/6NRj), and even JAX (as indicated by the "NJ" at the end of the name: C57BL/6NJ stock # 005304). In Europe, Inotiv distributes the C57BL/KaLwRijHsd substrain, a branch of the original B6/J but separated in 1947 (almost exclusively used as a model for multiple myeloma). In the 1970s and 1980s, other colonies stemming from C57BL/6J were created, like the Eicher substrain (C57BL/6JEiJ), separated in 1976. Other examples are the C57BL/6JOlaHsd substrain created in the UK (Harlan Olac) and the C57BL/6JRccHsd

Table 1.1 This table displays the spontaneous mutations found in and shared among the different substrains of C57BL/6 from the main international vendors (identified by their laboratory codes in boldface) offering this inbred strain. Please note that some genes have not been tested across all substrains and some mutations are suspected only based on the colony's history

Substrain	Vendor	Affected gene (mutant allele)						
		Nnt (*Nnt*^{B6J})	*Crb1* (*Crb1*^{rd8})	*Snca* (*Snca*^{del})^1	*Mmrn1* (*Mmrn1*^{del})^1	*Nlrp12* (*Nlrp12*^{B6J})	*Dock2* (*Dock2*^{Hsd})	*Cyfip2* (*Cyfip2*^{B6N})
C57BL/6**J**	JAX (C57BL/6**J**)	MUTATED	WT	WT	WT	MUTATED	WT	WT
	CRL (Europe) (C57BL/6**JCrl**)	MUTATED	WT	WT	WT	not tested	suspected WT	WT
	Janvier (C57BL/6**JRj**)	MUTATED	WT	WT	WT	not tested	suspected WT	WT
C57BL/6**N**	JAX (C57BL/6**NJ**)	WT	MUTATED	WT	WT	WT	WT	MUTATED
	CRL (C57BL/6**NCrl**)	WT	MUTATED	WT	WT	suspected WT	WT	MUTATED
	Envigo (C57BL/6**NHsd**)	WT	MUTATED	WT	WT	suspected WT	MUTATED	MUTATED
	Taconic (C57BL/6**NTac**)	WT	MUTATED	WT	WT	suspected WT	WT	MUTATED
	Janvier (C57BL/6**NRj**)	WT	MUTATED	WT	WT	suspected WT	suspected WT	MUTATED
C57BL/6**ByJ**	JAX	WT	WT	WT	WT	not tested	not tested	WT
C57BL/6**JEiJ**	JAX	WT	WT	WT	WT	not tested	not tested	WT
C57BL/6**JOlaHsd**	Envigo	WT	WT	MUTATED	MUTATED	not tested	not tested	not tested
C57BL/6**JRccHsd**	Envigo	WT	WT	WT	WT	not tested	not tested	not tested
C57BL/6**JBomTac**	Taconic^2	WT	WT	WT	WT	not tested	not tested	not tested

^{1} Del(6)*Snca* 1Slab

^{2} A mutation on the Y chromosome was recently identified in C57BL/6JBomTac

substrain in Switzerland (RCC), both distributed exclusively in Europe by Inotiv. The C57BL/6JBomTac substrain was started in Germany (Hannover) in the early 1970s and is now distributed by Taconic in Europe.

Table 1.1 lists the mutations present in the C57BL/6 substrains described above, potentially affecting several types of studies, including neurological and behavioral. Mice of the C57BL/6JOlaHsd substrain, for example, are homozygous for a deletion of the *Snca* locus (encoding for α-synuclein) [89]. This deletion has modest phenotypic effects but might interfere in an unpredictable manner with other mutations if this substrain is used as a background strain for making a KO [106]. Also, a few spontaneous mutations have been described to segregate differentially in the most popular substrains of C57BL/6 mice, including a retinal degeneration mutation (*Crb1*^{rd8}) present in the B6/N substrain [60] and a deletion in the *Nnt* gene present only in the B6/J substrain [36]. The most comprehensive comparative phenotypic and genomic analysis of these popular substrains has been published [84].

The 129 group of inbred mice is a rather heterogeneous collection of strains that shared a common origin starting in the 1920s. They present genetic variability, including different coat colors and distinctive behavior [10], and even, in some cases, traces of genetic contamination (accidental crosses). With some differences, 129 mice were recognized for having high incidence of spontaneous testicular teratomas (and these strains were the first to be used for the development of pluripotent ES cell lines). Today we distinguish four groups of 129 substrains: named with the letters P (derived from the original parental strains), S (derived from a line carrying the "Steel" mutation), T (derived from a line carrying the teratoma "ter" mutation), and X (derived from a line with a genetic contamination) [85]. Strains 129X1/SvJ, 129P1, 129P2 and 129P3, are albino or white-bellied, pink-eyed, light chinchilla, but those from the Steel group (129S1, 129S2, 129S3, 129S4, 129S5, 129S6, 129S7, and 129S8) are white- (or light)-bellied agouti. Some of the passenger mutations present in the 129 family are: (i) a *Cdt1* deletion in some substrains from the P and S group [24]; (ii) a missense mutation in the *Slc3a1* gene in 129S2/SvPas [58]; (iii) a *Disc1* deletion in 129S6/SvEv [54]; and (iv) a deletion in the *Casp4* gene in 129X1, 129S1, 129S2, 129S6 and 129P3 [53]. The nomenclature of the 129 family of strains was revised in 1997 (see http://www.informatics.jax.org/mgihome/nomen/strain_129.shtml).

BALB/c mice originated from an albino stock maintained by Halsey *Bagg* at Memorial Hospital in New York City (*B*agg's *ALB*ino). The three most common substrains, BALB/cAn (Andervont), BALB/cByJ (Bailey), and BALB/cJ were separated from the BALB/c maintained by Snell at JAX in the mid-1930s [71]. The BALB/cJ substrain has some phenotypic differences with the other BALB/c substrains, including persistent postnatal expression of genes that are normally expressed only in the fetal liver, resistance to atherosclerosis when placed on a high-fat diet, and reduced expression of major urinary proteins. All these features related with a mutation in the *Zhx2* gene [20]. BALB/cAnN originated from a group of BALB/cJ mice donated to Howard B. Andervont in 1935 and later transferred to the NIH in 1951. This substrain, with more than 80 years of divergence from the original BALB/c lineage is available from Charles River, Inotiv, and Taconic (in addition to the BALB/cJBomTac substrain). The BALB/cBy substrain originates from BALB/cAnN from which it was separated in 1961 and maintained originally by Dr. Bailey. This substrain is homozygous for the $Cdh23^{\mathrm{ahl}}$ mutant allele associated with age-related hearing loss (mutation also present in several other inbred strains). Both BALB/cAnN and BALB/cByJ substrains are a good choice to produce monoclonal antibodies (inducing plasmacytomas upon injection of mineral oil) [72].

Several C3H substrains have been produced over the years, but the two most common are C3H/HeJ (JAX) and C3H/HeN (NIH). The C3H/HeN branched out from the C3H/HeJ substrain in 1951. The C3H/HeJ substrain carries a spontaneous missense mutation in the *Tlr4* gene (toll-like receptor 4 gene, *Tlr4*) making these mice highly susceptible to infection by Gram-negative bacteria [70]. On the other hand, C3H/HeN mice carry a wild-type allele and responds well to bacterial lipopolysaccharides (LPS). Most of the vendors offer this NIH substrain.

As is the case with mice, many rat inbred strains have at least two substrains. For example, BN/Rj, BN/SsN, and BN/Crl, or SHR/Ola and SHR/NCrl. Substrain variability has been confirmed through sequencing, with WKY showing the highest degree of substrain variation [47]. It is important to note that different suppliers are likely to offer different substrains potentially carrying hidden mutations. For example, LEW/N (NIH), LEW/Crl (Charles River Laboratories), and LEW/Hsd (Inotiv) are available as LEW rats, while F344/N (NIH), F344/NTac (Taconic), F344/NCrl (Charles River Laboratories), and F344/NSlc (Japan SLC) are available as F344/N rats.

What we just reported concerning the insidious and unavoidable occurrence of new mutations in an inbred strain also explains and justifies the recommendation by the International Committee on Standardized Genetic Nomenclature for Mice that inbreeding should never be relaxed. Inbreeding is inefficient in preventing mutations from occurring, but it contributes to the elimination of a substantial proportion of the new mutant alleles, and accordingly, helps to preserve the genetic profile of a given strain in the long term.

1.2.4 Other Inbred Rodents

Besides mice and rats, there are a few inbred strains from other rodents, however, a quick search on PubMed shows that very few recent publications list these strains. As mentioned previously, Guinea pigs (*Cavia porcellus*) were the first rodents to be used in the development of pure lines, which included the creation of the 2/N and 13/N strains. For example, strain 13/N is still used as model for emerging and high-pathogenicity viruses. Syrian hamsters (*Mesocricetus auratus*), also known as golden hamsters, have been widely used as animal models in biomedical research. Various inbred strains have been developed through brother x sister mating to create genetically identical animals. Some examples of inbred Syrian hamsters include the CB, LHC, LSH, MHA, PD4, and the BIO14.6 (cardiomyopathic) strains. Laboratory Mongolian gerbils (*Meriones unguiculatus*) are descendants of the Tumblebrook Farm animals, which can be traced back to 20 pairs of wild-caught Mongolian gerbils that were sent to Japan in 1935 [68]. MON/Tum is an inbred strain developed in the late 1970s and used as a model for Rift Valley Fever viral encephalitis but is now rarely used. Recently, two new inbred strains of gerbils have been developed: LSAG (albino), which has a low susceptibility to seizures, and SPBG (black), which is prone to seizures [37].

1.3 F1 Hybrids

Resulting from the cross of two inbred strains, F1 hybrids are heterozygous at all loci for which the parental strains have different alleles, but they are genetically uniform (isogenic) like their parents (Fig. 1.5). Pairs of the same sex are equivalent to monozygotic (identical) twins or to cloned mice. They are also histocompatible and permanently accept tissue transplantations from either parental strain, from their littermates, and from all their offspring; however, the parental strains will not accept a graft from the F1 hybrids.

F1 mice also exhibit hybrid vigor (heterosis), the opposite of inbreeding depression, making them a choice in many experimental protocols. This is common, for example, in the protocols aimed at the production of genetically engineered animals, where F1 hybrids are often used because of their high production of pre-implantation embryos that are highly resistant to manipulation (e.g., transgene

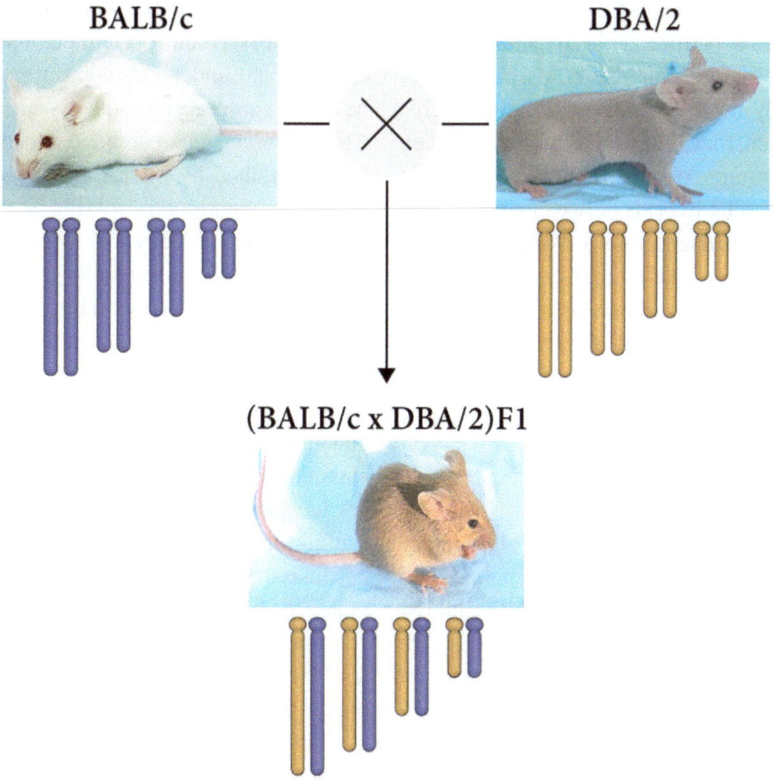

Fig. 1.5 *F1 hybrids.* This figure depicts the creation of F1 mice by intercrossing parental inbred BALB/c (albino) and DBA/2 (diluted brown). Below the mouse pictures, only four pair of chromosomes are shown as an example, with different colors representing the different backgrounds (although not all alleles will be polymorphic between the parental strains). Note that the F1 mouse obtained has the characteristic brown agouti (cinnamon) coat color. The standard nomenclature is (BALB/c x DBA/2)F1 (maternal strain listed first). Also acceptable is the abbreviated version CD2F1. Note that hybrid F1 mice are isogenic because they all receive the same maternal and paternal chromosomes. However, crossing F1 mice will generate hybrid F2 mice that are not isogenic because they will have recombinant chromosomes showing different patterns of BALB/c and DBA/2 alleles

or CRISPR/Cas pronuclear microinjection). However, a major drawback is that, when intercrossed, their progeny (F2) is genetically heterogeneous since the alleles at all polymorphic loci start segregating due to recombination events in the F1 gametes during meiosis. Examples of common, commercially available, hybrid mice are the agouti B6C3F1 (a cross between female C57BL/6 and male C3H), the black B6D2F1 (female C57BL/6 x male DBA/2), the agouti CB6F1 (female BALB/c x male C57BL/6), and the agouti NZBNZWF1 (female NZB x male NZW). Recently, FVBB6F1 (female FVB/NJ x male C57BL/6J) mice have been proposed as an improved mouse model for neurobehavioral research [86]. An example of a commercially available hybrid rat strain is the black hooded ZSF1, a cross between female ZDF and male SHHF (available as lean control and obese mutant) [92].

Interstrain hybrids can also be used to generate genetically heterogeneous populations. This is the case when, for example, F1 hybrids between hypothetical strains A and B (abbreviated ABF1) are crossed with F1 hybrids between strain C and strain D (CDF1) to generate a 4-way heterogeneous stock. In this case the basic ingredients of such a genetically heterogeneous stock (i.e., the original inbred strains A, B, C, and D) are perfectly identified.

1.4 Outbred Stocks

Outbred stocks are populations of laboratory rodents that are genetically heterogeneous, and therefore, significantly different from the inbred strains discussed above. Outbred stocks are closed populations bred to maintain maximum heterozygosity. Unlike inbred strains, F1 hybrids, and congenic strains (see below), the genetic makeup of a randomly selected animal from an outbred stock is not known beforehand. These stocks are bred according to a system that minimizes inbreeding and ensures a certain level of heterozygosity in the population [2, 12, 46]. The level of genetic heterogeneity present in outbred colonies is determined by the history of the colony [17]. This heterogeneity can be quite low, resulting from genetic drift or when the pool of breeders is intentionally or accidentally reduced to a small number of individuals, as is typical when initiating a new breeding program with a limited number of imported breeders (a bottleneck effect). In contrast, genetic heterogeneity can be very high when the colony has been recently outcrossed with individuals from an unrelated stock.

Because outbred colonies, like human populations, are heterogeneous, they are often considered the most appropriate category of laboratory animals for toxicology and pharmacology research. However, several geneticists have disputed this point and have even suggested that in many studies, outbred mice were used inappropriately, wasting animals' lives and research resources on suboptimal experiments [31]. Other investigators recently claimed that, contrary to conventional understanding, outbred mice might be better subjects for some research areas [101]. Also, outbred stock can be replaced with a population obtained by intercrossing classical inbred strains. As mentioned, crossing two inbred strains to produce F1 progeny followed by crossing two independent F1 individuals generates a 4-way polymorphic population. This population is heterogenic, in the sense that individuals are genetically different. In addition, the population often carries a greater number of allelic forms, which is generally considered an advantage compared to a classical outbred population. Recently, researchers have realized that outbred stocks might be useful for refining QTL mapping experiments, because these heterogeneous stocks accumulate many recombination breakpoints that over time split their chromosomes into "fine-grained mosaics," facilitating high-resolution mapping of complex traits [19, 93, 96, 108].

Outbred animals are typically bred to maintain a defined level of population heterozygosity and to avoid inbreeding [46]. Several breeding schemes for rigorous outbreeding have been developed and should be applied instead of experimenting with random mating. Frequently used systems include the so-called rotational breeding [29, 69, 75]. A key element of these schemes is dividing the colony into a fixed number of equally sized groups, determined by which females and males will be mated to each other. A constant number of progeny per parent of the breeding stock is selected to avoid unintended directional selection: the greater the number of breeders in a stock, the smaller the variations in allele frequencies (genetic drift) at each generation [69]. Typically, a large number of breeding pairs ($\geq$ 25). Software for generating random mating schemes have been also published [79].

Examples of classical outbred stocks of mice are ICR (CD-1), CFW, and NMRI (all derived from the original "Swiss" mice imported to the United States by Clara J. Lynch in 1926) and the (non-Swiss) CF-1 mice [17]. A new Swiss-derived albino outbred stock (related to CD-1) called J:ARC(S) is currently maintained at The Jackson Laboratory (JAX). It was established in 2020 by importing a group of 128 mice from the Australian Animal Resource Center (ARC), comprising 64 males and 64 females. This stock is ideal as a general purpose or for toxicology and pharmacology screens due to its high fecundity (average of 14 pups per female), docility, and low variability in body weight [23]. As we will explain below, other non-Swiss outbred stocks, like the multiparent J:DO from JAX, are now used for mapping complex traits.

Examples of outbred rat stocks are Sprague Dawley (SD), Wistar (WI), and Long-Evans (LE). These rat stocks are commonly used in toxicology and pharmacology studies, as well as physiology or general laboratory use. They are also known for their robust health and reproductive capabilities, making them ideal for long-term studies. Outbred mice and rats are available from various commercial vendors and breeders such as Charles River Laboratories, Inotiv, Taconic Biosciences, Janvier Labs, and others. They are also available from many academic institutions and research facilities that maintain their own breeding colonies. It is important to keep in mind that outbred stocks from different vendors may have different genetic profiles. For this reason, it is recommended to use the same outbred stock for a series of experiments and avoid switching to another vendor in the middle of a study. Standard nomenclature for outbred stocks was proposed as early as the 1970s [32] and will be addressed in Chap. 3.

There are also commercially available outbred stocks for guinea pigs [(Crl:HA (Hartley) and HsdDhl:DH (Dunkin Hartley)]; Syrian hamsters [HsdHan:AURA and Crl:LVG(SYR)]; Mongolian gerbils [Crl:MON(Tum)], and cotton rats (*Sigmodon hispidus*) (Hsd:Cotton Rat). These stocks are typically marketed as multi-purpose models, for example, Syrian hamster stocks have been extensively used in the development of vaccines and the study of SARS-CoV-2 pathogenesis. These "outbred" rodents are usually maintained as closed colonies without following specific outbred breeding programs, as is the case for rats and mice [12]. This implies that they are likely undergoing a gradual shift toward inbreeding, so using them for genetic studies is challenging due to their unclear and uncontrolled population structure.

1.5 Standardized Strains of Mice and Rats Used for Genetic Studies

These specialized laboratory strains are designed for use as backgrounds for GA lines or in genetic studies, specifically in crosses used for gene mapping and the positional cloning of spontaneous mutations. Additionally, these strains play a crucial role in the challenging effort to identify complex traits in these species. Complex traits, such as susceptibility to certain cancers, do not follow a simple Mendelian inheritance pattern and are typically the result of intricate interactions between various genetic and environmental components.

1.5.1 Strains Used to Maintain Spontaneous and Genetically Engineered Mutations

The strains described below are primarily used for maintaining and studying mutant alleles or chromosomal regions in a defined background. They can also serve to confirm the location of a disease-causing gene identified through linkage analysis. Therefore, they will also be included in Sect. 5.2.

1.5.1.1 Co-isogenic Strains

When a mutation arises within the breeding nucleus of an inbred strain, assuming the new mutant allele has replaced the original one, the inbred strain in question diverges from the original strain solely at this new mutant allele. Spontaneous and chemically induced mouse mutations will be discussed in Chap. 2. If the new mutation is viable and does not impair fertility, one can propagate the new strain by intercrossing mutant mice [44]. These two strains are said to be *co-isogenic*. Co-isogenic strains are extremely useful for gene annotation because they allow a comparison of the phenotypes of two allelic forms of a particular gene with no influence of the genetic background. Many such strains are stored worldwide in the major genetic repositories. For example, C57BL/6 has several co-isogenic "companion" strains segregating for a variety of allelic forms involved in the determinism of coat color. Mice of the C57BL/6-*Tyr*^{c} albino co-isogenic strain have been popular for the production of easily recognizable C57BL/6:C57BL6-*Tyr*^{c}/*Tyr*^{c} chimeric mice from C57BL/6 ES cells injected into albino C57BL/6-*Tyr*^{c}/*Tyr*^{c} blastocysts [80].

Co-isogenic strains, when developed in parallel to the background strain, may accumulate other genetic differences over time because of genetic drift. Thus, to minimize the effects of this drift, they must be periodically backcrossed to the original parental strain or be cryopreserved (see Chap. 6). Co-isogenic strains have two significant drawbacks inherent to their origin, restricting their utility. Firstly, they typically arise as a result of rare and fortuitous mutations. Secondly, while they may manifest in any inbred strain, it is generally not the strain of primary interest for researchers. Still, genetically engineered mice can also be considered co-isogenic strains when the genetic modification is done in a way such that the targeted or edited locus is the only difference from the wild-type animals. In the case of classical additive transgenesis or endonucleases-mediated gene editing, this can be achieved by performing the pronuclear microinjection using embryos derived from an inbred strain (e.g., FVB/N mice or F344 rats). A significant distinction between classical transgenics and co-isogenic strains lies in the dynamic nature of the transgenic insertions over time. This can manifest as alterations in structure, such as changes in copy number or DNA methylation (see Chap. 2).

1.5.1.2 Congenic Strains

Congenic strains are an alternative to co-isogenic strains with the advantage that any allele or segment of the genome may be moved (*introgressed*) into any inbred background. The disadvantage, as we will explain, is that the situation is not as pure, from the genetic point of view, as it is in the case of co-isogenic strains. Congenic strains are produced by crossing two strains: the first one carries the allele or chromosome region of interest (i.e., spontaneous, induced, or targeted mutations, as well as transgenes), and is referred as the *donor strain*; the second strain is referred to as the *recipient strain* or *background* strain [44]. The F1 offspring generated by crossing the above-mentioned two strains are backcrossed to the background strain, and the offspring that carry the allele of interest (i.e., the one originating from the donor strain) are crossed again to the background strain, typically for ten or more successive generations [76]. Incipient congenics typically refer to strains that have been backcrossed for around 5 to 9 generations. Once this process progresses to 10 or more generations of backcrossing, it is usually termed a "full congenic" (see Chap. 3 for standard nomenclature). During this succession of backcrosses, the chromosomes of the background strain progressively replace those of the donor strain, except for the one that carries the allele of interest (Fig. 1.6).

For this chromosome, the segment containing the selected or targeted allele is reduced in size only when a recombination event occurs that replaces a piece of chromosome of the donor strain for the homologous segment of the background strain. Since the occurrence of this sort of event depends upon the size of the segment, one understands then that the chromosome carrying the targeted allele is gradually "worn" on both sides, generation after generation, but in a nonlin-

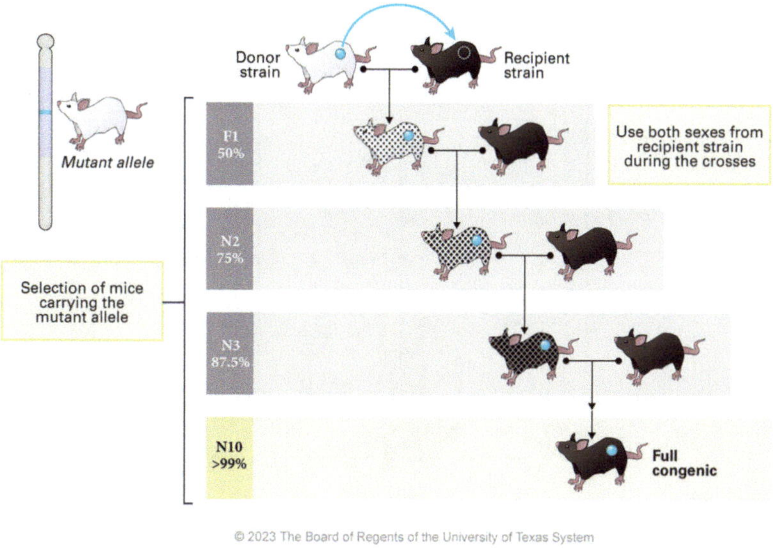

Fig. 1.6 *Congenic strains.* This scheme represents the successive steps in the establishment of a congenic strain. The initial step is a cross between two strains: a donor strain (white in the example) carrying the gene of interest (e.g., a mutant allele) and a recipient or background strain (black in the example). At each generation, a breeder carrying the allele of interest (blue dot) is backcrossed to a partner of the recipient strain. The degree of gray/black color indicates that, after each backcross generation, the offspring have an increased amount of the background genome. When the mutant allele has no recognizable phenotype, specific molecular genotyping is necessary (closely linked markers may also be used). After 10 generations, the animals are globally >99% recipient background and the line is called full congenic

ear manner. The chromosomal segments flanking the selected locus tend to remain associated with this locus, and this is the basic difference between congenic and co-isogenic strains. It must be kept in mind that the reduction in size of the introgressed chromosomal segment by definition depends upon the occurrence of crossover events. For this reason, the reduction in size is nonlinear.

Since, on average, at each generation, an equivalent proportion of the background strain replaces one half of the genome of the donor strain, the progression of genome substitution is given by the formula $1/2^N$, where N is the number of backcross generations. This means that, theoretically, after 10 backcross generations only $1/2^{10}$ (~ 1/1000) of the donor genome remains in the congenic strain. This assumption is, again, purely statistical and the actual percentage of donor genome is subject to variations at each generation. In addition, and as we already pointed out, this estimation stands only for the chromosomes that do not carry the allele of interest. In the case of the chromosome carrying the allele of interest, the reduction in size is a much slower process (Fig. 1.7).

Another important consideration is that for the congenic strain to possess both mitochondrial and Y chromosome recipient DNA, both sexes of the recipient strain must be utilized during the crossing process. If only recipient males are involved, the mitochondrial DNA would continue to originate from the donor strain. Conversely, if only recipient females are employed, the Y chromosome would retain its donor strain origin.

The development of congenic strains has become more efficient and precise with the use of polymorphic DNA markers, through a process known as marker-assisted breeding (or backcrossing). This method, also referred to as *speed congenics*, enables a more rapid and rigorous development of congenic strains [59, 82, 104]. The speed congenics process is based on the principle that breeders can be selected at each generation of backcrossing based on the percentage of

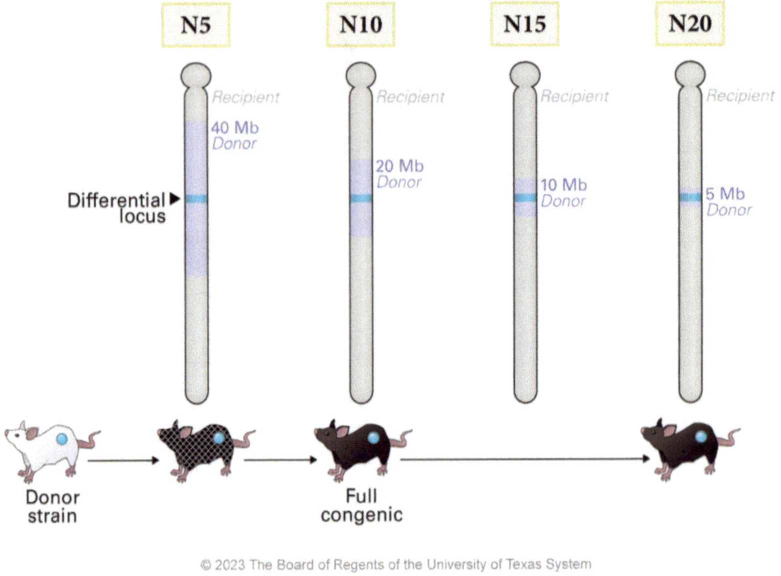

Fig. 1.7 *The flanking region in congenic strains.* The region flanking the gene of interest in congenic strains can pose a challenge as it may impact the phenotype of the strain under investigation. This is due to the potential presence of modifier genes within the donor strain's chromosomal region, which could interact with genes from the recipient strain, resulting in unintended genetic effects that might influence the interpretation of experimental outcomes. Consequently, minimizing the size of flanking regions in congenic mouse strains is crucial in order to mitigate the likelihood of confounding effects. As shown in the figure, even at backcross generation N10 the size of the flanking (donor) region (lavender color) could be as big as 20 megabases (Mb), potentially carrying hundreds of genes

donor genome they possess. This selection process is made possible using Single Nucleotide Polymorphisms (SNPs) to differentiate between the two parental strains. Obviously, the animal with the lowest percentage of donor DNA is the one to select as breeder for setting the next backcross. This procedure reduces the number of generations necessary to reach full congenicity from N10 to N5, and the strain development time, approximately by half (see Chap. 5).

1.5.2 Strains Used for Gene Mapping

Below, we describe special mouse and rat strains that are used for gene mapping, including complex traits such as diabetes, hypertension, or susceptibility to cancer. Many mouse and rat inbred strains exhibit features that can model specific human conditions and are extremely valuable for identifying and studying QTLs, especially when used in combination with SNP and the extensive genomic sequence information currently available.

1.5.2.1 Congenic Strains as Mapping Tool

Congenic strains, explained above, are commonly used in gene mapping studies to identify the specific genetic regions responsible for a particular phenotype or disease. When employing congenic strains in mapping projects, it becomes essential to possess prior data that identifies the specific region or regions that are to be transferred onto a distinct inbred background. The advantage of using congenic strains is that they allow researchers to isolate and study specific genetic regions of interest, while minimizing the potential confounding effects of other genetic variability. Mouse congenic strains have been extensively used since the early days of mouse genetics. They are particularly suited for the genetic analysis of phenotypes that are controlled by several genes, and it is precisely by developing such strains that George D. Snell and his colleagues from The Jackson Laboratory could elucidate the genetic determinism of histocompatibility.[2] Tissue transplantations performed between mice belonging to different inbred strains, are rejected. On the contrary, the same transplantations performed between any two mice of the same inbred strains (and the same sex) are permanently accepted. The problem is that, in the case of tissue transplantations, the rejection, which is the observed phenotype, is controlled by several loci, each of them independently triggering the same phenotype. To clarify the situation, Snell bred a series of strains, with the same C57BL/10 genetic background, but congenic for a single locus inducing tissue incompatibility. He called these mice *congenic resistant* and developed a very clever protocol to characterize each of these strains. By doing this, Snell succeeded in making a comprehensive inventory of the histocompatibility (*H*) loci segregating among the laboratory strains [88].

This strategy could be adapted with almost no change to the genetic analysis of any trait that is under polygenic control, for example, resistance to infectious diseases. When a congenic strain has been established, there is still a lot of work to do to finally characterize the gene involved in the phenotype, given that the chromosomal fragment is sometimes quite large. Nevertheless, congenic strains are certainly of great help in these investigations. To illustrate, congenic mouse lines have been used to study the genetic basis of autoimmune diseases such as lupus and rheumatoid arthritis. For instance, a study employed the congenic line B10.D1-H2^{q}/SgJ (B10.Q) to introgress a specific targeted deletion and map a gene linked to susceptibility to arthritis [4]. Congenic rat strains have also been used for gene mapping of complex traits. For example, reciprocal congenic strains derived from alcohol-preferring and non-preferring rats have been used to confirm a QTL [14]. Another example is the identification of blood pressure QTLs using congenic Dahl rats. These congenic strains were created by introgressing Lewis (LEW inbred) low-blood-pressure QTL alleles into the genetic background of Dahl salt-sensitive (SS inbred) rats [38].

[2]G.D. Snell, J. Dausset and B. Benacerraf were awarded the Nobel Prize in 1980 *"for their discoveries concerning genetically determined structures on the cell surface that regulate immunological reactions"*.

1.5.2.2 Consomic Strains

Consomic strains, also designated chromosome substitution strains (CSS), are a variation of the congenic strains concept in which the introgressed DNA is a complete chromosome, rather than a piece of chromosome flanking a given gene [64] (Fig. 1.8). These strains have been very useful for the rapid mapping of phenotypic traits to a specific chromosome. They are also useful for the detection of chromosomal regions having an influence in the determinism of a particular phenotype, for example the susceptibility to diabetes [3]. Using a marker-assisted protocol, consomic strains are easy to produce, however, one must keep in mind that tiny pieces of chromosomes of the donor strain might escape the marker-assisted selection process. To illustrate a real mouse example, C57BL/6J-Chr 1$^{A/J}$ is a consomic strain with C57BL/6J as background and A/J as donor, replacing the C57BL/6 chromosome 1. Sets of consomic strains involving different subspecies of *Mus musculus* have also been developed [41]. Consomic rat strains have been utilized to investigate the genetic foundation of hypertension and renal disease in SS rats. This was accomplished by transferring one chromosome at a time from the BN rat onto the genetic background of the SS rat [61].

1.5.2.3 Conplastic Strains

Conplastic strains have been created to study the effects of mtDNA variations on various traits. In these strains, the nuclear genome is that of the recipient strain, while the mitochondrial genome is derived from a donor strain [113]. This allows researchers to isolate the effects of mtDNA variations on phenotypic traits, as observed differences are primarily attributed to differences in mitochondrial genes. Conplastic strains are used to investigate the role of mitochondrial genetics

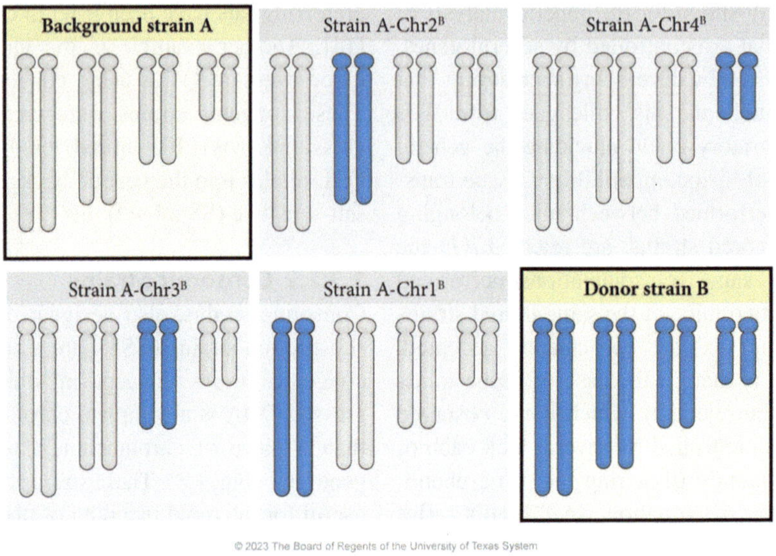

Fig. 1.8 *Consomic Strains.* A consomic strain is an inbred strain that has undergone a series of marker-assisted backcrosses, resulting in the replacement of one pair of chromosomes with the homologous chromosome pair from another strain. A complete panel of consomic strains consists of 21 strains, each derived from the same donor and recipient strains but having each a different chromosome pair (Chr 1-19, X or Y) of the recipient strain replaced by its homolog from the donor. A reciprocal panel can be produced by inverting the donor and host strains, respectively. The example illustrates only four pairs of chromosomes (i.e., 1 to 4) from a consomic strain where strain A (gray chromosomes) is the recipient and strain B the donor (blue chromosomes). The standard nomenclature for consomic strains is HOST STRAIN-Chr #$^{\text{DONOR STRAIN}}$ (e.g., A-Chr1^{B}, A-Chr2^{B}, etc.)

in various aspects, including cancer, aging, and neurodegenerative diseases [13]. As expected, the mitochondrial donor is always the female during the backcrosses onto the recipient background. For example, C57BL/6J-mt^{FVB} is a strain with the nuclear genome of C57BL/6J and the mitochondrial genome of FVB/N.

1.5.2.4 Recombinant Inbred Strains and the Collaborative Cross

Recombinant inbred strains (RIS) are developed by crossing two parental inbred strains to generate F1 hybrids and then by intercrossing these F1 to generate F2s. Then, randomly chosen F2 animals are brother x sister mated for 20 or more generations to develop a group of related inbred strains [5]. RIS are commonly grouped into sets, also known as panels. A panel is a collection of RIS that are derived from the same parental strains, for example mouse strains DBA/2 and AKR (Fig. 1.9).

The C57BL/6 × DBA/2 (BXD) is today the largest mouse RI panel with ~90 strains. These are true inbred strains, meaning that they are homozygous at all loci but have the additional characteristic that each RIS has a unique fixed combination of the parental alleles in a 50%:50% proportion (on average). For example, each strain of the set of 33 AXB-BXA strains, derived from the initial cross of a C57BL/6 mouse with a A/J mouse, carries either the B6 allele or the A allele at each locus of its genome; by typing all of these allelic forms, one can establish a strain distribution pattern (SDP) for each of the strains, which lists the collection of alleles inherited from either the parental strain A or the parental strain B6. Of course, this SDP is fixed forever in each strain (not considering the rare mutations that inevitably occur), and new data are constantly added to it, allowing correlations to be made between genotypes and phenotypes simply by scanning, generally with the help of a simple computer program, the co-segregation of a new phenotype (or genotype) with the existing SDP. RIS have proven helpful when used for mapping chromosomal regions involved in the genetic determinism of behavioral characteristics or

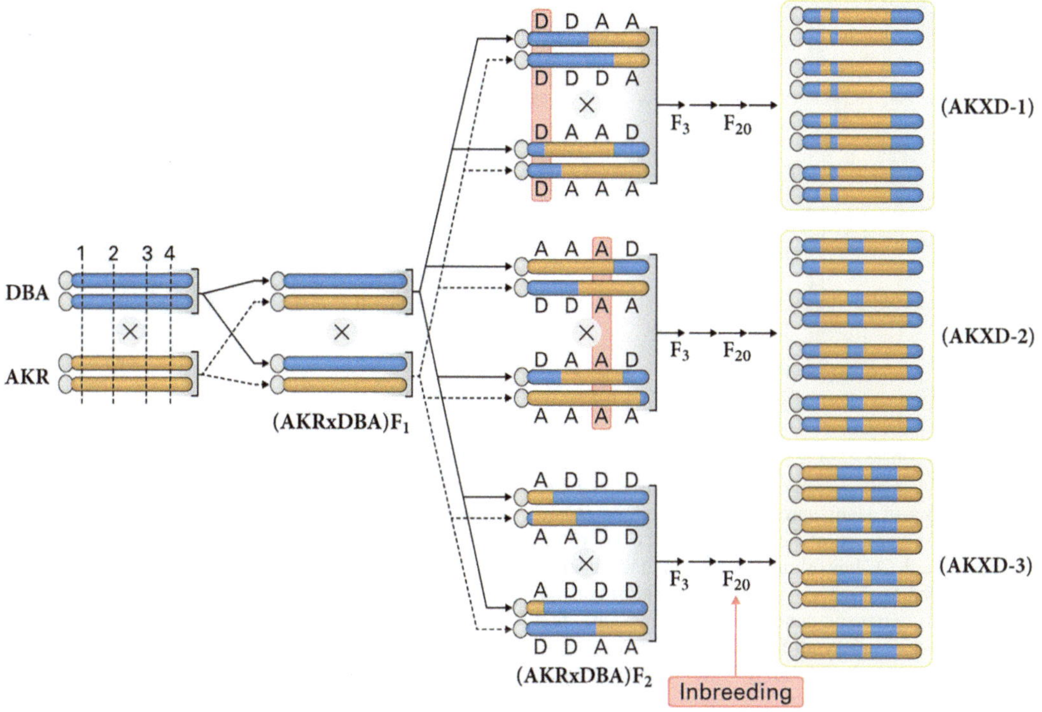

Fig. 1.9 *Recombinant Inbred Strains.* This diagram represents the creation of a set of three recombinant inbred strains (RIS) originated by intercrossing parental inbred strains DBA/2 (D) and AKR (AK) (only one pair of chromosomes is shown as an example). The positions of four hypothetical loci are indicated with dotted lines in the parental chromosomes (numbers 1–4). The pink rectangles show alleles that are already fixed (D or AK) in some breeders at the F2 generation. After >20 generations of inbreeding, these truly inbred strains carry, on average, 50% of alleles from each parental strain. The boxes on the right represent the same chromosome pair showing identical patterns in four random mice from three different RIS (AKXD-1, AKXD-2, and AKXD-3). Individual RISs have a unique combination of loci derived by recombination of the alleles present in the original parental strains. Since RISs are inbred, each strain has a unique genotype

immunological responses. They will very likely still be of great help in many other experiments where the phenotype is measured on a group of animals rather than on individuals [114]. RIS offer advantages, such as pre-existing genotyping for numerous markers, enabling cost and time savings. Phenotyping is enhanced by the ability to assess groups of mice from the same RI strain, benefiting from their identical genotypes. Additionally, being inbred strains allows for cryopreservation, facilitating sharing among diverse research institutions.

The Collaborative Cross is a variation of the recombinant inbred strain concept with a much higher power of resolution and a much higher level of genetic diversity segregating in the panel [16, 18, 78]. The Collaborative Cross is a randomized cross of eight inbred mouse strains that have been carefully selected by a panel of mouse geneticists (The Complex Trait Consortium) [22]. These strains consist of *i*) three classical inbred strains (A/J, C57BL/6J, 129S1/SvImJ); *ii*) two inbred strains affected by a genetically complex pathology (diabetes/obesity: NOD/ShiLtJ, NZO/HILtJ) and *iii*) three strains recently derived from wild progenitors (CAST/EiJ—from *Mus m. castaneus*; PWK/PhJ—from *Mus m. musculus*; and WSB/EiJ—from *Mus m. domesticus*) (Fig. 1.10). These special RISs are produced via a funnel breeding scheme that combined the eight founder strains (otherwise mitochondrial, and Y chromosomes

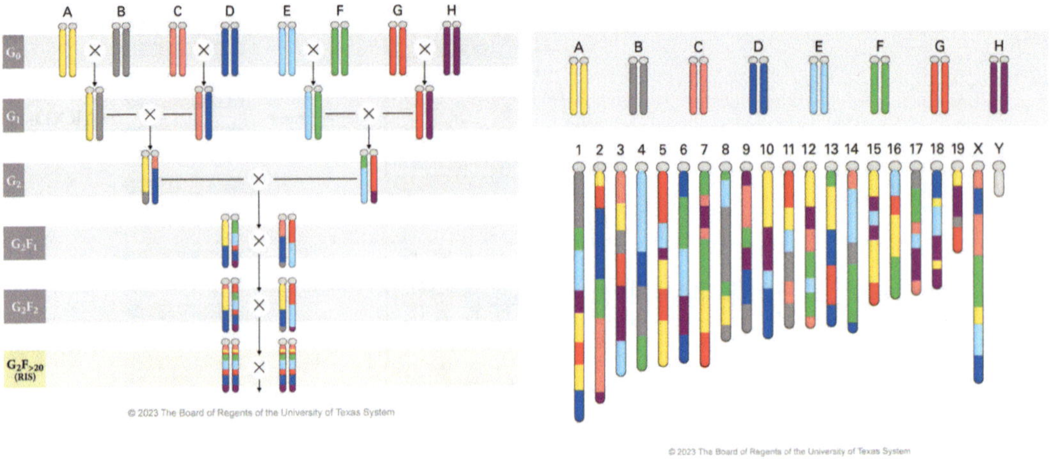

Fig. 1.10 *The Collaborative Cross (CC).* This is a randomized cross of eight unrelated mouse inbred strains selected by the members of the Complex Trait Consortium. *Left panel*: Each CC line originates from an independently breeding funnel so that every recombination site in the CC population is uniquely generated. The lines are first crossed pairwise to make all 56 possible G1 parents. A set of possible 4-way crosses is performed, keeping Y-chromosome and mitochondrial balance. Finally, all eight genomes are brought together in G2:F1, and the offspring of this cross are inbred (G2:F2, G2:F3, etc.). *Right panel*: An example of a set of chromosomes from a hypothetical mouse belonging to a CC line showing chromosomal regions (multicolor component) of the original eight parental strains above

would always have the same origin). The eight strains are first crossed pairwise to make all (8 × 7 = 56) possible G1 parents, then all eight genomes are brought together in a series of F1, and the offspring of this cross are inbred for several generations [66]. Several tens of these new RIS are progressively becoming available. Such panels represents a valuable community resource; despite the concerns related with the facilities needed to host all these strains, their distribution, and the funding to preserve them [99].

1.5.2.5 The J:DO Outbred Mouse Stock

Recently, there has been a suggestion by researchers that outbred stocks could enhance the identification of complex genetic traits [108]. As mentioned, outbred mice prove to be valuable in gene mapping experiments due to the accumulation of numerous recombination events over time. This fact facilitates the high-resolution mapping of genes that are responsible for complex traits [33, 67, 78, 93, 108]. The Diversity Outbred (J:DO) stock was designed to be the most genetically diverse mouse resource, allowing for refined gene mapping resolution [19]. The J:DO heterogeneous stock shares the same eight parental strains as the Collaborative Cross. It was developed by randomly outcrossing 160 G2F1s from the CC (see above), and the colony is maintained through continued random matings using 175 breeding funnels. This method is employed to prevent mating between siblings and thereby reduce the risk of inbreeding and unintentional phenotypic selection [23, 78, 93]. Due to its genetic diversity, the J:DO stock allows for the identification of genetic loci that may be associated with complex traits and may be used for toxicology and drug resistance studies.

References

1. Aitman TJ, Critser JK, Cuppen E, Dominiczak A, Fernandez-Suarez XM, Flint J, Gauguier D, Geurts AM, Gould M, Harris PC, Holmdahl R, Hubner N, Izsvak Z, Jacob HJ, Kuramoto T, Kwitek AE, Marrone A, Mashimo T, Moreno C, Mullins J, Mullins L, Olsson T, Pravenec M, Riley L, Saar K, Serikawa T, Shull JD, Szpirer C, Twigger SN, Voigt B, Worley K. Progress and

prospects in rat genetics: a community view. Nat Genet. 2008;40:516–22.
2. Aldinger KA, Sokoloff G, Rosenberg DM, Palmer AA, Millen KJ. Genetic variation and population substructure in outbred CD-1 mice: implications for genome-wide association studies. PLoS One. 2009;4:e4729.
3. Babaya N, Ueda H, Noso S, Hiromine Y, Itoi-Babaya M, Kobayashi M, Fujisawa T, Ikegami H. Genetic dissection of susceptibility genes for diabetes and related phenotypes on mouse chromosome 14 by means of congenic strains. BMC Genet. 2014;15:93.
4. Backlund A, Holmdahl M, Mattsson R, Hakansson K, Lindstrom V, Nandakumar KS, Grubb A, Holmdahl R. Cystatin C influences the autoimmune but not inflammatory response to cartilage type II collagen leading to chronic arthritis development. Arthritis Res Ther. 2011;13:R54.
5. Bailey DW. Recombinant-inbred strains. An aid to finding identity, linkage, and function of histocompatibility and other genes. Transplantation. 1971;11:325–7.
6. Beck JA, Lloyd S, Hafezparast M, Lennon-Pierce M, Eppig JT, Festing MF, Fisher EM. Genealogies of mouse inbred strains. Nat Genet. 2000;24:23–5.
7. Benavides F, Rulicke T, Prins JB, Bussell J, Scavizzi F, Cinelli P, Herault Y, Wedekind D. Genetic quality assurance and genetic monitoring of laboratory mice and rats: FELASA Working Group Report. Lab Anim. 2019;54:0023677219867719.
8. Bogue MA, Grubb SC, Walton DO, Philip VM, Kolishovski G, Stearns T, Dunn MH, Skelly DA, Kadakkuzha B, Tehennepe G, Kunde-Ramamoorthy G, Chesler EJ. Mouse Phenome Database: an integrative database and analysis suite for curated empirical phenotype data from laboratory mice. Nucleic Acids Res. 2018;46:D843–50.
9. Bogue MA, Ball RL, Walton DO, Dunn MH, Kolishovski G, Berger A, Lamoureux A, Grubb SC, Gerring M, Kim M, Liang H, Emerson J, Stearns T, He H, Mukherjee G, Bluis J, Davis S, Desai S, Sundberg B, Kadakkuzha B, Kunde-Ramamoorthy G, Philip VM, Chesler EJ. Mouse phenome database: curated data repository with interactive multi-population and multi-trait analyses. Mamm Genome. 2023;34:509–19.
10. Boleij H, Salomons AR, Van Sprundel M, Arndt SS, Ohl F. Not all mice are equal: welfare implications of behavioural habituation profiles in four 129 mouse substrains. PLoS One. 2012;7:e42544.
11. Brayton CF, Boyd KL, Everitt JL, Meyerholz DK, Treuting PM, Bolon B. An introduction to pathology in biomedical research: a mission-critical specialty for reproducibility and rigor in translational research. ILAR J. 2018;59:1–3.
12. Brekke TD, Steele KA, Mulley JF. Inbred or outbred? Genetic diversity in laboratory rodent colonies. G3 (Bethesda). 2018;8:679–86.
13. Bussard KM, Siracusa LD. Understanding mitochondrial polymorphisms in cancer. Cancer Res. 2017;77:6051–9.
14. Carr LG, Habegger K, Spence JP, Liu L, Lumeng L, Foroud T. Development of congenic rat strains for alcohol consumption derived from the alcohol-preferring and nonpreferring rats. Behav Genet. 2006;36:285–90.
15. Chang PL, Kopania E, Keeble S, Sarver BAJ, Larson E, Orth A, Belkhir K, Boursot P, Bonhomme F, Good JM, Dean MD. Whole exome sequencing of wild-derived inbred strains of mice improves power to link phenotype and genotype. Mamm Genome. 2017;28:416–25.
16. Chesler EJ, Miller DR, Branstetter LR, Galloway LD, Jackson BL, Philip VM, Voy BH, Culiat CT, Threadgill DW, Williams RW, Churchill GA, Johnson DK, Manly KF. The Collaborative Cross at Oak Ridge National Laboratory: developing a powerful resource for systems genetics. Mamm Genome. 2008;19:382–9.
17. Chia R, Achilli F, Festing MF, Fisher EM. The origins and uses of mouse outbred stocks. Nat Genet. 2005;37:1181–6.
18. Churchill GA, Airey DC, Allayee H, Angel JM, Attie AD, Beatty J, Beavis WD, Belknap JK, Bennett B, Berrettini W, Bleich A, Bogue M, Broman KW, Buck KJ, Buckler E, Burmeister M, Chesler EJ, Cheverud JM, Clapcote S, Cook MN, Cox RD, Crabbe JC, Crusio WE, Darvasi A, Deschepper CF, Doerge RW, Farber CR, Forejt J, Gaile D, Garlow SJ, Geiger H, Gershenfeld H, Gordon T, Gu J, Gu W, De Haan G, Hayes NL, Heller C, Himmelbauer H, Hitzemann R, Hunter K, Hsu HC, Iraqi FA, Ivandic B, Jacob HJ, Jansen RC, Jepsen KJ, Johnson DK, Johnson TE, Kempermann G, Kendziorski C, Kotb M, Kooy RF, Llamas B, Lammert F, Lassalle JM, Lowenstein PR, Lu L, Lusis A, Manly KF, Marcucio R, Matthews D, Medrano JF, Miller DR, Mittleman G, Mock BA, Mogil JS, Montagutelli X, Morahan G, Morris DG, Mott R, Nadeau JH, Nagase H, Nowakowski RS, O'Hara BF, Osadchuk AV, Page GP, Paigen B, Paigen K, Palmer AA, Pan HJ, Peltonen-Palotie L, Peirce J, Pomp D, Pravenec M, Prows DR, Qi Z, Reeves RH, Roder J, Rosen GD, Schadt EE, Schalkwyk LC, Seltzer Z, Shimomura K, Shou S, Sillanpaa MJ, Siracusa LD, Snoeck HW, Spearow JL, Svenson K, et al. The Collaborative Cross, a community resource for the genetic analysis of complex traits. Nat Genet. 2004;36:1133–7.
19. Churchill GA, Gatti DM, Munger SC, Svenson KL. The diversity outbred mouse population. Mamm Genome. 2012;23:713–8.
20. Clinkenbeard EL, Turpin C, Jiang J, Peterson ML, Spear BT. Liver size and lipid content differences between BALB/c and BALB/cJ mice on a high-fat diet are due, in part, to Zhx2. Mamm Genome. 2019;30(7–8):226–36. https://doi.org/10.1007/

s00335-019-09811-6. Epub 2019 Jul 18. PMID: 31321500. PMCID: PMC6761023.
21. Coleman DL, Hummel KP. The influence of genetic background on the expression of the obese (Ob) gene in the mouse. Diabetologia. 1973;9:287–93.
22. Collaborative Cross C. The genome architecture of the Collaborative Cross mouse genetic reference population. Genetics. 2012;190:389–401.
23. Cornes BK, Paisie C, Swanzey E, Fields PD, Schile A, Brackett K, Reinholdt LG, Srivastava A. Protein coding variation in the J:ARC and J:DO outbred laboratory mouse stocks provides a molecular basis for distinct research applications. G3 (Bethesda). 2023;13:jkad015.
24. Coulombe P, Gregoire D, Tsanov N, Mechali M. A spontaneous Cdt1 mutation in 129 mouse strains reveals a regulatory domain restraining replication licensing. Nat Commun. 2013;4:2065.
25. Dejager L, Libert C, Montagutelli X. Thirty years of Mus spretus: a promising future. Trends Genet. 2009;25:234–41.
26. Doetschman T. Influence of genetic background on genetically engineered mouse phenotypes. Methods Mol Biol. 2009;530:423–33.
27. Dwinell MR, Worthey EA, Shimoyama M, Bakir-Gungor B, Depons J, Laulederkind S, Lowry T, Nigram R, Petri V, Smith J, Stoddard A, Twigger SN, Jacob HJ, Team, R. G. D. The Rat Genome Database 2009: variation, ontologies and pathways. Nucleic Acids Res. 2009;37:D744–9.
28. Erickson RP. Twentieth century mouse genetics: a historical and scientific review. London: Academic, an imprint of Elsevier; 2022.
29. Falconer DS. Genetic aspects of breeding methods. In: Lane-Petter W, editor. UFAW handbook on the care and management of laboratory animals. 3rd ed. Edinburgh: Livingstone; 1967.
30. Ferris SD, Sage RD, Wilson AC. Evidence from mtDNA sequences that common laboratory strains of inbred mice are descended from a single female. Nature. 1982;295:163–5.
31. Festing MF. Inbred strains should replace outbred stocks in toxicology, safety testing, and drug development. Toxicol Pathol. 2010;38:681–90.
32. Festing M, Kondo K, Loosli R, Poiley SM, Spiegel A. International standardized nomenclature for outbred stocks of laboratory animals. ILAR J. 2021;62:362–6.
33. Flint J, Valdar W, Shifman S, Mott R. Strategies for mapping and cloning quantitative trait genes in rodents. Nat Rev Genet. 2005;6:271–86.
34. Frazer KA, Eskin E, Kang HM, Bogue MA, Hinds DA, Beilharz EJ, Gupta RV, Montgomery J, Morenzoni MM, Nilsen GB, Pethiyagoda CL, Stuve LL, Johnson FM, Daly MJ, Wade CM, Cox DR. A sequence-based variation map of 8.27 million SNPs in inbred mouse strains. Nature. 2007;448:1050–3.
35. Freeman D, Lesche R, Kertesz N, Wang S, Li G, Gao J, Groszer M, Martinez-Diaz H, Rozengurt N, Thomas G, Liu X, Wu H. Genetic background controls tumor development in PTEN-deficient mice. Cancer Res. 2006a;66:6492–6.
36. Freeman HC, Hugill A, Dear NT, Ashcroft FM, Cox RD. Deletion of nicotinamide nucleotide transhydrogenase: a new quantitive trait locus accounting for glucose intolerance in C57BL/6J mice. Diabetes. 2006b;55:2153–6.
37. Fujisawa N, Maeda Y, Yamamoto Y, Sato NL, Niimura S. Newly established low seizure susceptible and seizure-prone inbred strains of Mongolian gerbil. Exp Anim. 2003;52:169–72.
38. Garrett MR, Dene H, Walder R, Zhang QY, Cicila GT, Assadnia S, Deng AY, Rapp JP. Genome scan and congenic strains for blood pressure QTL using dahl salt-sensitive rats. Genome Res. 1998;8:711–23.
39. Gibbs RA, Weinstock GM, Metzker ML, Muzny DM, Sodergren EJ, Scherer S, Scott G, Steffen D, Worley KC, Burch PE, Okwuonu G, Hines S, Lewis L, Deramo C, Delgado O, Dugan-Rocha S, Miner G, Morgan M, Hawes A, Gill R, Celera, Holt RA, Adams MD, Amanatides PG, Baden-Tillson H, Barnstead M, Chin S, Evans CA, Ferriera S, Fosler C, Glodek A, Gu Z, Jennings D, Kraft CL, Nguyen T, Pfannkoch CM, Sitter C, Sutton GG, Venter JC, Woodage T, Smith D, Lee HM, Gustafson E, Cahill P, Kana A, Doucette-Stamm L, Weinstock K, Fechtel K, Weiss RB, Dunn DM, Green ED, Blakesley RW, Bouffard GG, De Jong PJ, Osoegawa K, Zhu B, Marra M, Schein J, Bosdet I, Fjell C, Jones S, Krzywinski M, Mathewson C, Siddiqui A, Wye N, McPherson J, Zhao S, Fraser CM, Shetty J, Shatsman S, Geer K, Chen Y, Abramzon S, Nierman WC, Havlak PH, Chen R, Durbin KJ, Egan A, Ren Y, Song XZ, Li B, Liu Y, Qin X, Cawley S, Cooney AJ, D'Souza LM, Martin K, Wu JQ, Gonzalez-Garay ML, Jackson AR, Kalafus KJ, McLeod MP, Milosavljevic A, Virk D, Volkov A, Wheeler DA, Zhang Z, Bailey JA, Eichler EE, Tuzun E, et al. Genome sequence of the Brown Norway rat yields insights into mammalian evolution. Nature. 2004;428:493–521.
40. Goios A, Pereira L, Bogue M, Macaulay V, Amorim A. mtDNA phylogeny and evolution of laboratory mouse strains. Genome Res. 2007;17:293–8.
41. Gregorova S, Divina P, Storchova R, Trachtulec Z, Fotopulosova V, Svenson KL, Donahue LR, Paigen B, Forejt J. Mouse consomic strains: exploiting genetic divergence between Mus m. musculus and Mus m. domesticus subspecies. Genome Res. 2008;18:509–15.
42. Guenet JL, Benavides FJ. Mouse strains and genetic nomenclature. Curr Protoc Mouse Biol. 2011;1:213–38.
43. Guenet JL, Bonhomme F. Wild mice: an ever-increasing contribution to a popular mammalian model. Trends Genet. 2003;19:24–31.
44. Guenet JL, Benavides F, Panthier J, Montagutelli X. Genetics of the mouse. Berlin: Springer; 2015.

45. Guo X, Brenner M, Zhang X, Laragione T, Tai S, Li Y, Bu J, Yin Y, Shah AA, Kwan K, Li Y, Jun W, Gulko PS. Whole-genome sequences of DA and F344 rats with different susceptibilities to arthritis, autoimmunity, inflammation and cancer. Genetics. 2013;194:1017–28.
46. Hartl DL. Genetic management of outbred laboratory rodent populations. Wilmington: Charles River Genetic Literature; 2001.
47. Hermsen R, De Ligt J, Spee W, Blokzijl F, Schafer S, Adami E, Boymans S, Flink S, Van Boxtel R, Van Der Weide RH, Aitman T, Hubner N, Simonis M, Tabakoff B, Guryev V, Cuppen E. Genomic landscape of rat strain and substrain variation. BMC Genomics. 2015;16:357.
48. Howe K, Dwinell M, Shimoyama M, Corton C, Betteridge E, Dove A, Quail MA, Smith M, Saba L, Williams RW, Chen H, Kwitek AE, McCarthy SA, Uliano-Silva M, Chow W, Tracey A, Torrance J, Sims Y, Challis R, Threlfall J, Blaxter M. The genome sequence of the Norway rat, Rattus norvegicus Berkenhout 1769. Wellcome Open Res. 2021;6:118.
49. Hulme-Beaman A, Orton D, Cucchi T. The origins of the domesticate brown rat (Rattus norvegicus) and its pathways to domestication. Anim Front. 2021;11:78–86.
50. Jacob HJ, Kwitek AE. Rat genetics: attaching physiology and pharmacology to the genome. Nat Rev Genet. 2002;3:33–42.
51. Kalbfleisch TS, Hussien Abouel Ela NA, Li K, Brashear WA, Kochan KJ, Hillhouse AE, Zhu Y, Dhande IS, Kline EJ, Hudson EA, Murphy TD, Thibaud-Nissen F, Smith ML, Doris PA. The assembled genome of the stroke-prone spontaneously hypertensive rat. Hypertension. 2023;80:138–46.
52. Kaldunski ML, Smith JR, Hayman GT, Brodie K, De Pons JL, Demos WM, Gibson AC, Hill ML, Hoffman MJ, Lamers L, Laulederkind SJF, Nalabolu HS, Thorat K, Thota J, Tutaj M, Tutaj MA, Vedi M, Wang SJ, Zacher S, Dwinell MR, Kwitek AE. The Rat Genome Database (RGD) facilitates genomic and phenotypic data integration across multiple species for biomedical research. Mamm Genome. 2022;33:66–80.
53. Kenneth NS, Younger JM, Hughes ED, Marcotte D, Barker PA, Saunders TL, Duckett CS. An inactivating caspase 11 passenger mutation originating from the 129 murine strain in mice targeted for c-IAP1. Biochem J. 2012;443:355–9.
54. Koike H, Arguello PA, Kvajo M, Karayiorgou M, Gogos JA. Disc1 is mutated in the 129S6/SvEv strain and modulates working memory in mice. Proc Natl Acad Sci USA. 2006;103:3693–7.
55. Kuramoto T, Inoue S, Neoda Y, Yamasaki K, Hashimoto R, Mashimo T, Oda S, Serikawa T. Genetic and phenotypic characterization of a Japanese wild-derived DOB/Oda rat strain. Mamm Genome. 2013;24:303–8.
56. Lazar J, Moreno C, Jacob HJ, Kwitek AE. Impact of genomics on research in the rat. Genome Res. 2005;15:1717–28.
57. Linder CC. Genetic variables that influence phenotype. ILAR J. 2006;47:132–40.
58. Livrozet M, Vandermeersch S, Mesnard L, Thioulouse E, Jaubert J, Boffa JJ, Haymann JP, Baud L, Bazin D, Daudon M, Letavernier E. An animal model of type A cystinuria due to spontaneous mutation in 129S2/SvPasCrl mice. PLoS One. 2014;9:e102700.
59. Markel P, Shu P, Ebeling C, Carlson GA, Nagle DL, Smutko JS, Moore KJ. Theoretical and empirical issues for marker-assisted breeding of congenic mouse strains. Nat Genet. 1997;17:280–4.
60. Mattapallil MJ, Wawrousek EF, Chan CC, Zhao H, Roychoudhury J, Ferguson TA, Caspi RR. The Rd8 mutation of the Crb1 gene is present in vendor lines of C57BL/6N mice and embryonic stem cells, and confounds ocular induced mutant phenotypes. Invest Ophthalmol Vis Sci. 2012;53:2921–7.
61. Mattson DL, Dwinell MR, Greene AS, Kwitek AE, Roman RJ, Jacob HJ, Cowley AW Jr. Chromosome substitution reveals the genetic basis of Dahl salt-sensitive hypertension and renal disease. Am J Physiol Renal Physiol. 2008;295:F837–42.
62. Moriwaki K, Shiroishi T, Yonekawa H. Genetics in wild mice: its application to biomedical research. Tokyo: Japan Scientific Societies Press; 1994.
63. Morse HC 3rd. Origins of inbred mice. New York: Academic; 1978.
64. Nadeau JH, Singer JB, Matin A, Lander ES. Analysing complex genetic traits with chromosome substitution strains. Nat Genet. 2000;24:221–5.
65. Paigen K, Eppig JT. A mouse phenome project. Mamm Genome. 2000;11:715–7.
66. Panthier JJ, Montagutelli X. [The Collaborative Cross, a groundbreaking tool to tackle complex traits]. Med Sci (Paris). 2012;28:103–8.
67. Peissel B, Zaffaroni D, Pazzaglia S, Manenti G, Zanesi N, Zedda I, Rebessi S, Covelli V, Dragani TA, Saran A. Use of intercross outbred mice and single nucleotide polymorphisms to map skin cancer modifier loci. Mamm Genome. 2001;12:291–4.
68. Petrij F, Van Veen K, Mettler M, Bruckmann V. A second acromelanistic allelomorph at the albino locus of the Mongolian gerbil (*Meriones unguiculatus*). J Hered. 2001;92:74–8.
69. Poiley SM. A systematic method of breeder rotation for non-inbred laboratory animals colonies. Proc Anim Care Panel. 1960;10:159.
70. Poltorak A, He X, Smirnova I, Liu MY, Van Huffel C, Du X, Birdwell D, Alejos E, Silva M, Galanos C, Freudenberg M, Ricciardi-Castagnoli P, Layton B, Beutler B. Defective LPS signaling in C3H/HeJ and C57BL/10ScCr mice: mutations in Tlr4 gene. Science. 1998;282:2085–8.
71. Potter M. History of the BALB/c family. Curr Top Microbiol Immunol. 1985;122:1–5.

72. Potter M, Wax JS, Blankenhorn E. BALB/c subline differences in susceptibility to plasmacytoma induction. Curr Top Microbiol Immunol. 1985;122:234–41.
73. Radaelli E, Santagostino SF, Sellers RS, Brayton CF. Immune relevant and immune deficient mice: options and opportunities in translational research. ILAR J. 2018;59:211–46.
74. Rader K. Making mice: standardizing animals for American Biomedical Research, 1900–1955. Princeton: Princeton University Press; 2004.
75. Rapp KG. HAN-rotation, a new system for rigorous outbreeding. Z Versuchstierkd. 1972;14:133–42.
76. Rogner UC, Avner P. Congenic mice: cutting tools for complex immune disorders. Nat Rev Immunol. 2003;3:243–52.
77. Rulicke T, Montagutelli X, Pintado B, Thon R, Hedrich HJ. FELASA guidelines for the production and nomenclature of transgenic rodents. Lab Anim. 2007;41:301–11.
78. Saul MC, Philip VM, Reinholdt LG, Center for Systems Neurogenetics of, A, Chesler EJ. High-diversity mouse populations for complex traits. Trends Genet. 2019;35:501–14.
79. Schmitt AO, Bortfeldt R, Neuschl C, Brockmann GA. RandoMate: a program for the generation of random mating schemes for small laboratory animals. Mamm Genome. 2009;20:321–5.
80. Schuster-Gossler K, Lee AW, Lerner CP, Parker HJ, Dyer VW, Scott VE, Gossler A, Conover JC. Use of coisogenic host blastocysts for efficient establishment of germline chimeras with C57BL/6J ES cell lines. BioTechniques. 2001;31:1022–4, 1026.
81. Serikawa T, Mashimo T, Takizawa A, Okajima R, Maedomari N, Kumafuji K, Tagami F, Neoda Y, Otsuki M, Nakanishi S, Yamasaki K, Voigt B, Kuramoto T. National BioResource Project-Rat and related activities. Exp Anim. 2009;58:333–41.
82. Serreze DV, Chapman HD, Varnum DS, Hanson MS, Reifsnyder PC, Richard SD, Fleming SA, Leiter EH, Shultz LD. B lymphocytes are essential for the initiation of T cell-mediated autoimmune diabetes: analysis of a new "speed congenic" stock of NOD.Ig mu null mice. J Exp Med. 1996;184:2049–53.
83. Shimoyama M, Smith JR, Bryda E, Kuramoto T, Saba L, Dwinell M. Rat genome and model resources. ILAR J. 2017;58:42–58.
84. Simon MM, Greenaway S, White JK, Fuchs H, Gailus-Durner V, Wells S, Sorg T, Wong K, Bedu E, Cartwright EJ, Dacquin R, Djebali S, Estabel J, Graw J, Ingham NJ, Jackson IJ, Lengeling A, Mandillo S, Marvel J, Meziane H, Preitner F, Puk O, Roux M, Adams DJ, Atkins S, Ayadi A, Becker L, Blake A, Brooker D, Cater H, Champy MF, Combe R, Danecek P, Di Fenza A, Gates H, Gerdin AK, Golini E, Hancock JM, Hans W, Holter SM, Hough T, Jurdic P, Keane TM, Morgan H, Muller W, Neff F, Nicholson G, Pasche B, Roberson LA, Rozman J, Sanderson M, Santos L, Selloum M, Shannon C, Southwell A, Tocchini-Valentini GP, Vancollie VE, Westerberg H, Wurst W, Zi M, Yalcin B, Ramirez-Solis R, Steel KP, Mallon AM, De Angelis MH, Herault Y, Brown SD. A comparative phenotypic and genomic analysis of C57BL/6J and C57BL/6N mouse strains. Genome Biol. 2013;14:R82.
85. Simpson EM, Linder CC, Sargent EE, Davisson MT, Mobraaten LE, Sharp JJ. Genetic variation among 129 substrains and its importance for targeted mutagenesis in mice. Nat Genet. 1997;16:19–27.
86. Sloin HE, Bikovski L, Levi A, Amber-Vitos O, Katz T, Spivak L, Someck S, Gattegno R, Sivroni S, Sjulson L, Stark E. Hybrid offspring of C57BL/6J mice exhibit improved properties for neurobehavioral research. eNeuro. 2022;9:ENEURO.0221-22.2022.
87. Smits BM, Cuppen E. Rat genetics: the next episode. Trends Genet. 2006;22:232–40.
88. Snell GD. Methods for the study of histocompatibility genes. J Genet. 1948;49:87–108.
89. Specht CG, Schoepfer R. Deletion of the alpha-synuclein locus in a subpopulation of C57BL/6J inbred mice. BMC Neurosci. 2001;2:11.
90. Steensma DP, Kyle RA, Shampo MA. Abbie Lathrop, the "mouse woman of Granby": rodent fancier and accidental genetics pioneer. Mayo Clin Proc. 2010;85:e83.
91. Stevens JC, Banks GT, Festing MF, Fisher EM. Quiet mutations in inbred strains of mice. Trends Mol Med. 2007;13:512–9.
92. Stolina M, Luo X, Dwyer D, Han CY, Chen R, Zhang Y, Xiong Y, Chen Y, Yin J, Shkumatov A, Ason B, Hale C, Veniant MM. The evolving systemic biomarker milieu in obese ZSF1 rat model of human cardiometabolic syndrome: characterization of the model and cardioprotective effect of GDF15. PLoS One. 2020;15:e0231234.
93. Svenson KL, Gatti DM, Valdar W, Welsh CE, Cheng R, Chesler EJ, Palmer AA, McMillan L, Churchill GA. High-resolution genetic mapping using the Mouse Diversity outbred population. Genetics. 2012;190:437–47.
94. Takada T, Ebata T, Noguchi H, Keane TM, Adams DJ, Narita T, Shin IT, Fujisawa H, Toyoda A, Abe K, Obata Y, Sakaki Y, Moriwaki K, Fujiyama A, Kohara Y, Shiroishi T. The ancestor of extant Japanese fancy mice contributed to the mosaic genomes of classical inbred strains. Genome Res. 2013;23:1329–38.
95. Takada T, Fukuta K, Usuda D, Kushida T, Kondo S, Kawamoto S, Yoshiki A, Obata Y, Fujiyama A, Toyoda A, Noguchi H, Shiroishi T, Masuya H. MoG+: a database of genomic variations across three mouse subspecies for biomedical research. Mamm Genome. 2022;33:31–43.
96. Talbot CJ, Nicod A, Cherny SS, Fulker DW, Collins AC, Flint J. High-resolution mapping of quantitative trait loci in outbred mice. Nat Genet. 1999;21:305–8.
97. Tam WY, Cheung KK. Phenotypic characteristics of commonly used inbred mouse strains. J Mol Med (Berl). 2020;98:1215–34.

98. Tanave A, Imai Y, Koide T. Nested retrotransposition in the east Asian mouse genome causes the classical nonagouti mutation. Commun Biol. 2019;2:283.
99. Threadgill DW, Churchill GA. Ten years of the collaborative cross. G3 (Bethesda). 2012;2:153–6.
100. Tucker PK, Lee BK, Lundrigan BL, Eicher EM. Geographic origin of the Y chromosomes in "old" inbred strains of mice. Mamm Genome. 1992;3:254–61.
101. Tuttle AH, Philip VM, Chesler EJ, Mogil JS. Comparing phenotypic variation between inbred and outbred mice. Nat Methods. 2018;15:994–6.
102. Twigger SN, Pruitt KD, Fernandez-Suarez XM, Karolchik D, Worley KC, Maglott DR, Brown G, Weinstock G, Gibbs RA, Kent J, Birney E, Jacob HJ. What everybody should know about the rat genome and its online resources. Nat Genet. 2008;40:523–7.
103. Vedi M, Smith JR, Thomas Hayman G, Tutaj M, Brodie KC, De Pons JL, Demos WM, Gibson AC, Kaldunski ML, Lamers L, Laulederkind SJF, Thota J, Thorat K, Tutaj MA, Wang SJ, Zacher S, Dwinell MR, Kwitek AE. 2022 updates to the rat genome database: a findable, accessible, interoperable, and reusable (FAIR) resource. Genetics. 2023;224:iyad042.
104. Wakeland E, Morel L, Achey K, Yui M, Longmate J. Speed congenics: a classic technique in the fast lane (relatively speaking). Immunol Today. 1997;18:472–7.
105. Waterston RH, Lindblad-Toh K, Birney E, Rogers J, Abril JF, Agarwal P, Agarwala R, Ainscough R, Alexandersson M, An P, Antonarakis SE, Attwood J, Baertsch R, Bailey J, Barlow K, Beck S, Berry E, Birren B, Bloom T, Bork P, Botcherby M, Bray N, Brent MR, Brown DG, Brown SD, Bult C, Burton J, Butler J, Campbell RD, Carninci P, Cawley S, Chiaromonte F, Chinwalla AT, Church DM, Clamp M, Clee C, Collins FS, Cook LL, Copley RR, Coulson A, Couronne O, Cuff J, Curwen V, Cutts T, Daly M, David R, Davies J, Delehaunty KD, Deri J, Dermitzakis ET, Dewey C, Dickens NJ, Diekhans M, Dodge S, Dubchak I, Dunn DM, Eddy SR, Elnitski L, Emes RD, Eswara P, Eyras E, Felsenfeld A, Fewell GA, Flicek P, Foley K, Frankel WN, Fulton LA, Fulton RS, Furey TS, Gage D, Gibbs RA, Glusman G, Gnerre S, Goldman N, Goodstadt L, Grafham D, Graves TA, Green ED, Gregory S, Guigo R, Guyer M, Hardison RC, Haussler D, Hayashizaki Y, Hillier LW, Hinrichs A, Hlavina W, Holzer T, Hsu F, Hua A, Hubbard T, Hunt A, Jackson I, Jaffe DB, Johnson LS, Jones M, Jones TA, Joy A, Kamal M, Karlsson EK, et al. Initial sequencing and comparative analysis of the mouse genome. Nature. 2002;420:520–62.
106. Wotjak CT. C57BLack/BOX? The importance of exact mouse strain nomenclature. Trends Genet. 2003;19:183–4.
107. Yalcin B, Fullerton J, Miller S, Keays DA, Brady S, Bhomra A, Jefferson A, Volpi E, Copley RR, Flint J, Mott R. Unexpected complexity in the haplotypes of commonly used inbred strains of laboratory mice. Proc Natl Acad Sci USA. 2004;101:9734–9.
108. Yalcin B, Nicod J, Bhomra A, Davidson S, Cleak J, Farinelli L, Osteras M, Whitley A, Yuan W, Gan X, Goodson M, Klenerman P, Satpathy A, Mathis D, Benoist C, Adams DJ, Mott R, Flint J. Commercially available outbred mice for genome-wide association studies. PLoS Genet. 2010;6:e1001085.
109. Yang H, Bell TA, Churchill GA, Pardo-Manuel De Villena F. On the subspecific origin of the laboratory mouse. Nat Genet. 2007;39:1100–7.
110. Yang H, Wang JR, Didion JP, Buus RJ, Bell TA, Welsh CE, Bonhomme F, Yu AH, Nachman MW, Pialek J, Tucker P, Boursot P, McMillan L, Churchill GA, De Villena FP. Subspecific origin and haplotype diversity in the laboratory mouse. Nat Genet. 2011;43:648–55.
111. Yoshiki A, Moriwaki K. Mouse phenome research: implications of genetic background. ILAR J. 2006;47:94–102.
112. Yoshiki A, Ike F, Mekada K, Kitaura Y, Nakata H, Hiraiwa N, Mochida K, Ijuin M, Kadota M, Murakami A, Ogura A, Abe K, Moriwaki K, Obata Y. The mouse resources at the RIKEN BioResource center. Exp Anim. 2009;58:85–96.
113. Yu X, Gimsa U, Wester-Rosenlof L, Kanitz E, Otten W, Kunz M, Ibrahim SM. Dissecting the effects of mtDNA variations on complex traits using mouse conplastic strains. Genome Res. 2009;19:159–65.
114. Zou F, Gelfond JA, Airey DC, Lu L, Manly KF, Williams RW, Threadgill DW. Quantitative trait locus analysis using recombinant inbred intercrosses: theoretical and empirical considerations. Genetics. 2005;170:1299–311.

2 Genetically Altered Rodents

Martina Crispo, María Noel Meikle, Fabien Delerue, and Fernando Benavides

Abstract

Spontaneous mutations, along with those induced by chemical or physical mutagens, remain crucial as mouse and rat models. Despite this, numerous technologies for generating genetically modified animals have emerged over the years, with mice and rats standing out as premier models. Initial methods involved pronuclear microinjection of DNA fragments into one-cell embryos, resulting in approximately 10% of the offspring carrying the transgene. More recently, advanced techniques like homologous recombination in embryonic stem cells, predominantly in mice, have been developed to yield knockout or knockin animals. These approaches, while more intricate, offer even lower efficiencies in genetic modification. Lentivirus, transposons, and sperm mediated gene transfer were also available during the following years to develop new animal models with higher efficiencies, but it was not until the arrival of endonucleases and more specifically the CRISPR/Cas system that the field of genome editing grew exponentially. This technology allows the researchers to generate genetically modified mice and rats as never imagined before due to its high efficiency, low cost, and ease of design. In this chapter, we will outline both spontaneous and induced mutations, alongside the primary technologies employed for creating genetically modified mice and rats. These methodologies have been in continuous development over the past four decades. In this chapter, the term "genetically altered" (GA) rodents is utilized to encompass genetically engineered, spontaneous, and chemically induced mutations.

Keywords

Spontaneous mutations · Chemical mutagenesis · Transgenesis · Genetic engineering · Genome editing · Repositories · Phenotyping

M. Crispo · M. N. Meikle
Laboratory Animal Biotechnology Unit, Institut Pasteur de Montevideo, Montevideo, Uruguay
e-mail: crispo@pasteur.edu.uy; manomeikle@pasteur.edu.uy

F. Delerue (✉)
Department of Genetics, The University of Texas MD Anderson Cancer Center, Houston, TX, USA
e-mail: fdelerue@mdanderson.org

F. Benavides
Department of Epigenetics and Molecular Carcinogenesis, The University of Texas MD Anderson Cancer Center, Houston, TX, USA
e-mail: fbenavid@mdanderson.org

F. Benavides, A. K. Hansen (eds.), *Rodent Quality Control: Genes and Bugs*, Laboratory Animal Science and Medicine 2, https://doi.org/10.1007/978-3-031-59103-7_2

2.1 Introduction

In this chapter, we present the fascinating evolution of mutant mice and rats, unraveling the variety of genetic modifications that have revolutionized the field of biomedical research. From the serendipitous discovery of spontaneous mutations to the precision of genetically engineered alterations using embryonic stem cells and the cutting-edge CRISPR technology, the genetic landscape of rodent models has been dynamically shaped over the years. The advent of CRISPR technology further transformed the landscape, providing unprecedented precision and efficiency in genetic engineering. We provide an overview of commonly used methods, ranging from the classic random transgenesis to chemically induced mutations, and the state-of-the-art genome editing with nucleases. This chapter not only traces the historical milestones in the production of GA lines but also elucidates the implications of these advancements for laboratory animal veterinarians, technicians, and researchers using these types of models.

2.2 Spontaneous and Chemically Induced Mutations

Mutations take place within the genomes of all living organisms, spanning from minor alterations involving a single base pair to substantial chromosomal reorganizations. These mutations can emerge in either somatic or germ cells at various developmental stages, and get passed down to descendant cells, unless they result in lethality or produce a detrimental selective trait. In instances where mutations arise in the germline, they possess the potential to be inherited by the subsequent generation. Germinal mutations, by definition, generate new alleles that enter the gene pool of the species and contribute to an increase in polymorphism. Most of these new alleles have no effects or effects that do not influence the fitness of the affected individuals and are called "*neutral mutations.*" A small proportion of these new mutations may result in a better adaptation of the animals to their environment. Finally, some mutations have deleterious effects, frequently leading to pathological conditions. In this context, given that most genes in mouse and rat genomes have counterparts in the human genome, it becomes evident that many of the identified mutant alleles are valuable models for investigating human genetic diseases [29, 125].

2.2.1 Types and Examples of Spontaneous Mutations

Mutations can be dominant, semi-dominant (where the trait is milder in heterozygotes than in homozygous mutants), co-dominant (both alleles equally contributing), or recessive. The Nobel laureate Hermann Muller, who did pioneer research on experimental mutagenesis in Drosophila flies using X-rays, proposed a classification of the mutations into five categories based on the effect of the genetic change on gene activity. The first category, the *amorphic* mutations, consisted of those mutations that completely abolish the activity of the gene and were equivalent to *null* or *loss-of-function alleles. Hypomorphic* mutations were associated with reduced activity compared to the wild type allele, while *hypermorphic* mutations showed increased activity. *Neomorphic* mutations were the group of mutations exhibiting a new function, and *antimorphic* alleles were mutations with dominant-negative effects.

Synonymous mutations are those that occur in a DNA sequence and do not result in a change in the corresponding amino acid in the protein encoded by the gene. This is because the genetic code is degenerate, meaning that multiple codons can code for the same amino acid. Therefore, even if the DNA sequence changes, the codon will still specify the same amino acid. Synonymous mutations usually have little to no effect on the function of the protein. They are often referred to as "silent mutations" because they don't alter the protein's primary structure or function [122]. Synonymous mutations are often observed, mainly as single nucleotide variants (SNVs) or single nucleotide polymorphisms (SNPs), in the genomes of inbred mouse and rat strains [126, 139] (See Chap. 1). In essence, all SNPs are SNVs, but not all SNVs are necessarily SNPs. SNPs are single nucleotide changes that are common in a population and are present in at least 1% of that population.

Non-synonymous mutations, on the other hand, are mutations that result in a change in the corresponding amino acid or protein. These mutations can be further divided into *missense* and *nonsense* mutations. Missense mutations occur where a single DNA base pair change leads to the substitution of one amino acid for another. Depending on the specific amino acid change and the location within the protein, missense mutations can have varying effects on the protein's structure and function. Some missense mutations might lead to a loss of function, while others might lead to a gain of function or altered protein behavior. Nonsense mutations are those that introduce a premature stop codon in the DNA sequence. As a result, translation of the mRNA is terminated prematurely, leading to the production of a truncated, usually non-functional protein. Typically, inactive alleles resulting from a stop codon have no phenotypic manifestation when heterozygous, except in the case of haploinsufficiency or parental imprinting of the normal allele. The mRNAs with a premature stop codon are in general rapidly degraded by specific exonucleases, process known as nonsense mediated mRNA decay. Nonetheless, certain nonsense mutations can yield dominant-negative effects, where the truncated protein persists without degradation and disrupts the normal function of the wild type protein [107].

Insertions are mutations resulting from the intercalation of a DNA sequence of variable size into the genome. Instead, alterations where a sequence is missing are called *deletions*. Insertions or deletions can be as small as a single nucleotide or they can expand over several kilobases of DNA, affecting a variable number of genes. *Indels*, a combination of insertion and deletion, is the term generally used to refer to these mutations. Indels have consequences when they occur in an Open Reading Frame (ORF), in the close vicinity of splicing sites, or at DNA binding sites. They can also have phenotypic consequences when happening in non-coding parts of the genome, however, these mutations are more difficult to analyze because the role and importance of non-coding RNAs only emerged recently [96]. When mutations occur in an ORF and have a size greater or lesser than three nucleotides (i.e., a number not divisible by three), they result in frameshift mutations, whose effects are similar to those of the mutations occurring in the splicing sites and are transcribed, in general, into aberrant mRNA molecules. When indels have a size of three or a multiple of three nucleotides, they result in the incorporation of additional amino acids into the protein chain, and their effects are difficult to predict. Duplications are another type of mutation whose effects and consequences are similar to insertions.

The gene encoding the leptin receptor (*Lepr*), along with its numerous alleles, serves as an illustrative case demonstrating point mutations, indels, as well as duplications in both mice and rats. The mouse *Lepr* gene has 16 documented spontaneous mutations in the Mouse Genome Informatics (MGI) database, including the well-known *diabetes* allele ($Lepr^{db}$) [65] (Fig. 2.1). Similarly, in the rat, mutant alleles of the leptin receptor include $Lepr^{cp}$, associated with corpulence, and $Lepr^{fa}$, linked to adiposity [110]. These instances collectively underscore the intricate variations and modifications within the *Lepr* gene across these rodent species.

Spontaneous mutations resulting from the insertion of mobile elements are also observed in mice and rats. For example, the mouse dilute ($Myo5a^{d}$) mutation, a very ancient mutation with several alleles, is the result of the integration of the ecotropic murine leukemia virus *Emv-3* into the myosin VA (*Myo5a*) gene. The *a* (non-agouti) mouse mutation is also the consequence of the insertion of a 5.5 kb virus-like element (VL30) into the first intron of the agouti gene (*ASIP*, agouti signaling protein in humans), which interferes with the transcription process. Finally, the hairless (Hr^{hr}) mutant allele in mice was caused by the insertion of a murine leukemia virus into intron 6 that results in aberrant splicing of the *Hr* gene [1]. Some strategies have been designed to make use of the capacity of transposons to move

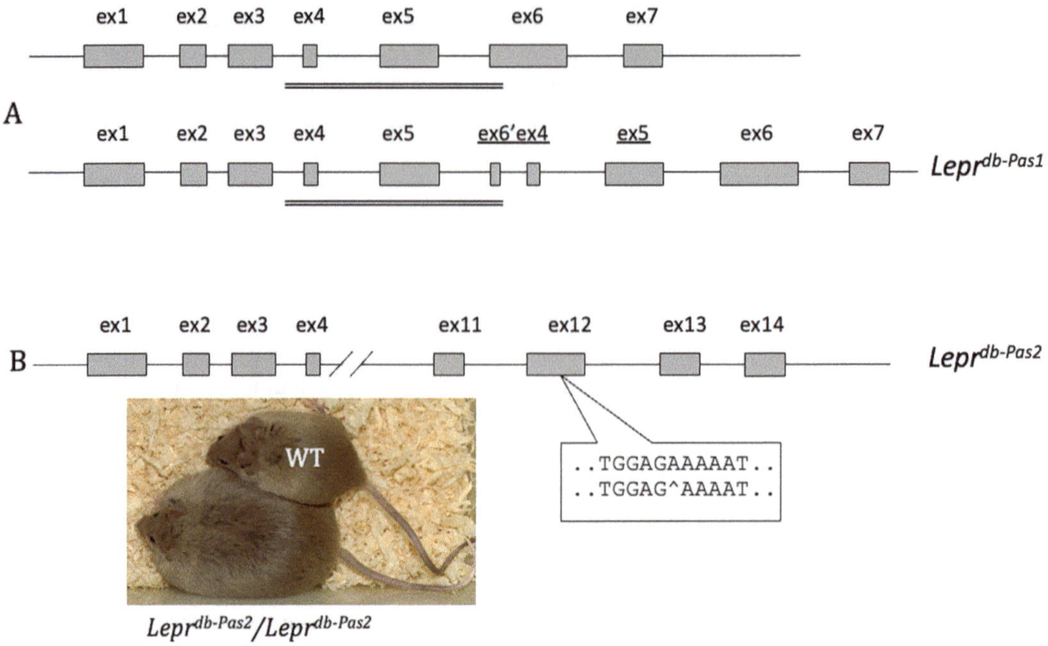

Fig. 2.1 Mutations resulting from duplications and deletions. In the mouse, 16 spontaneous mutations have been reported, so far, for the gene encoding the leptin receptor (*Lepr*). This gene normally consists of 18 exons and has multiple splice variants, comprising at least five isoforms. (**a**) Among these mutant alleles, *Lepr*$^{db\text{-}Pas1}$ is the consequence of a partial duplication that spans the entirety of exons 4 and 5, plus 21 bp of coding exon 6 (as well as the two introns between exons 4 and 6). This produces a null allele that is unable to encode a functional receptor. (**b**) Another spontaneous allele, *Lepr*$^{db\text{-}Pas2}$, is the consequence of a 1-bp deletion producing a frameshift in exon 12, altering a domain of the protein. The mutant allele is inactive, and the mouse becomes obese and diabetic when homozygous (picture) [59]. (Reprinted from Genetics of the Mouse, with permission from Springer Nature)

in the mammalian genome, for the production of new mutations in the mouse and in the rat [11].

A meticulous examination of the mutations (both recessive and dominant) gathered by the scientists at The Jackson Laboratory in the USA has revealed significant differences in the mutation rates at various loci in mice. This phenomenon arises from the fact that numerous mutant alleles go undetected either due to their subtle or severe phenotypes or their late-onset phenotypes. These variations may also be attributed to differences in the sizes of the loci or the presence of multiple exons, offering a broader target for mutagenic events. Nonetheless, these two explanations alone are inadequate to account for certain observed differences. Currently, it is well established that certain genes exhibit an unexpectedly higher mutation rate than the average. For instance, this holds true for the mouse gene encoding the Kit receptor tyrosine kinase (*Kit*), where the MGI database lists 74 instances of spontaneous mutant alleles. Curiously, only one spontaneous mutation of this gene has been documented in rats (*Kit*^{Ws}).

Interestingly, several spontaneous mutations result in marked immunodeficiency, allowing them to accept allogeneic (same species) and xenogeneic (different species) grafts. For example, mice with the *nude* phenotype [50] fail to develop a functional thymus and are thus, deficient in those T cell subsets that mature in the thymus. Several strains of mice homozygous for this *Foxn1*^{nu} mutation (athymic) are available. Similar to mice, rats with mutations in the *Foxn1* gene have been identified, as is the case of the athymic Rowett nude rat (*Foxn1*^{rnu} allele) [62]. Severe combined immunodeficient (SCID) mice are homozygous for the *Prkdc*^{scid} allele, a spontaneous mutation that arose in BALB/c-*Igh*^{b} (C.B-17) mice and lack function of the DNA-activated protein kinase essential for recombination of immunoglobulin and T cell receptor genes [19]. Thus, they produce no mature T and B cells, and lack both humoral and cell-mediated immunity (they do possess NK cells, however). The *Prkdc*^{scid} mutation has been placed on the non-obese diabetic (NOD) inbred background (technically, a congenic strain), which has very low NK cell activity, multiple defects in myeloid development and function, and an absence of C5, which renders NOD mice unable to activate complement. These NOD.CB17-*Prkdc*^{scid} (NOD-scid) mice support long-term engraftment of xenotransplants and show greatly reduced leakiness.

Highlighting the significance of spontaneous mutations in mice, The Jackson Laboratory oversees the Mouse Mutant Resource (https://www.jax.org/research-and-faculty/resources/mouse-mutant-resource) where they employ high-throughput DNA sequencing to investigate mice exhibiting deviant phenotypes that emerge spontaneously within their extensive mouse colonies [42, 43]. In conclusion, spontaneous mutations, which are generally identified through the observation of an abnormal phenotype, present several advantages. The first and probably the most important is that they are produced at virtually no cost and are in general freely available. Another advantage is that they have, in general, an obvious phenotype given that they are identified based on observation. Also, spontaneous mutations represent a great variety of molecular events, such as deletions, insertions, and point mutations, generating not only loss-of-function alleles but also hypomorphic and hypermorphic alleles. An excellent example of this situation is a spontaneous model of human citrullinemia type I [106], where two hypomorphic alleles appeared to be much better models of the human syndrome (OMIM 215700) than the KO mice.

2.2.2 Chemical Mutagenesis in the Mouse

Experimental mutagenesis can be "phenotype-driven," when unknown genes are identified based on the phenotypic changes associated with at least one mutant allele [103]. In this case, the structure of the gene affected by the mutation is elucidated afterwards, by positional cloning. Experimental mutagenesis can also be "gene-driven," where mutations are massively and randomly induced and then sought only in pre-selected genes or DNA regions, for example, for the purpose of genome annotation [35, 58].

Chemical mutagenesis consists of exposing progenitors of either sex to a mutagenic agent, with the aim of increasing the occurrence of novel mutations in the progenies of the treated animals. For practical reasons, only male progenitors are exposed to the mutagens because spermatogenesis is a continuous process, starting at puberty and lasting several months or even years. Because spermatogenesis is a continuous and precisely timed process, one can calculate the precise stage of development of a specific germ cell, at the time of exposure to a mutagen, depending on the time elapsed between the treatment and the fertile mating. When the stem cells of spermatogenesis (i.e., the spermatogonia) are successfully mutagenized, the male becomes a permanent provider of mutations. When males receive a mutagenic treatment, the number of affected stem cells depends on the dose. If the dose is too high, most spermatogonia are killed and the male becomes permanently sterile. Conversely, if the dose is too low, the lethal effect is limited but the mutation rate is low, and the experiment might not be successful. Selecting the best dose is very important and may require preliminary experiments [56].

In 1979, William Russell from the Oak Ridge National Laboratory reported that a simple alkylating agent, N-ethyl-N-nitroso-urea (ENU), had considerable mutagenic power [116]. These observations had a major impact on genetic research and are considered as an important milestone in the history of mouse genetics. The mutagenic activity of ENU results from its capacity to transfer an ethyl group to oxygen or nitrogen radicals in the DNA molecule, inducing mispairing and ultimately leading to base pair substitutions or deletions [69, 123]. In the vast majority of cases, adenine (A) is the primary target of ENU activity, resulting in either A-T to T-A transversions or A-T to G-C transitions as the main genetic alterations [69]. In their initial paper, W. Russell and colleagues [116] reported a mutation rate 90 times higher than the spontaneous rate.

ENU mutagenesis enables the creation of a diverse range of novel mutant alleles for protein-coding genes, each proving invaluable data for genome annotation. The only drawback to consider is that mutagens act randomly, necessitating an extensive and costly identification process. Several large ENU mutagenesis programs have been conducted in Germany, England, and the USA [63, 68, 89, 103], using BALB/c, C3H/He, or C57BL/6 strains. In the USA, *Mutagenetix* (https://mutagenetix.utsouthwestern.edu/) is a database of phenotypes and mutations produced through random germline mutagenesis with ENU [137]. ENU mutagenesis is also possible in the rat [87], however, was not extensively used in this species.

2.2.2.1 Phenotype-Driven (Genome-Wide) ENU Mutagenesis

Bode and colleagues [16] were among the first to use a phenotype-driven ENU mutagenesis program to produce relevant animal models of phenylketonuria (PKU—OMIM 261640). A phenotype-driven mutagenesis program consists of four successive steps. First, males (G0) are treated with the mutagen and, once they have recovered from the sterility period, mated with females. Unusual phenotypes (often called *phenodeviants*) that could result from dominant mutations are then screened by careful examination of the offspring from this first cross (G1 population). In the second step, G1 males, which are all potential carriers of recessive mutations at a number of unknown loci, are gathered for the establishment of individual micro-pedigrees. For this, each G1 male is mated to a few females, either of the same or from a different strain, and a sample of G2 females, offspring of this cross, is selected and backcrossed to their father to produce a G3 population. This G3 generation is then carefully examined for the detection of possible recessive mutations (Fig. 2.2). The rigorous and systematic examination of the G3 progeny is part of the phenotyping process.

Many interesting mutations have been discovered using these protocols [5]. Among the numerous mutations identified are *Clock*^{m1Jt}, a semidominant mutation altering the circadian rhythm in affected

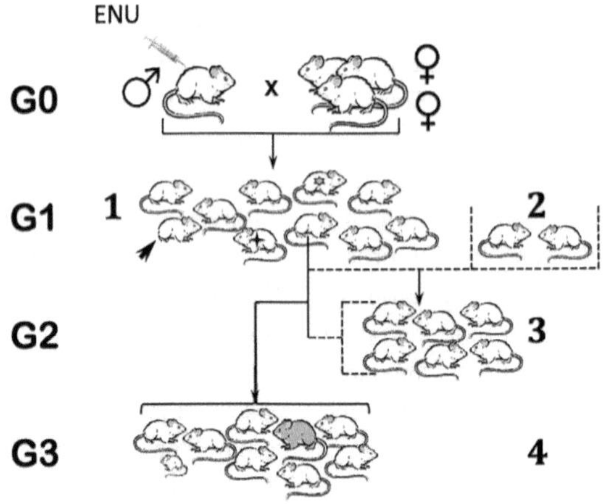

Fig. 2.2 Phenotype-driven genome-wide mutagenesis. Phenotype-driven mutagenesis consists of four successive steps. In the first step, males are treated with the powerful mutagen ENU and mated with 2–3 females after recovery from a 10 to 13-week sterility period (G0 generation). The entire G1 progeny is then carefully scrutinized, looking for possible dominant mutations (arrow). In the second step, males of the G1 generation (which are potential heterozygous carriers of recessive mutations of all kinds) are selected for the establishment of micro-pedigrees. First, they are mated with females of either the same or a different strain, and 4–6 female offspring (G2) are backcrossed to their G1 father. Finally, the progenies of the G1 male crossed with the G2 female offspring (G3) are subjected to careful phenotypic examination. Micro-pedigrees producing mutant phenotypes are then isolated for in-depth analysis. The number of G2 females and their G3 offspring are established after statistical computation to optimize the possibility of detection of new phenotypes [59]. (Reprinted from Genetics of the Mouse, with permission from Springer Nature)

mice [132] and the codominant *Ticam1*^{Lps2}, a mutation leading to compromised defense mechanisms against viral and bacterial diseases [61]. In a European project focusing on deafness syndromes, no less than 13 new independent genes involved in inner ear differentiation and pathology were identified by ENU mutagenesis [78].

Forward genetics is the bottom-up methodology used for these studies, progressing from the phenotype to the genotype [59]. Within this strategy, animals showcasing phenotypes of interest due to spontaneous or induced mutations serve as the foundational material. Subsequent to the identification of these individuals, the process necessitates meticulous breeding and the construction of a genetic map. Positional cloning, a specific method within the realm of forward genetics, is the intricate procedure of pinpointing a gene's location within the genome, devoid of any prior knowledge about its functional characteristics [7]. It involves narrowing down the location of a gene responsible for a particular phenotype by studying the inheritance of genetic markers in specific mouse or rat crosses. Once the region containing the gene is identified, researchers can isolate and study the candidate genes in that region to determine which one is causing the observed phenotype. An illustrative historical instance of positional cloning involves the identification of leptin as the gene responsible for the obese mutation (*Lep*^{ob}) [140].

2.2.2.2 Gene-Driven Chemical Mutagenesis

As previously mentioned, gene-driven ENU mutagenesis is an experimental genetic method employed to induce mutations in specific genes, aiming to study their functions and the resulting impact on phenotype [52]. The strategy generally consists of four steps. First, adult males of an inbred strain are treated with an appropriate dose of ENU and then mated with female of the same inbred strain for the production of a large G1 population. In the second step, sperm samples are collected from adult G1 offspring of this initial cross and cryopreserved for performing future in vitro fertilization (IVF). Simultaneously, genomic DNA samples from the same G1 males are prepared, cross-referenced with the sperm samples, and stored. The third step consists of the analysis of the DNA sequence of all G1 mice, looking for any structural changes that may have occurred in a selected gene or a well-delimited region of the genome. This can be accomplished through next-generation sequencing, which involves identifying all nucleotide mismatches following the pooling of DNA samples. Once a mutation is identified and marked as potentially significant (nonsense or missense mutations with predicted deleterious effects), the fourth and final step is initiated. The sperm cell sample from the potentially intriguing mutant mouse is defrosted, and oocytes of the identical strain are fertilized in vitro. These fertilized oocytes are then implanted into pseudo-pregnant females. Upon birth, the potentially heterozygous offspring are bred and subjected to crosses to generate homozygous progeny, whose resulting phenotype is subsequently observed. In this micro-pedigree, the molecular analysis of the progeny stands as a crucial step (Fig. 2.3).

This gene-driven protocol allows the production of all types of mutations (not limited to loss-of-function) in all regions of the genome (coding and non-coding). A drawback is the difficulty and time required for identifying the mutations in the targeted regions. Once identified in the batch of frozen sperm cells, a mutation can be retrieved and made available as heterozygous adult mice. Several laboratories have published interesting results using a gene-driven approach in mice [51, 113] and rats [88].

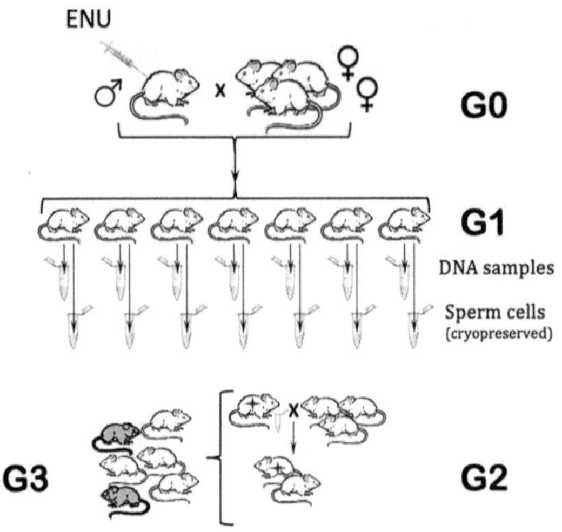

Fig. 2.3 Genotype-driven mutagenesis. Male mice are treated with ENU and mated to females (preferably of the same inbred strain) once they have recovered from the sterile period (G0). A large number of G1 males, which all are heterozygous carriers of a great number of random independent point mutations (mostly base pair changes), are then bred. Sperm samples from each G1 mouse are collected and cryopreserved, while DNA samples from the same mice are processed and stored with the same reference. Identification of the mutations generated by the ENU treatment in a specific target (a gene or any other specific sequence) is carried out by molecular techniques to identify DNA mismatches, or directly by sequencing. Once the base pair changes are identified and considered potentially interesting (stop codons, missense, etc.), the corresponding sperm cells are thawed, and heterozygous mice are produced by IVF with oocytes of the same background strain. A major advantage of this method is that it produces all types of point mutations, not only knockouts. A drawback is the difficulty and time required for identifying the mutations in the targeted region. With the rapid expansion of new sequencing techniques, the identification step should be somewhat easier [59]. (Reprinted from Genetics of the Mouse, with permission from Springer Nature)

2.3 Genetically Engineered Mutations

Since the first reports of genetically modified mice through pronuclear microinjection of DNA fragments in the early 80s [53, 54], thousands of models have been created worldwide. The first knockout (KO) mouse models were developed in the late 1980s and early 1990s using embryonic stem (ES) cells [75]. Subsequently, knockin (KI) models were introduced, allowing the insertion of specific sequences or mutations into the gene of interest rather than deactivating the gene. The low efficiency (i.e., 10%) and random integration of the transgene as well as the complexity of ES cells culture for homologous recombination and generation of chimeric mice made these technologies challenging. These hurdles include prolonged generation times and the requirement for highly skilled technicians to achieve satisfactory results. The emergence of genome editing tools, particularly the CRISPR/Cas systems, made the technology rapidly available to more facilities. This has pushed various fields, including biomedicine, to unprecedented advancements [37, 67]. Nowadays it is easier, faster, cheaper, and more efficient to obtain sophisticated rodent models. Continuous enhancements in technology are consistently reported, contributing to the ongoing evolution of these methods on a weekly basis [97, 136]. In addition, the possibility to electroporate the CRISPR reagents into multiple zygotes at once instead of microinjecting each zygote individually makes the system even more accessible to facilities and research laboratories [111]. The CRISPR discovery was deemed worthy of the 2020 Nobel Prize to Jennifer Doudna and Emmanuelle Charpentier "for the development of a method for genome editing" and is considered one of the major breakthroughs of the decade in this field, similar to somatic cell nuclear transfer in the 90s, democratizing the science to the whole scientific community.

2.3.1 Transgenesis by Pronuclear Microinjection

The first approaches to generate mutant mice were published in the early 1980s when pioneer researchers explored the direct microinjection of exogenous DNA (i.e., the *transgene*) into the pronuclei of early stage mouse embryos, with very low success rates [21, 22, 54]. Several subsequent attempts refined the technique, culminating in the birth of the first transgenic mice. Subsequently, transgenic mice displaying a straight phenotype were observed [23, 53, 105]. These were considered milestones for this technology, and thousands of mice have since been generated. The generation of transgenic rats has been achieved for the first time in 1990 [99]. Since then, many transgenic rats have been generated [49].

The technique consists basically in the microinjection of a small DNA fragment or plasmid into the pronucleus of 1-cell embryos. This transgene will be integrated randomly into the genome of the developing embryo and generate transgenic offspring. Typically, specific DNA sequences called promoters are used to drive the expression of the transgene in a particular tissue or organ as a *gain of function* model. Although conceptually simple, the technology is difficult to master and must be complemented by other assisted reproductive techniques such as hormonal donor superovulation, embryo harvest, in vitro culture, and transfer of microinjected zygotes into synchronized recipients to achieve acceptable results both in mice and rats. The technique has been described elsewhere for both species [33, 38, 49, 93, 102], requiring expensive equipment and highly skilled technicians (Fig. 2.4).

The recommended generic symbol for a transgenic insertion is Tg. The founder transgenic ani-

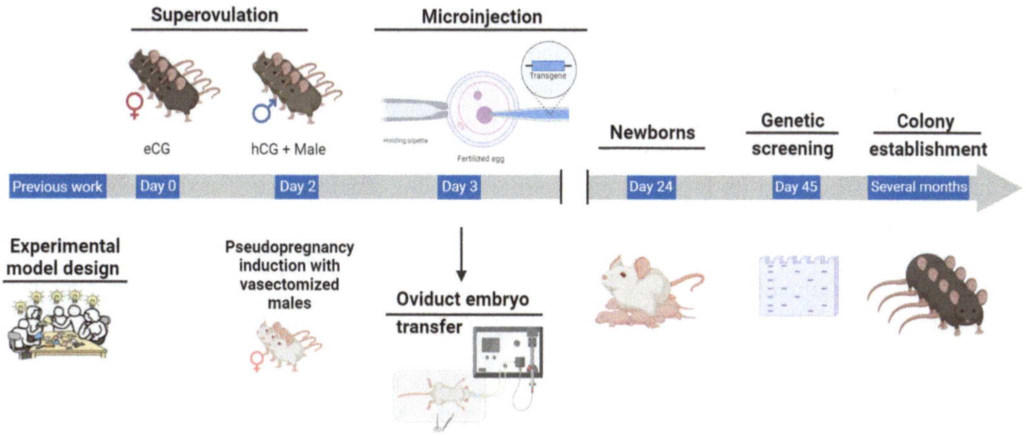

Fig. 2.4 Producing transgenic mice. The flowchart represents the steps for the production of transgenic mice by pronuclear microinjection. One-cell embryos are recovered from the oviduct and the transgene is microinjected with a glass micropipette into one of the pronucleus (typically the male pronucleus). Once injected, the embryos are kept in vitro for a few hours and then transplanted into pseudo-pregnant females (previously mated with vasectomized males). Genotyping of the G0 (presumptive) transgenic mice can be achieved after weaning. Every pup genotyped as positive by PCR (i.e., hemizygous Tg/0 carrier) or expressing a reporter protein (e.g., GFP) should be considered a "founder", and independent lines should be developed from each founder. *eCG* equine chorionic hormone, *hCG* human chorionic gonadotropin. (Created with BioRender.com)

mals are hemizygous for the DNA segment and are designated Tg/0.[1] Each transgenic line generated via random integration creates a unique animal model and each putative founder must be bred independently. Establishing a transgenic line, in which the transgene is propagated by mating, requires genotyping each generation to detect to which offspring the transgene was transmitted, unless the carriers have an obvious phenotype. Lines are normally kept by backcrossing transgenic carriers (hemizygous Tg/0) with wild type animals from the inbred background strain and by selecting carriers at each generation. When viability and fertility are unaffected, a transgene may be maintained by keeping transgenic lines in the homozygous state.

Traditionally, to distinguish between homozygous (Tg/Tg) and hemizygous (Tg/0) mice, the mouse of interest is crossed to a non-transgenic partner and the progeny is statistically analyzed for Mendelian segregation of the transgene. Quantitative real-time PCR (qPCR) can be used to distinguish hemizygous from homozygous transgenic mice [9]. In order to achieve a pure genetic background (recommended), the transgene must be introduced into embryos derived from an inbred strain, such as FVB/N, which is widely used because its zygotes possess large and prominent male pronuclei, and the females are excellent breeders that produce large litters [128]. However, care should be taken when selecting the background strain and its intended research area. If the background utilized during production is inappropriate, a costly process of backcrossing may be required (e.g., FVB/N mice are not appropriate for many behavioral studies because they are blind). As often in science, the C57BL/6J background serendipitously became the most commonly used mouse inbred strain, yet this strain is also the most challenging to produce transgenic mice [6].

[1]Transgenes are extra segments of DNA that have no corresponding "wild type" sequence in the unmodified homologous chromosome in hemizygous animals, which is why it is recommended to use of "0" instead of "+" (typically used to denote wild type alleles).

A later improvement on the constructs (i.e., transgenes) used in the transgenesis approach was the introduction of *inducible systems* in which transgene expression can be turned on and off. Examples of this strategy are the Tet-on and Tet-off expression systems. In these systems, transcription of a given transgene is placed under the control of a tetracycline-controlled transactivator protein, which can be regulated, both reversibly and quantitatively, by exposing the transgenic mice to either *Tetracycline* or one of its derivatives, such as *Doxycycline*. Both Tet-on and Tet-off are binary systems that require the generation of double transgenic (*bigenic*) mice. These mice carry both a responder construct consisting of a tetracycline response element regulated transgene and an effector construct (tTA or rtTA) containing a tetracycline-controlled transactivator. This technology was developed by Bujard and colleagues at the University of Heidelberg [10, 55].

Although a technology of choice for several decades and still in use, transgenesis is an additive approach that has several major drawbacks including low efficiency, random integration, a high degree of mosaicism, and variable transgene expression patterns. Several of these limitations have been improved upon throughout the years, such as including better transcription factors or lowering DNA toxicity among others. Nevertheless, alternative methodologies for improving efficiency such as the use of lentivirus [85] or transposons [90] were developed, yet the randomness of the integration into the genome remained a major limitation that has since been overcome by targeted mutagenesis.

2.3.2 Targeted Mutagenesis by Homologous Recombination Using ES Cells

Homologous recombination is a natural biological process that plays a crucial role in the maintenance and repair of DNA in living organisms. In the context of ES cells and chimeric mice, homologous recombination has been extensively utilized in genetic engineering and gene targeting

[8]. During the early 1980s, and almost in parallel with the development of the transgenesis technology, a group of pioneer researchers discovered the existence of pluripotent cells in the inner cell mass of mouse embryos at the blastocyst stage [46]. These ES cells have the remarkable ability to self-renew indefinitely and differentiate into various cell types representing all three germ layers (endoderm, mesoderm, and ectoderm).

These pluripotent mouse ES cells were then isolated and cultured in vitro, and the first experiments to perform gene targeting in a controlled environment were developed. The possibility of introducing specific mutations into cultured ES cells through homologous recombination encouraged the researchers to create genetically modified lines with the desired genetic alterations, making possible to obtain KO (*loss-of-function* model) or KI (change in protein sequence/gain of function) animals in contrast to the classical random insertion transgenesis [20, 115]. To underscore the significance of this technology in the field of biomedicine, Mario Capecchi, Martin Evans, and Oliver Smithies were awarded the 2007 Nobel Prize in Physiology for their groundbreaking contributions to "introducing specific gene modifications in mice by the use of embryonic stem cells" [133].

The modified ES cells have to be selected in vitro and then injected into blastocysts to promote their contribution for the formation of various tissues, resulting in chimeric mice. Chimeric mice are composed of cells derived from two different embryos: the wild type cells from the host embryo (blastocyst) and the genetically modified cells from the injected ES cells [60]. Chimeric mice harboring the modified ES cells in their germline will eventually pass on the genetic modifications to their offspring, providing that the ES cells contributed to the germline (Fig. 2.5). This process is crucial for establishing stable mutant lines with precise genetic alterations. Such lines serve as valuable models for human diseases [40].

Historically, the first mouse ES cells were derived from embryos of the 129 family (129S2, 129P3, etc.). Today ES cell lines are available from many mouse strains, with those originating from C57BL/6 N being widely adopted for multinational projects (e.g., EUCOMM). On the

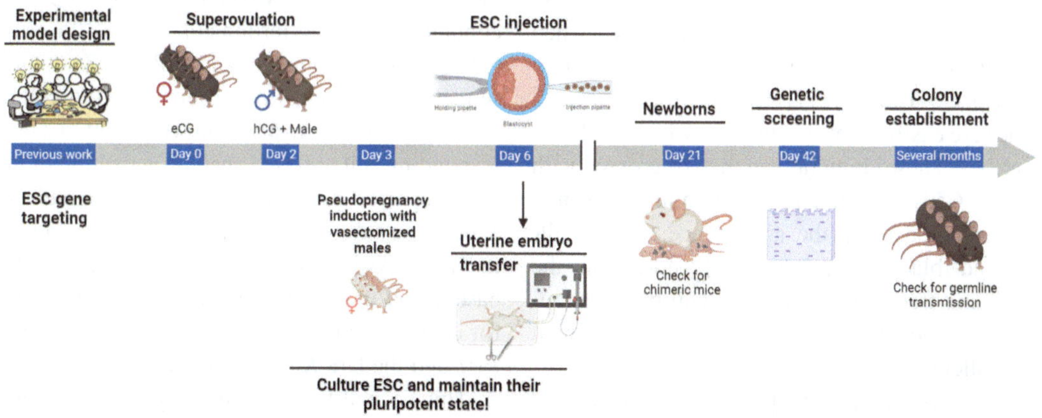

Fig. 2.5 Targeted mutagenesis in the mouse using engineered ES cells. The flowchart represents the different steps for the production of targeted mutants (KO and KI) using genetically modified ES cells. ES cells can be purchase from repositories with the desired mutation or made in the laboratory. Pluripotent ES cells can be cultured in vitro, for several generations, remaining in an undifferentiated state. While in vitro, the ES cells can be manipulated and selected on the basis of specific criteria. Then, 10–15 ES cells are typically injected per blastocyst where they merge with the inner cell mass. After embryo transfer into the uterus of a pseudo-pregnant female, and provided that the ES cells are still pluripotent, fertile chimeric mice can result from these reconstructed blastocysts. The chimeras with the best level of chimerism are then crossed with wild type mice in order to confirm germ line transmission, producing mice carrying the targeted allele as heterozygous (one extra generation is then necessary to obtain the alteration in the homozygous state). (Created with BioRender.com)

other hand, it was not until the late 2000s that the isolation of rat ES cells could be achieved and few KO rats were created using this technology [28, 71, 80].

Chimeric mice, created by combining engineered ES cells with cells from the inner cell mass of a recipient blastocyst, can be recognized by their variegated coat color shortly after birth. This distinction is particularly noticeable when the ES cells originate from C57BL/6N (non-agouti a/a, displaying a black appearance), and the recipient blastocyst is derived from either a wild type (agouti *A/A*) or albino (*Tyr^c/Tyr^c*) strain. In these conditions, the chimeras exhibit a mixture of black and agouti or albino spots. Using coat color as a reference, one can estimate the degree of chimerism, but a high level of chimerism does not necessarily parallel with a high rate of germline transmission. The majority of ES cell lines are XY, therefore, although chimeras can be from either sex, the number of males is usually unproportionally high within high chimeric litters and chimeric males are generally the only sex with germline transmission. In order to avoid mixed background lines, co-isogenic KO/KI lines should be generated by crossing the chimeras with wild type mice from the same background as the ES cells. For example, when C57BL/6-derived ES cells are injected into albino C57BL/6-*Tyr^c/Tyr^c* blastocysts, chimeric mice are easily identified by their white and black patches. These chimeras can then be crossed with albino C57BL/6 mice to test germline transmission, validated by the appearance of ES cell-derived black offspring [119] (note that the germline transmitted offspring will be heterozygous for the targeted allele as well as the mutant *Tyr^c* allele).

In cases where constitutive KO (null) alleles lead to early death (i.e., essential genes), *conditional alleles* may be generated, allowing one to control the time and tissue where a gene is turned off, typically using the Cre/LoxP system [117]. Production of *conditional KOs* requires two independent lines: one providing a source of Cre recombinase, an enzyme derived from bacteriophage P1, in the tissue under study, and the other containing LoxP (*l*ocus *o*f *X*-ing over *P*1) sites flanking the DNA segment of interest. These two lines need to be crossed to generate double mutant mice in order to generate a conditional model. When a mouse carries a copy of the Cre sequence and is homozygous for the gene of interest, which is "floxed" (flanked by LoxP sites), the Cre enzyme will cut and recombine the floxed sequence at the LoxP sites in the tissue where Cre is expressed. This process results in the generation of a tissue-specific KO model [129].

The Cre transgene can be made inducible, adding more sophistication to the system. The tamoxifen-inducible Cre^{ERT2}, which can be activated in a spatio-temporal manner by administration of tamoxifen, is widely used [47]. However, tamoxifen treatment during early pregnancy can impair placental development, resulting in embryonic death and abortion [41]. Many Cre-expressing lines are being produced as KI mice that incorporate the Cre sequence directly into the gene of interest rather than by creating transgenic lines by pronuclear microinjection. The Cre-LoxP strategy can also be used to regulate the expression of reporter genes. For example, the *lac*Z gene can be driven by a ubiquitous promoter (e.g., *Rosa 26*) with a floxed stop sequence, containing several terminator codons inserted between the promoter and the *lac*Z coding sequence (Lox-Stop-Lox system) [121]. A cell or tissue-specific Cre expression will result in a corresponding cell or tissue-specific activation of the *lac*Z reporter gene [134]. Many available Cre expressing strains (Cre-deleters) have been added to The Jackson Laboratory Cre Resources database (http://www.creportal.org/). Other strategies for creating conditional mutants include the Flp-*FRT* and the Dre-*Rox* systems [27, 118].

This ES cell-based technology was a standard approach until the arrival of genome editing through endonucleases in mid-2010s. Nevertheless, homologous recombination is still valid for the introduction or substitution of large DNA fragments into the genome and is still applied in several facilities around the world. Today, the generation of chimeric mice using ES cells remains an optional tool in the field of genetics and biomedical research. For a comprehensive overview of ES cell generation, microinjection, and related assisted reproductive technologies (ART) techniques in mice, please consult [12, 64, 102, 138]. Similarly, for rats, please refer to [44, 79].

2.3.3 Gene Editing Using Nucleases

Although incredibly powerful to generate genetically modified animals, both pronuclear injection and gene targeting using ES cells suffer multiple limitations, both at the technical and the genetic levels. Both technologies require specialized equipment and highly skilled technicians to perform microinjection of zygotes (pronuclear) or blastocysts (ES cells). Besides, transgenesis by pronuclear injection suffers from "position effect." Because the integration of the transgene occurs randomly in the genome, a microinjected transgene is likely to land in a "silent" part of the genome. Indeed, a vast majority (~98%) of the mammalian genome does not encode proteins. As such, a transgene integrated in such "silent" part of the genome may not express the protein for which it encodes. Even when the transgene is expressed, the genetic landscape surrounding it influences the expression level. Hence, researchers need to generate several transgenic lines and check for expression of the transgene, which is time consuming and expensive. On the other hand, gene targeting using ES cells has an intrinsic limitation related to the availability of the cell lines. Although quite common in mice, stable and pluripotent ES cells in other species are not readily available. Besides, germline transmission is not warranted when creating chimeric mice, pushing researchers to continually improve this technology [74].

Engineered endonucleases are a type of modified nucleases that revolutionized the field of gene targeting. Such disruptive technology is based on the ability of such nucleases to scan the entire genome and modify a specific DNA sequence in this genome. The mammalian genome is made of approximately 3 billion base pairs (3 Gbp), and such nucleases have the ability to modify one (or more) with unprecedented accuracy and ease, similar to a text editor able to pinpoint and correct a single misspelled word in a 23-volume encyclopedia in a fraction of second. The basic mechanisms of endonucleases reside in their ability to hybridize (using base pair complementarity) with a short genomic sequence (usually 20–30 bases) and induce a double stranded break (DSB) at this precise location. The DSB stimulates the repair mechanisms of the cell to protect it from potentially deleterious mutations. Taking advantage of these repair mechanisms, endonucleases can induce small mutations (e.g., indels) if the repair processes fail to properly resolve the DSB. These indels have the potential to result in the KO of a specific gene. Alternatively, when a DNA template containing sequences homologous to those located on either side of the DSB (referred to as "homologous arms") is delivered, the endonuclease-induced DSB can be repaired using such template by homologous recombination (similar to gene targeting using ES cells), generating a KI (i.e., the targeted integration of a DNA sequence). Historically, the first generation of endonucleases that have been engineered and used to generate genetically modified mice and rats included meganucleases [95], Zinc-Finger nucleases (ZFN) [39] and transcription activator-like effector nucleases (TALEN) [124]. Although efficient, these nucleases rely on modular recognition elements that render their design, synthesis, and assembly quite cumbersome, time consuming and expensive.

2.3.3.1 The CRISPR/Cas System

The Clustered Regularly Interspaced Short Palindromic Repeats (CRISPR) system emerged in the 2010s as a new tool for genome editing. This bacterial immune system has been repurposed for gene targeting in cells, and the whole process was coined "gene editing." Unlike its predecessors, CRISPR relies on a short RNA strand (called "single guide" or sgRNA) that hybridizes with its complementary genomic sequence. The CRISPR associated nuclease (e.g., Cas9) linked to the sgRNA is guided to the matching genomic sequence and induces a DSB, which in turns induces either indels or homologous directed repair (HDR). The fact that CRISPR is an RNA-guided nuclease (unlike its predecessors) renders this technology easy to design and cheap to synthesize quickly. As such, CRISPR has been successfully used in mice for the first time in 2013 with unprecedented efficiency [135]. Since then, thousands of GA mouse and rat lines have been generated during the last decade using CRISPR, and large consortiums are well on their way to knocking out every single coding gene of the mouse genome (see Sect. 2.3 below).

Using CRISPR, researchers around the world are able to generate very subtle (e.g., single base changes) or large genetic modifications (e.g., Kbp insertions, Mbp chromosomal translocations etc.) directly in the mouse zygotes [14, 34, 101]. However, the targeted integration of very large sequences in zygotes remains one of the rare (together with the high level of mosaicism) limitations attached to the CRISPR system, and this limit can only be overcome using ES cells-based gene targeting. Nevertheless, because CRISPR is amenable to the zygote for most genetic modifications, the advent of CRISPR/Cas9 was accompanied with improvements in assisted reproductive technologies (ART) and gene transfer techniques in rodents in the 2010s. Indeed, Qin et al. showed for the first time that CRISPR could be electroporated (rather than microinjected) in mouse zygotes [111] (Fig. 2.6), while Kaneko et al. showed that it was also achievable in rat embryos [70]. This improvement streamlines the process, since it does not require highly skilled technicians and is amenable to many zygotes at once (unlike microinjection, which requires manipulation of the zygotes one at a time). Furthermore, Takahashi et al. showed that GA mice could be generated using CRISPR by electroporation of zygotes in-utero (rather than ex vivo), simplifying further the process [127]. Finally, technologies such as adeno-associated-virus (AAV) have been coupled with electroporation (the so-called CRISPR-READI method) to bring the DNA template inside the zygote and generate large KI in a seamless manner [31].

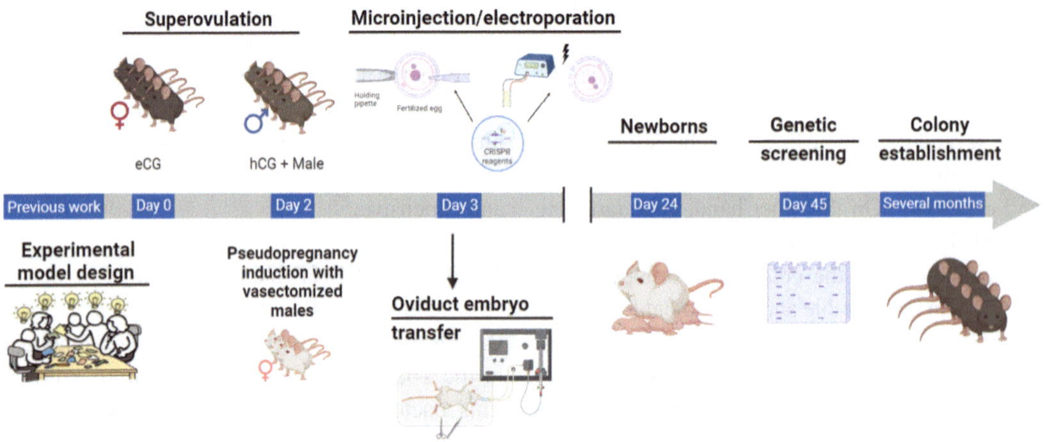

Fig. 2.6 Genome editing using site-specific RNA-guided DNA endonuclease (CRISPR/Cas system). With the CRISPR strategy, Cas9 unwinds the DNA duplex and performs a double strand break after recognition of a specific target by the sgRNA, provided that the correct protospacer adjacent motif (PAM) is present. DSBs are repaired through non-homologous end joining (NHEJ) or through homology directed repair (HDR). In the case of DSBs repaired by NHEJ, the mechanism will induce indels and potentially produce KO alleles. For HDR to occur, a DNA molecule or a single stranded synthetic DNA is to be added as a template. DNA can be electroporated (with size restrictions) or injected into either the cytoplasm or pronuclei of 1-cell, or 2-cell stage embryos, thus avoiding the use of ES cells and chimeras. However, as each engineered animal is unique, this technology requires extensive sequence analysis to characterize multiple putative founders (see Chap. 5) to ensure the presence of the desired mutation and the absence of undesired on- and off-target mutations or unpredictable larger genome alterations, while also identifying mosaic founders (G0). Once identified, the selected founder should be bred with wild type animals to evaluate transmission of the mutation to their offspring. *Created with BioRender.com*

2.3.3.2 Base Editing and Prime Editing

Since its inception, a primary concern to the use of CRISPR is its putative ability to generate "off-target" mutations. Because the recognition sequence is short (usually 20–30 bp), similar sequences containing one (or more) mismatches are frequently found in the genome of the targeted species. The propension of CRISPR to cut at these off-target locations have been assessed in mice. However, it is important to note that CRISPR-induced off-target mutations are rare in mice [66], and the latest reports put the rate of off-target mutations within the context of naturally occurring mutations [108]. Moreover, potential off-target mutations can easily be diluted out by means of breeding in rodents. Yet, the concerns around CRISPR-induced off-target mutations pushed researchers to develop and use alternative technologies, such as base editing and prime editing. These technologies were primarily developed by the laboratory of David Liu at the Broad Institute (Harvard University and Massachusetts Institute of Technology). These third-generation endonucleases differ from CRISPR/Cas9 essentially by their inability to generate a DSB. Instead, both base editing (BE) and prime editing (PE) nucleases induce a single stranded break, referred to as a "nick." Nicking the genomic DNA is considered safer, as it does not involve the natural cell repair mechanisms.

BE nucleases consist of a guide RNA coupled to a "dead" Cas9 (dCas9) (a mutation in the Cas9 coding sequence prevents it from inducing DSB) and "converting" enzymes (e.g., deaminase) able to convert one base to another (e.g., replacing a G with an A) [76]. This "rewriting" of the genomic DNA was further developed to allow for not only any type of conversions, but also small modifications, such as the insertion of LoxP sites (34 bp). PE nucleases are also built with a dCas9; however, they are fused to an engineered reverse transcriptase which uses the guide RNA to produce a piece of DNA that will be synthesized at the nicking site [4]. Both BE and PE have been successfully used to generate GA mice [72, 81, 82]. However, the efficiency and accuracy of the third-generation editors proved lower than the traditional CRISPR/Cas9 system in zygotes [141], and efforts continue to improve these technologies before a wide uptake of these editors be seen in transgenic laboratories around the world [30].

2.3.3.3 Gene Editing in Rats

Although the mouse is the most commonly used animal for genetic modifications (see https://www.understandinganimalresearch.org.uk/), the rat has intrinsic advantages over the mouse. For instance, their cognitive processes and strategies are more similar to those of humans than their mouse counterparts. Consequently, they are extensively utilized in the field of neuroscience, as well as in immunology and cancer research. Many GA rats have been generated as a result [49]. However, as previously mentioned, pluripotent rat ES cells have only been isolated years after the mouse ES cells, allowing for the first KO rats to be generated only in 2008 [80]. Since then, the first-generation endonucleases such as ZFN [39] and TALEN [94, 131], as well as CRISPR/Cas [114] have allowed for direct gene editing in rat zygotes. In fact, the Rat Resource and Research Center (RRRC) generates and archives thousands of GA rat lines using CRISPR.

One of the main restrictions to the use of GA rats is the limited development of ART such as IVF and cryopreservation (see Chap. 6), further preventing the wide uptake of this rodent species for genetic modifications. For instance, although GA rats have recently been produced in utero, similar to mice [73], the rat embryo culture medium usually does not sustain the development to the blastocyst stage, despite recent improvements [92]. Consequently, electroporation of CRISPR into rat zygotes is often used to generate GA rats, as it is easier and faster than microinjection [114].

Besides mice and rats, other GA rodents are slowly been generated using CRISPR technology, for example, Syrian hamsters [130] and prairie voles [13]. Beside rodents, the achievements that the CRISPR technology brought to the field of gene editing are unprecedented. Indeed, there has been no report to date of a single living species that is not amenable to CRISPR gene editing [48].

2.4 International Phenotyping Consortiums, Mutant Repositories, and Databases

As we discussed in this chapter, researchers now have all the means and tools to create a great variety of alterations in the mouse and rat genomes. Many of these alterations are expected to result in changes in phenotype, and the careful analysis of these phenotypic changes is fundamental for the process of genome annotation and model creation. The collection of physical and biochemical traits of an animal is known as the *phenome*, and *phenomics* is the discipline that deals with the measurement of these traits. *Phenotyping* of rodent models has become a main concern over the last decades. Therefore, many laboratories and institutions have developed highly standardized phenotyping protocols. The range of phenotyping platforms, including dual-energy X-ray absorptiometry, electrocardiography, high-resolution imaging, and FACS, ensures the recovery of phenotype data across multiple systems and disease states. In most cases the basic protocols include behavior, neurology, clinical chemistry, development, immunology, energy metabolism, vision and hearing, pain perception, and cardiovascular and gross pathology assessments. The use of standard procedures and defined protocols allows data to be comparable and shareable, even across species, which may help identify mouse and rat models of human diseases [24]. However, phenotyping loss-of-function mutations cannot predict the relevance of these alleles (and their phenotypes) to complex human diseases that are likely driven by several alleles of modest effect [26].

One of the earliest collaborative projects using standard phenotyping procedures was the European Eumorphia project. This program also developed the Europhenome data repository and the European Mouse Phenotyping Resource for Standardized Screens [86]. The Jackson Laboratory has developed a program to collect baseline phenotypic data on the most commonly used inbred strains of mice through a coordinated international effort. Information collected through the *Mouse Phenome Database* (MPD) is freely available to the community through the Internet (http://phenome.jax.org/) [17, 18, 104].

The International Mouse Phenotyping Consortium (IMPC) was established in 2011 with several goals: (i) to maintain and expand a worldwide consortium of institutions with capacity and expertise to produce germ line transmission of targeted KO mutations in ES cells; (ii) to test each mutant mouse line through a broad based primary phenotyping pipeline; (iii) to systematically aim to discover and ascribe biological function to each gene; (iv) to maintain and expand collaborative networks with specialist phenotyping consortia or laboratories; and (v) to provide a centralized data center and portal for free, unrestricted access to primary and secondary data from the scientific community [25, 29, 57, 83, 84, 100, 109]. Some European members of IMPC are the Medical Research Council (Harwell), the Wellcome Trust Sanger Institute (Cambridge), and the European Bioinformatics Institute (Hinxton) in the UK; the Helmholtz-Zentrum Muenchen in Germany; the PHENOMIN (Strasbourg) in France; the CNR (Montorotondo) in Italy; the Czech Centre for Phenogenomics in the Czech Republic; and the Universitat Autònoma de Barcelona in Spain [77]. IMPC also has members in Australia, Canada, China, India, Japan, South Africa, South Korea, Taiwan, and the USA. Phenotyping data are accessible on the IMPC website (http://www.mousephenotype.org/).

Using both gene trapping and gene targeting approaches, the IMPC has developed mutant ES cells (many with conditional mutations) for thousands of genes representing more than 90% of the mouse protein-coding genes [26]. Since its inception in 2011, IMPC has published 17 data releases describing over 8000 genes and their associations to 90,000 significant phenotypes, in addition to 750,000 images and over 1000 human disease models [57].

The IMPC uses the International Mouse Phenotyping Resource of Standardized Screens (IMPReSS) phenotyping protocols, which are essential for the characterization of mouse phenotypes (see https://www.mousephenotype.org/impress). In this case, homozygous (or heterozygous in the case of embryonic lethal mutations) adult mutant mice enter a standardized pipeline [91] where cohorts of males and females undergo a wide range of phenotyping tests from 9 to 16 weeks, followed by a variety of terminal tests. The phenotyping of both male and female cohorts has allowed an in-depth analysis of the extent of sexual dimorphism. More importantly, approximately 90% of the gene–phenotype annotations described by the IMPC have not been previously reported [26]. Data from the IMPC also shows that around 24% of genes will not produce homozygous KO (null allele) offspring because they are homozygous lethal.

Recently, the Rat Genome Database (RGD) has initiated the PhenoMiner project to integrate quantitative phenotype data from the PhysGen Program for Genomic Applications and the National BioResource Project in Japan as well as manual annotations from biomedical literature [36, 112].

To conclude, it is important to note that numerous repositories and databases for mouse and rat mutants have been established over time. These repositories play a vital role in scientific research by serving as centralized hubs for storing and disseminating GA mouse and rat strains featuring particular mutations.

The Mutant Mouse Resource and Research Center (MMRRC) (https://www.mmrrc.org/), is supported by the National Institutes of Health of the United States and distributes and cryopreserves scientifically valuable, GA mouse strains and mouse ES cell [3].

The Rat Resource and Research Center (RRRC) (https://www.rrrc.us/) is a resource that provides services related to the supply, distribution, and maintenance of rat models for research purposes. The RRRC primarily focuses on rat strains that are genetically modified or have specific traits relevant to biomedical research [120].

The INFRAFRONTIER consortium (https://www.infrafrontier.eu/) was created in 2006 to investigate the link between gene function and human diseases by the use of mouse and rat models. This network consists of 22 partners from 15 countries and is continuously enhancing and broadening its portfolio of resources and services that are offered to the research community on a non-profit basis [2]. The INFRAFRONTIER database was designed to be a public platform for the biomedical community in order to provide up-to-date curated information. As an example, The INFRAFRONTIER Rare Disease Resource presently lists more than 1600 rare diseases where mouse models are available for distribution from the European Mouse Mutant Archive (EMMA) (https://www.infrafrontier.eu/emma/).

The RIKEN BioResource Research Center (BRC) (https://mus.brc.riken.jp/en/) was established in 2001 as a comprehensive biological resource center in Japan. The Experimental Animal Division, one of the BRC infrastructure divisions, has been designated as the core facility for mouse resources within the National BioResource Project (NBRP). This center also performs collection, preservation, quality control, and distribution of mouse resources [98].

The Asian Mouse Mutagenesis Resource Association (AMMRA) is a non-profit organization consisting of major resource and research institutions with rodent expertise from within the Asia Pacific region [32].

Finally, the Mouse Genome Database (MGD) is a core resource within the MGI (http://www.informatics.jax.org) that incorporates strain data from multiple mouse mutant repositories including JAX, MMRRC, EMMA, RIKEN, and many others via the International Mouse Strain Resource (IMSR, http://www.findmice.org). IMSR is a searchable online catalog of mouse strains, stocks, and mutant ES cell lines available worldwide [15, 45].

References

1. Ahmad W, Panteleyev AA, Sundberg JP, Christiano AM. Molecular basis for the rhino (hrrh-8J) phenotype: a nonsense mutation in the mouse hairless gene. Genomics. 1998;53:383–6.
2. Ali Khan A, Valera Vazquez G, Gustems M, Matteoni R, Song F, Gormanns P, Fessele S, Raess M, Hrabe De Angelis M, INFRAFRONTIER Consortium. INFRAFRONTIER: mouse model resources for modelling human diseases. Mamm Genome. 2023;34:408–17.
3. Amos-Landgraf J, Franklin C, Godfrey V, Grieder F, Grimsrud K, Korf I, Lutz C, Magnuson T,

Mirochnitchenko O, Patel S, Reinholdt L, Lloyd KCK. The mutant mouse resource and research center (MMRRC): the NIH-supported national public repository and distribution archive of mutant mouse models in the USA. Mamm Genome. 2022;33:203–12.
4. Anzalone AV, Randolph PB, Davis JR, Sousa AA, Koblan LW, Levy JM, Chen PJ, Wilson C, Newby GA, Raguram A, Liu DR. Search-and-replace genome editing without double-strand breaks or donor DNA. Nature. 2019;576:149–57.
5. Arnold CN, Barnes MJ, Berger M, Blasius AL, Brandl K, Croker B, Crozat K, Du X, Eidenschenk C, Georgel P, Hoebe K, Huang H, Jiang Z, Krebs P, La Vine D, Li X, Lyon S, Moresco EM, Murray AR, Popkin DL, Rutschmann S, Siggs OM, Smart NG, Sun L, Tabeta K, Webster V, Tomisato W, Won S, Xia Y, Xiao N, Beutler B. ENU-induced phenovariance in mice: inferences from 587 mutations. BMC Res Notes. 2012;5:577.
6. Auerbach AB, Norinsky R, Ho W, Losos K, Guo Q, Chatterjee S, Joyner AL. Strain-dependent differences in the efficiency of transgenic mouse production. Transgenic Res. 2003;12:59–69.
7. Avraham KB. Positional-candidate cloning of genes from mouse mutants. Methods Mol Biol. 2001;158:369–79.
8. Babinet C, Cohen-Tannoudji M. Genome engineering via homologous recombination in mouse embryonic stem (ES) cells: an amazingly versatile tool for the study of mammalian biology. An Acad Bras Cienc. 2001;73:365–83.
9. Ballester M, Castello A, Ibanez E, Sanchez A, Folch JM. Real-time quantitative PCR-based system for determining transgene copy number in transgenic animals. BioTechniques. 2004;37:610–3.
10. Baron U, Bujard H. Tet repressor-based system for regulated gene expression in eukaryotic cells: principles and advances. Methods Enzymol. 2000;327:401–21.
11. Beckmann PJ, Largaespada DA. Transposon insertion mutagenesis in mice for Modeling human cancers: critical insights gained and new opportunities. Int J Mol Sci. 2020;21:1172.
12. Behringer R, Gertsenstein M, Vintersten Nagy K, Nagy A. Manipulating the mouse embryo-A laboratory manual. Cold Spring Harbor Laboratory Press; 2014.
13. Berendzen KM, Sharma R, Mandujano MA, Wei Y, Rogers FD, Simmons TC, Seelke AMH, Bond JM, Larios R, Goodwin NL, Sherman M, Parthasarthy S, Espineda I, Knoedler JR, Beery A, Bales KL, Shah NM, Manoli DS. Oxytocin receptor is not required for social attachment in prairie voles. Neuron. 2023;111:787–96.e4.
14. Birling MC, Schaeffer L, Andre P, Lindner L, Marechal D, Ayadi A, Sorg T, Pavlovic G, Herault Y. Efficient and rapid generation of large genomic variants in rats and mice using CRISMERE. Sci Rep. 2017;7:43331.
15. Blake JA, Baldarelli R, Kadin JA, Richardson JE, Smith CL, Bult CJ, Mouse Genome Database, G. Mouse Genome Database (MGD): knowledgebase for mouse-human comparative biology. Nucleic Acids Res. 2021;49:D981–7.
16. Bode VC, Mcdonald JD, Guenet JL, Simon D. hph-1: a mouse mutant with hereditary hyperphenylalaninemia induced by ethylnitrosourea mutagenesis. Genetics. 1988;118:299–305.
17. Bogue MA, Ball RL, Walton DO, Dunn MH, Kolishovski G, Berger A, Lamoureux A, Grubb SC, Gerring M, Kim M, Liang H, Emerson J, Stearns T, He H, Mukherjee G, Bluis J, Davis S, Desai S, Sundberg B, Kadakkuzha B, Kunde-Ramamoorthy G, Philip VM, Chesler EJ. Mouse phenome database: curated data repository with interactive multipopulation and multi-trait analyses. Mamm Genome. 2023;34(4):509–19.
18. Bogue MA, Grubb SC, Walton DO, Philip VM, Kolishovski G, Stearns T, Dunn MH, Skelly DA, Kadakkuzha B, Tehennepe G, Kunde-Ramamoorthy G, Chesler EJ. Mouse Phenome Database: an integrative database and analysis suite for curated empirical phenotype data from laboratory mice. Nucleic Acids Res. 2018;46:D843–50.
19. Bosma MJ. The scid mutation: occurrence and effect. Curr Top Microbiol Immunol. 1989;152:3–9.
20. Bradley A, Evans M, Kaufman MH, Robertson E. Formation of germ-line chimaeras from embryo-derived teratocarcinoma cell lines. Nature. 1984;309:255–6.
21. Brinster RL, Chen HY, Trumbauer ME. Mouse oocytes transcribe injected Xenopus 5S RNA gene. Science. 1981;211:396–8.
22. Brinster RL, Chen HY, Trumbauer ME, Avarbock MR. Translation of globin messenger RNA by the mouse ovum. Nature. 1980;283:499–501.
23. Brinster RL, Chen HY, Trumbauer ME, Yagle MK, Palmiter RD. Factors affecting the efficiency of introducing foreign DNA into mice by microinjecting eggs. Proc Natl Acad Sci USA. 1985;82:4438–42.
24. Brommage R, Liu J, Hansen GM, Kirkpatrick LL, Potter DG, Sands AT, Zambrowicz B, Powell DR, Vogel P. High-throughput screening of mouse gene knockouts identifies established and novel skeletal phenotypes. Bone Res. 2014;2:14034.
25. Brown SD, Moore MW. The international mouse phenotyping consortium: past and future perspectives on mouse phenotyping. Mamm Genome. 2012;23:632–40.
26. Brown SDM, Holmes CC, Mallon AM, Meehan TF, Smedley D, Wells S. High-throughput mouse phenomics for characterizing mammalian gene function. Nat Rev Genet. 2018;19:357–70.
27. Buchholz F, Angrand PO, Stewart AF. Improved properties of FLP recombinase evolved by cycling mutagenesis. Nat Biotechnol. 1998;16:657–62.
28. Buehr M, Meek S, Blair K, Yang J, Ure J, Silva J, Mclay R, Hall J, Ying QL, Smith A. Capture of

authentic embryonic stem cells from rat blastocysts. Cell. 2008;135:1287–98.
29. Cacheiro P, Haendel MA, Smedley D, International Mouse Phenotyping Consortium and the Monarch Initiative. New models for human disease from the International Mouse Phenotyping Consortium. Mamm Genome. 2019;30:143–50.
30. Caso F, Davies B. Base editing and prime editing in laboratory animals. Lab Anim. 2022;56:35–49.
31. Chen S, Sun S, Moonen D, Lee C, Lee AY, Schaffer DV, He L. CRISPR-READI: efficient generation of Knockin mice by CRISPR RNP electroporation and AAV donor infection. Cell Rep. 2019;27(3780–3789):e4.
32. Chin HJ, Dobbie MS, Gao X, Hennessy JE, Nam KH, Seong JK, Shiroishi T, Takeo T, Yoshiki A, Zao J, Wang CL. Asian Mouse Mutagenesis Resource Association (AMMRA): mouse genetics and laboratory animal resources in the Asia Pacific. Mamm Genome. 2022;33:192–202.
33. Cho A, Haruyama N, Kulkarni AB. Generation of transgenic mice. Curr Protoc Cell Biol. 2009;42:19.11.1–19.11.22.
34. Chu VT, Weber T, Graf R, Sommermann T, Petsch K, Sack U, Volchkov P, Rajewsky K, Kuhn R. Efficient generation of Rosa26 knock-in mice using CRISPR/Cas9 in C57BL/6 zygotes. BMC Biotechnol. 2016;16:4.
35. Coghill EL, Hugill A, Parkinson N, Davison C, Glenister P, Clements S, Hunter J, Cox RD, Brown SD. A gene-driven approach to the identification of ENU mutants in the mouse. Nat Genet. 2002;30:255–6.
36. Collier N, Groza T, Smedley D, Robinson PN, Oellrich A, Rebholz-Schuhmann D. PhenoMiner: from text to a database of phenotypes associated with OMIM diseases. Database (Oxford). 2015;2015:bav104.
37. Cong L, Ran FA, Cox D, Lin S, Barretto R, Habib N, Hsu PD, Wu X, Jiang W, Marraffini LA, Zhang F. Multiplex genome engineering using CRISPR/Cas systems. Science. 2013;339:819–23.
38. Cozzi J, Anegon I, Braun V, Gross AC, Merrouche C, Cherifi Y. Pronuclear DNA injection for the production of transgenic rats. Methods Mol Biol. 2009;561:73–88.
39. Cui X, Ji D, Fisher DA, Wu Y, Briner DM, Weinstein EJ. Targeted integration in rat and mouse embryos with zinc-finger nucleases. Nat Biotechnol. 2011;29:64–7.
40. Da Silva-Buttkus P, Spielmann N, Klein-Rodewald T, Schutt C, Aguilar-Pimentel A, Amarie OV, Becker L, Calzada-Wack J, Garrett L, Gerlini R, Kraiger M, Leuchtenberger S, Ostereicher MA, Rathkolb B, Sanz-Moreno A, Stoger C, Holter SM, Seisenberger C, Marschall S, Fuchs H, Gailus-Durner V, Hrabe De Angelis M. Knockout mouse models as a resource for the study of rare diseases. Mamm Genome. 2023;34:244–61.
41. Danielian PS, Muccino D, Rowitch DH, Michael SK, Mcmahon AP. Modification of gene activity in mouse embryos in utero by a tamoxifen-inducible form of Cre recombinase. Curr Biol. 1998;8:1323–6.
42. Davisson MT, Bergstrom DE, Reinholdt LG, Donahue LR. Discovery genetics – the history and future of spontaneous mutation research. Curr Protoc Mouse Biol. 2012;2:103–18.
43. Davisson MT, Taft RA. Strategies for managing an ever increasing mutant mouse repository. Brain Res. 2006;1091:255–7.
44. Demers S-P, Smith LC. Derivation, culture, and in vivo developmental capacity of embryonic cell lines from rat blastocysts. In: Anegon I, editor. Rat genomics: methods and protocols. Totowa: Humana Press; 2010.
45. Eppig JT, Motenko H, Richardson JE, Richards-Smith B, Smith CL. The International Mouse Strain Resource (IMSR): cataloging worldwide mouse and ES cell line resources. Mamm Genome. 2015;26:448–55.
46. Evans MJ, Kaufman MH. Establishment in culture of pluripotential cells from mouse embryos. Nature. 1981;292:154–6.
47. Feil S, Valtcheva N, Feil R. Inducible Cre mice. Methods Mol Biol. 2009;530:343–63.
48. Fernandez A, Josa S, Montoliu L. A history of genome editing in mammals. Mamm Genome. 2017;28:237–46.
49. Filipiak WE, Saunders TL. Advances in transgenic rat production. Transgenic Res. 2006;15:673–86.
50. Flanagan SP. 'Nude', a new hairless gene with pleiotropic effects in the mouse. Genet Res. 1966;8:295–309.
51. Gondo Y. Trends in large-scale mouse mutagenesis: from genetics to functional genomics. Nat Rev Genet. 2008;9:803–10.
52. Gondo Y, Fukumura R, Murata T, Makino S. ENU-based gene-driven mutagenesis in the mouse: a next-generation gene-targeting system. Exp Anim. 2010;59:537–48.
53. Gordon JW, Ruddle FH. Integration and stable germ line transmission of genes injected into mouse pronuclei. Science. 1981;214:1244–6.
54. Gordon JW, Scangos GA, Plotkin DJ, Barbosa JA, Ruddle FH. Genetic transformation of mouse embryos by microinjection of purified DNA. Proc Natl Acad Sci USA. 1980;77:7380–4.
55. Gossen M, Bujard H. Tight control of gene expression in mammalian cells by tetracycline-responsive promoters. Proc Natl Acad Sci USA. 1992;89:5547–51.
56. Gridley T, Murray SA. Mouse mutagenesis and phenotyping to generate models of development and disease. Curr Top Dev Biol. 2022;148:1–12.
57. Groza T, Gomez FL, Mashhadi HH, Munoz-Fuentes V, Gunes O, Wilson R, Cacheiro P, Frost A, Keskivali-Bond P, Vardal B, Mccoy A, Cheng TK, Santos L, Wells S, Smedley D, Mallon AM,

Parkinson H. The International Mouse Phenotyping Consortium: comprehensive knockout phenotyping underpinning the study of human disease. Nucleic Acids Res. 2023;51:D1038–45.
58. Guenet JL. Chemical mutagenesis of the mouse genome: an overview. Genetica. 2004;122:9–24.
59. Guenet JL, Benavides F, Panthier J, Montagutelli X. Genetics of the mouse. Springer; 2015.
60. Hall B, Limaye A, Kulkarni AB. Overview: generation of gene knockout mice. Curr Protoc Cell Biol. 2009;Chapter 19:Unit 19 12 19 12 1–17.
61. Hoebe K, Du X, Georgel P, Janssen E, Tabeta K, Kim SO, Goode J, Lin P, Mann N, Mudd S, Crozat K, Sovath S, Han J, Beutler B. Identification of Lps2 as a key transducer of MyD88-independent TIR signalling. Nature. 2003;424:743–8.
62. Hougen HP. The athymic nude rat. Immunobiological characteristics with special reference to establishment of non-antigen-specific T-cell reactivity and induction of antigen-specific immunity. APMIS Suppl. 1991;21:1–39.
63. Hrabe De Angelis MH, Flaswinkel H, Fuchs H, Rathkolb B, Soewarto D, Marschall S, Heffner S, Pargent W, Wuensch K, Jung M, Reis A, Richter T, Alessandrini F, Jakob T, Fuchs E, Kolb H, Kremmer E, Schaeble K, Rollinski B, Roscher A, Peters C, Meitinger T, Strom T, Steckler T, Holsboer F, Klopstock T, Gekeler F, Schindewolf C, Jung T, Avraham K, Behrendt H, Ring J, Zimmer A, Schughart K, Pfeffer K, Wolf E, Balling R. Genome-wide, large-scale production of mutant mice by ENU mutagenesis. Nat Genet. 2000;25:444–7.
64. Hughes ED, Saunders TL. Gene targeting in embryonic stem cells. In: Pease S, Saunders TL, editors. Advanced protocols for animal transgenesis: an ISTT manual. Berlin, Heidelberg: Springer; 2011.
65. Hummel KP, Dickie MM, Coleman DL. Diabetes, a new mutation in the mouse. Science. 1966;153:1127–8.
66. Iyer V, Shen B, Zhang W, Hodgkins A, Keane T, Huang X, Skarnes WC. Off-target mutations are rare in Cas9-modified mice. Nat Methods. 2015;12:479.
67. Jinek M, Chylinski K, Fonfara I, Hauer M, Doudna JA, Charpentier E. A programmable dual-RNA-guided DNA endonuclease in adaptive bacterial immunity. Science. 2012;337:816–21.
68. Justice MJ. Capitalizing on large-scale mouse mutagenesis screens. Nat Rev Genet. 2000;1:109–15.
69. Justice MJ, Noveroske JK, Weber JS, Zheng B, Bradley A. Mouse ENU mutagenesis. Hum Mol Genet. 1999;8:1955–63.
70. Kaneko T, Sakuma T, Yamamoto T, Mashimo T. Simple knockout by electroporation of engineered endonucleases into intact rat embryos. Sci Rep. 2014;4:6382.
71. Kawamata M, Ochiya T. Establishment of embryonic stem cells from rat blastocysts. Methods Mol Biol. 2010;597:169–77.
72. Kim K, Ryu SM, Kim ST, Baek G, Kim D, Lim K, Chung E, Kim S, Kim JS. Highly efficient RNA-guided base editing in mouse embryos. Nat Biotechnol. 2017;35:435–7.
73. Kobayashi T, Namba M, Koyano T, Fukushima M, Sato M, Ohtsuka M, Matsuyama M. Successful production of genome-edited rats by the rGONAD method. BMC Biotechnol. 2018;18:19.
74. Koentgen F, Lin J, Katidou M, Chang I, Khan M, Watts J, Mombaerts P. Exclusive transmission of the embryonic stem cell-derived genome through the mouse germline. Genesis. 2016;54:326–33.
75. Koller BH, Hagemann LJ, Doetschman T, Hagaman JR, Huang S, Williams PJ, First NL, Maeda N, Smithies O. Germ-line transmission of a planned alteration made in a hypoxanthine phosphoribosyl transferase gene by homologous recombination in embryonic stem cells. Proc Natl Acad Sci USA. 1989;86:8927–31.
76. Komor AC, Kim YB, Packer MS, Zuris JA, Liu DR. Programmable editing of a target base in genomic DNA without double-stranded DNA cleavage. Nature. 2016;533:420–4.
77. Koscielny G, Yaikhom G, Iyer V, Meehan TF, Morgan H, Atienza-Herrero J, Blake A, Chen CK, Easty R, Di Fenza A, Fiegel T, Grifiths M, Horne A, Karp NA, Kurbatova N, Mason JC, Matthews P, Oakley DJ, Qazi A, Regnart J, Retha A, Santos LA, Sneddon DJ, Warren J, Westerberg H, Wilson RJ, Melvin DG, Smedley D, Brown SD, Flicek P, Skarnes WC, Mallon AM, Parkinson H. The International Mouse Phenotyping Consortium web portal, a unified point of access for knockout mice and related phenotyping data. Nucleic Acids Res. 2014;42:D802–9.
78. Lewis MA, Quint E, Glazier AM, Fuchs H, De Angelis MH, Langford C, Van Dongen S, Abreu-Goodger C, Piipari M, Redshaw N, Dalmay T, Moreno-Pelayo MA, Enright AJ, Steel KP. An ENU-induced mutation of miR-96 associated with progressive hearing loss in mice. Nat Genet. 2009;41:614–8.
79. Li P, Schulze EN, Tong C, Ying Q-L. Rat embryonic stem cell derivation and propagation. In: Pease S, Saunders TL, editors. Advanced protocols for animal transgenesis: an ISTT manual. Berlin, Heidelberg: Springer; 2011.
80. Li P, Tong C, Mehrian-Shai R, Jia L, Wu N, Yan Y, Maxson RE, Schulze EN, Song H, Hsieh CL, Pera MF, Ying QL. Germline competent embryonic stem cells derived from rat blastocysts. Cell. 2008;135:1299–310.
81. Liang P, Sun H, Sun Y, Zhang X, Xie X, Zhang J, Zhang Z, Chen Y, Ding C, Xiong Y, Ma W, Liu D, Huang J, Songyang Z. Effective gene editing by high-fidelity base editor 2 in mouse zygotes. Protein Cell. 2017;8:601–11.
82. Liu Y, Li X, He S, Huang S, Li C, Chen Y, Liu Z, Huang X, Wang X. Efficient generation of mouse models with the prime editing system. Cell Discov. 2020;6:27.
83. Lloyd KC. A knockout mouse resource for the biomedical research community. Ann N Y Acad Sci. 2011;1245:24–6.

84. Lloyd KCK, Adams DJ, Baynam G, Beaudet AL, Bosch F, Boycott KM, Braun RE, Caulfield M, Cohn R, Dickinson ME, Dobbie MS, Flenniken AM, Flicek P, Galande S, Gao X, Grobler A, Heaney JD, Herault Y, De Angelis MH, Lupski JR, Lyonnet S, Mallon AM, Mammano F, Macrae CA, Mcinnes R, Mckerlie C, Meehan TF, Murray SA, Nutter LMJ, Obata Y, Parkinson H, Pepper MS, Sedlacek R, Seong JK, Shiroishi T, Smedley D, Tocchini-Valentini G, Valle D, Wang CL, Wells S, White J, Wurst W, Xu Y, Brown SDM. The deep genome project. Genome Biol. 2020;21:18.
85. Lois C, Hong EJ, Pease S, Brown EJ, Baltimore D. Germline transmission and tissue-specific expression of transgenes delivered by lentiviral vectors. Science. 2002;295:868–72.
86. Mallon AM, Blake A, Hancock JM. EuroPhenome and EMPReSS: online mouse phenotyping resource. Nucleic Acids Res. 2008;36:D715–8.
87. Mashimo T, Ohmori I, Ouchida M, Ohno Y, Tsurumi T, Miki T, Wakamori M, Ishihara S, Yoshida T, Takizawa A, Kato M, Hirabayashi M, Sasa M, Mori Y, Serikawa T. A missense mutation of the gene encoding voltage-dependent sodium channel (Nav1.1) confers susceptibility to febrile seizures in rats. J Neurosci. 2010;30:5744–53.
88. Mashimo T, Yanagihara K, Tokuda S, Voigt B, Takizawa A, Nakajima R, Kato M, Hirabayashi M, Kuramoto T, Serikawa T. An ENU-induced mutant archive for gene targeting in rats. Nat Genet. 2008;40:514–5.
89. Masuya H, Inoue M, Wada Y, Shimizu A, Nagano J, Kawai A, Inoue A, Kagami T, Hirayama T, Yamaga A, Kaneda H, Kobayashi K, Minowa O, Miura I, Gondo Y, Noda T, Wakana S, Shiroishi T. Implementation of the modified-SHIRPA protocol for screening of dominant phenotypes in a large-scale ENU mutagenesis program. Mamm Genome. 2005;16:829–37.
90. Mates L, Chuah MK, Belay E, Jerchow B, Manoj N, Acosta-Sanchez A, Grzela DP, Schmitt A, Becker K, Matrai J, Ma L, Samara-Kuko E, Gysemans C, Pryputniewicz D, Miskey C, Fletcher B, Vandendriessche T, Ivics Z, Izsvak Z. Molecular evolution of a novel hyperactive sleeping beauty transposase enables robust stable gene transfer in vertebrates. Nat Genet. 2009;41:753–61.
91. Meehan TF, Conte N, West DB, Jacobsen JO, Mason J, Warren J, Chen CK, Tudose I, Relac M, Matthews P, Karp N, Santos L, Fiegel T, Ring N, Westerberg H, Greenaway S, Sneddon D, Morgan H, Codner GF, Stewart ME, Brown J, Horner N, Haendel M, Washington N, Mungall CJ, Reynolds CL, Gallegos J, Gailus-Durner V, Sorg T, Pavlovic G, Bower LR, Moore M, Morse I, Gao X, Tocchini-Valentini GP, Obata Y, Cho SY, Seong JK, Seavitt J, Beaudet AL, Dickinson ME, Herault Y, Wurst W, De Angelis MH, Lloyd KCK, Flenniken AM, Nutter LMJ, Newbigging S, Mckerlie C, Justice MJ, Murray SA, Svenson KL, Braun RE, White JK, Bradley A, Flicek P, Wells S, Skarnes WC, Adams DJ, Parkinson H, Mallon AM, Brown SDM, Smedley D, International Mouse Phenotyping Consortium. Disease model discovery from 3,328 gene knockouts by the International Mouse Phenotyping Consortium. Nat Genet. 2017;49:1231–8.
92. Men H, Stone BJ, Bryda EC. Media optimization to promote rat embryonic development to the blastocyst stage in vitro. Theriogenology. 2020;151:81–5.
93. Menchaca A, Schlapp G, Noel Meikle M, Crispo M. Transgenesis and gene edition in mammals. In: Reference module in life sciences. Oxford: Elsevier; 2017.
94. Menoret S, De Cian A, Tesson L, Remy S, Usal C, Boule JB, Boix C, Fontaniere S, Creneguy A, Nguyen TH, Brusselle L, Thinard R, Gauguier D, Concordet JP, Cherifi Y, Fraichard A, Giovannangeli C, Anegon I. Homology-directed repair in rodent zygotes using Cas9 and TALEN engineered proteins. Sci Rep. 2015;5:14410.
95. Menoret S, Fontaniere S, Jantz D, Tesson L, Thinard R, Remy S, Usal C, Ouisse LH, Fraichard A, Anegon I. Generation of Rag1-knockout immunodeficient rats and mice using engineered meganucleases. FASEB J. 2013;27:703–11.
96. Miano JM, Long X, Lyu Q. Crispr links to long noncoding RNA function in mice: a practical approach. Vasc Pharmacol. 2019;114:1–12.
97. Mingarro G, Del Olmo ML. Improvements in the genetic editing technologies: Crispr-Cas and beyond. Gene. 2023;852:147064.
98. Mizuno-Iijima S, Nakashiba T, Ayabe S, Nakata H, Ike F, Hiraiwa N, Mochida K, Ogura A, Masuya H, Kawamoto S, Tamura M, Obata Y, Shiroishi T, Yoshiki A. Mouse resources at the RIKEN BioResource Research Center and the National BioResource Project core facility in Japan. Mamm Genome. 2022;33:181–91.
99. Mullins JJ, Peters J, Ganten D. Fulminant hypertension in transgenic rats harbouring the mouse Ren-2 gene. Nature. 1990;344:541–4.
100. Munoz-Fuentes V, Cacheiro P, Meehan TF, Aguilar-Pimentel JA, Brown SDM, Flenniken AM, Flicek P, Galli A, Mashhadi HH, Hrabe De Angelis M, Kim JK, Lloyd KCK, Mckerlie C, Morgan H, Murray SA, Nutter LMJ, Reilly PT, Seavitt JR, Seong JK, Simon M, Wardle-Jones H, Mallon AM, Smedley D, Parkinson HE, IMPC Consortium. The International Mouse Phenotyping Consortium (IMPC): a functional catalogue of the mammalian genome that informs conservation. Conserv Genet. 2018;19:995–1005.
101. Munoz-Santos D, Montoliu L, Fernandez A. Generation of genetically modified mice using CRISPR/Cas9. Methods Mol Biol. 2020;2110:129–38.
102. Nagy A, Gertsenstein M, Vintersten K, Behringer R. Manipulating the mouse embryo, a laboratory manual. 3rd ed. New York: Cold Spring Harbor Press; 2003.

103. Nolan PM, Peters J, Strivens M, Rogers D, Hagan J, Spurr N, Gray IC, Vizor L, Brooker D, Whitehill E, Washbourne R, Hough T, Greenaway S, Hewitt M, Liu X, Mccormack S, Pickford K, Selley R, Wells C, Tymowska-Lalanne Z, Roby P, Glenister P, Thornton C, Thaung C, Stevenson JA, Arkell R, Mburu P, Hardisty R, Kiernan A, Erven A, Steel KP, Voegeling S, Guenet JL, Nickols C, Sadri R, Nasse M, Isaacs A, Davies K, Browne M, Fisher EM, Martin J, Rastan S, Brown SD, Hunter J. A systematic, genome-wide, phenotype-driven mutagenesis programme for gene function studies in the mouse. Nat Genet. 2000;25:440–3.
104. Paigen K, Eppig JT. A mouse phenome project. Mamm Genome. 2000;11:715–7.
105. Palmiter RD, Brinster RL, Hammer RE, Trumbauer ME, Rosenfeld MG, Birnberg NC, Evans RM. Dramatic growth of mice that develop from eggs microinjected with metallothionein-growth hormone fusion genes. Nature. 1982;300:611–5.
106. Perez CJ, Jaubert J, Guenet JL, Barnhart KF, Ross-Inta CM, Quintanilla VC, Aubin I, Brandon JL, Otto NW, Digiovanni J, Gimenez-Conti I, Giulivi C, Kusewitt DF, Conti CJ, Benavides F. Two hypomorphic alleles of mouse Ass1 as a new animal model of citrullinemia type I and other hyperammonemic syndromes. Am J Pathol. 2010;177:1958–68.
107. Perez CJ, Mecklenburg L, Fernandez A, Cantero M, De Souza TA, Lin K, Dent SYR, Montoliu L, Awgulewitsch A, Benavides F. Naked (N) mutant mice carry a nonsense mutation in the homeobox of Hoxc13. Exp Dermatol. 2022;31:330–40.
108. Peterson KA, Khalouei S, Hanafi N, Wood JA, Lanza DG, Lintott LG, Willis BJ, Seavitt JR, Braun RE, Dickinson ME, White JK, Lloyd KCK, Heaney JD, Murray SA, Ramani A, Nutter LMJ. Whole genome analysis for 163 gRNAs in Cas9-edited mice reveals minimal off-target activity. Commun Biol. 2023;6:626.
109. Peterson KA, Murray SA. Progress towards completing the mutant mouse null resource. Mamm Genome. 2022;33:123–34.
110. Peterson RG, Little LA, Neel MA. Wky fatty rat as a model of obesity and non-insulin-dependent diabetes mellitus. Ilar News. 1990;32:13–5.
111. Qin W, Dion SL, Kutny PM, Zhang Y, Cheng AW, Jillette NL, Malhotra A, Geurts AM, Chen YG, Wang H. Efficient CRISPR/Cas9-mediated genome editing in mice by zygote electroporation of nuclease. Genetics. 2015;200:423–30.
112. Qiu W, Lei M, Tang H, Yan H, Wen X, Zhang W, Tan R, Wang D, Wu J. Hoxc13 is a crucial regulator of murine hair cycle. Cell Tissue Res. 2016;364:149–58.
113. Quwailid MM, Hugill A, Dear N, Vizor L, Wells S, Horner E, Fuller S, Weedon J, Mcmath H, Woodman P, Edwards D, Campbell D, Rodger S, Carey J, Roberts A, Glenister P, Lalanne Z, Parkinson N, Coghill EL, Mckeone R, Cox S, Willan J, Greenfield A, Keays D, Brady S, Spurr N, Gray I, Hunter J, Brown SD, Cox RD. A gene-driven ENU-based approach to generating an allelic series in any gene. Mamm Genome. 2004;15:585–91.
114. Remy S, Chenouard V, Tesson L, Usal C, Menoret S, Brusselle L, Heslan JM, Nguyen TH, Bellien J, Merot J, De Cian A, Giovannangeli C, Concordet JP, Anegon I. Generation of gene-edited rats by delivery of CRISPR/Cas9 protein and donor DNA into intact zygotes using electroporation. Sci Rep. 2017;7:16554.
115. Robertson E, Bradley A, Kuehn M, Evans M. Germline transmission of genes introduced into cultured pluripotential cells by retroviral vector. Nature. 1986;323:445–8.
116. Russell WL, Kelly EM, Hunsicker PR, Bangham JW, Maddux SC, Phipps EL. Specific-locus test shows ethylnitrosourea to be the most potent mutagen in the mouse. Proc Natl Acad Sci USA. 1979;76:5818–9.
117. Sauer B, Henderson N. Site-specific DNA recombination in mammalian cells by the Cre recombinase of bacteriophage P1. Proc Natl Acad Sci USA. 1988;85:5166–70.
118. Sauer B, Mcdermott J. DNA recombination with a heterospecific Cre homolog identified from comparison of the pac-c1 regions of P1-related phages. Nucleic Acids Res. 2004;32:6086–95.
119. Schuster-Gossler K, Lee AW, Lerner CP, Parker HJ, Dyer VW, Scott VE, Gossler A, Conover JC. Use of coisogenic host blastocysts for efficient establishment of germline chimeras with C57BL/6J ES cell lines. BioTechniques. 2001;31(1022–4):1026.
120. Shimoyama M, Smith JR, Bryda E, Kuramoto T, Saba L, Dwinell M. Rat genome and model resources. ILAR J. 2017;58:42–58.
121. Soriano P. Generalized lacZ expression with the ROSA26 Cre reporter strain. Nat Genet. 1999;21:70–1.
122. Stevens JC, Banks GT, Festing MF, Fisher EM. Quiet mutations in inbred strains of mice. Trends Mol Med. 2007;13:512–9.
123. Stottmann R, Beier D. ENU mutagenesis in the mouse. Curr Protoc Mouse Biol. 2014;4:25–35.
124. Sung YH, Baek IJ, Kim DH, Jeon J, Lee J, Lee K, Jeong D, Kim JS, Lee HW. Knockout mice created by TALEN-mediated gene targeting. Nat Biotechnol. 2013;31:23–4.
125. Szpirer C. Rat models of human diseases and related phenotypes: a novel inventory of causative genes. Mamm Genome. 2022;33:88–90.
126. Takada T, Fukuta K, Usuda D, Kushida T, Kondo S, Kawamoto S, Yoshiki A, Obata Y, Fujiyama A, Toyoda A, Noguchi H, Shiroishi T, Masuya H. MoG+: a database of genomic variations across three mouse subspecies for biomedical research. Mamm Genome. 2022;33:31–43.
127. Takahashi G, Gurumurthy CB, Wada K, Miura H, Sato M, Ohtsuka M. GONAD: genome-editing via oviductal nucleic acids delivery system: a novel microinjection independent genome engineering method in mice. Sci Rep. 2015;5:11406.

128. Taketo M, Schroeder AC, Mobraaten LE, Gunning KB, Hanten G, Fox RR, Roderick TH, Stewart CL, Lilly F, Hansen CT, et al. FVB/N: an inbred mouse strain preferable for transgenic analyses. Proc Natl Acad Sci USA. 1991;88:2065–9.
129. Taylor CA, Shawlot W, Ren JX, Mukhopadhyay S. Generation and validation of tissue-specific knockout strains for toxicology research. Curr Protoc Toxicol. 2019;81:e86.
130. Taylor JH, Walton JC, Mccann KE, Norvelle A, Liu Q, Vander Velden JW, Borland JM, Hart M, Jin C, Huhman KL, Cox DN, Albers HE. CRISPR-Cas9 editing of the arginine-vasopressin V1a receptor produces paradoxical changes in social behavior in Syrian hamsters. Proc Natl Acad Sci USA. 2022;119:e2121037119.
131. Tesson L, Remy S, Menoret S, Usal C, Thinard R, Savignard C, De Cian A, Giovannangeli C, Concordet JP, Anegon I. Genome editing in rats using TALE nucleases. Methods Mol Biol. 2016;1338:245–59.
132. Vitaterna MH, King DP, Chang AM, Kornhauser JM, Lowrey PL, Mcdonald JD, Dove WF, Pinto LH, Turek FW, Takahashi JS. Mutagenesis and mapping of a mouse gene, clock, essential for circadian behavior. Science. 1994;264:719–25.
133. Vogel G. Nobel Prizes. A knockout award in medicine. Science. 2007;318:178–9.
134. Wagner KU, Mcallister K, Ward T, Davis B, Wiseman R, Hennighausen L. Spatial and temporal expression of the Cre gene under the control of the MMTV-LTR in different lines of transgenic mice. Transgenic Res. 2001;10:545–53.
135. Wang H, Yang H, Shivalila CS, Dawlaty MM, Cheng AW, Zhang F, Jaenisch R. One-step generation of mice carrying mutations in multiple genes by CRISPR/Cas-mediated genome engineering. Cell. 2013;153:910–8.
136. Wang JY, Doudna JA. CRISPR technology: a decade of genome editing is only the beginning. Science. 2023;379:eadd8643.
137. Wang T, Bu CH, Hildebrand S, Jia G, Siggs OM, Lyon S, Pratt D, Scott L, Russell J, Ludwig S, Murray AR, Moresco EMY, Beutler B. Probability of phenotypically detectable protein damage by ENU-induced mutations in the Mutagenetix database. Nat Commun. 2018;9:441.
138. Williams E, Auerbach W, Dechiara TM, Gertsenstein M. Combining ES cells with embryos. In: Pease S, Saunders TL, editors. Advanced protocols for animal Transgenesis: an ISTT manual. Berlin/Heidelberg: Springer; 2011.
139. Yang H, Wang JR, Didion JP, Buus RJ, Bell TA, Welsh CE, Bonhomme F, Yu AH, Nachman MW, Pialek J, Tucker P, Boursot P, Mcmillan L, Churchill GA, De Villena FP. Subspecific origin and haplotype diversity in the laboratory mouse. Nat Genet. 2011;43:648–55.
140. Zhang Y, Proenca R, Maffei M, Barone M, Leopold L, Friedman JM. Positional cloning of the mouse obese gene and its human homologue. Nature. 1994;372:425–32.
141. Zuo E, Sun Y, Wei W, Yuan T, Ying W, Sun H, Yuan L, Steinmetz LM, Li Y, Yang H. Cytosine base editor generates substantial off-target single-nucleotide variants in mouse embryos. Science. 2019;364:289–92.

3 Standard Nomenclature of Mouse and Rat Strains

Fernando Benavides

Abstract

This chapter explores the fundamentals of standard mouse and rat nomenclature, highlighting the guidelines established by the international committees on standardized genetic nomenclature for mice and rats. The accurate designation and communication among researchers working with laboratory strains center on these rules, establishing connections between strains and their origin and history. Emphasizing fundamental principles, this chapter underscores the importance of adopting standard genetic nomenclature to foster clarity, reproducibility, and efficient data handling in scientific research.

Keywords

Rodents · Standard nomenclature · Laboratory codes · Inbred strains · Outbred stocks

F. Benavides (✉)
Department of Epigenetics and Molecular Carcinogenesis, The University of Texas MD Anderson Cancer Center, Houston, TX, USA
e-mail: fbenavid@mdanderson.org

3.1 Introduction

Nomenclature rules are of paramount importance and require meticulous consideration, as they serve as the foundation for accurately designating laboratory strains and stocks and connecting with their origin and history [10]. The *International Committee on Standardized Genetic Nomenclature for Mice* and the *Rat Genome and Nomenclature Committee* establish these rules, consistently updating them to align with advancements in the genetics of these species. Detailed insights, along with illustrative examples, are available on the MGI webpage under the title "Guidelines for Nomenclature of Mouse and Rat Strains," last revised in July 2022 (https://www.informatics.jax.org/mgihome/nomen/strains.shtml). Consequently, this chapter will focus on presenting only the fundamental principles of these rules.

The use of standard genetic nomenclature in laboratory strains is of utmost importance for several reasons: (i) Ensures clear and effective communication among researchers working with mouse and rat strains. It provides a common language that allows for accurate and unambiguous identification and description of strains and stocks, as well as genes, alleles, and mutations. (ii) Maintains consistency in naming strains, genes, and alleles across different laboratories and publications. (iii) Facilitates the integration and sharing of data across databases and reposi-

F. Benavides, A. K. Hansen (eds.), *Rodent Quality Control: Genes and Bugs*, Laboratory Animal Science and Medicine 2, https://doi.org/10.1007/978-3-031-59103-7_3

tories. (iv) Helps prevent the duplication of names, minimizing confusion. (v) Enables researchers to track and trace the history and lineage of mouse and rat strains, facilitating the organization and maintenance of breeding records, pedigrees, and genotyping information. In summary, adhering to standard genetic nomenclature for laboratory mice and rats promotes clarity, reproducibility, accessibility, and efficient data management [1, 7].

Finally, many scientific journals now require the use of standardized nomenclature. A complete list of these journals can be found on the MGI website under Journals Enforcing Standard Nomenclature (https://www.informatics.jax.org/mgihome/nomen/journals.shtml). For example, in this context, the use of incomplete or incorrect mouse strain designations such as "C3H," "C57," or "Balb" in place of C3H/HeNHsd, C57BL/6NCrl, or BALB/cByJ, would be flagged for correction. To assist authors with their nomenclature concerns, the MGI nomenclature group can be contacted via email at nomen@jax.org. The Jackson Laboratory provides a comprehensive and didactic online tutorial that explains the details which all researchers should be familiar with concerning the employed strains. (https://www.jax.org/jax-mice-and-services/customer-support/technical-support/genetics-and-nomenclature/nomenclature-tutorial).

Note that the inbred strains, outbred stocks, and genetically altered lines mentioned in this chapter are thoroughly described in Chaps. 1 and 2.

3.2 Nomenclature for Inbred Strains and Substrains

An inbred strain should be designated by a unique brief symbol made up of uppercase, roman letters, or a combination of letters and numbers beginning with a letter (e.g., DBA, FVB, and C57BL mice or BN, F344, and ACI rats) [8, 11]. However, some pre-existing strains do not follow this convention, like the 129 mouse strains, or the use of lower case "c" in BALB/c. In this case, the name derived by combining the name of the investigator (Bagg) with the albino color of the mouse (c/c genotype). Care should be taken that mouse and rat strains do not overlap in strain names. Also, the symbols used for inbred strains that have a common origin should reflect this relationship (for example, the "C57" or "NZ" family of mouse strains).

Use of approved abbreviations for common strains (and some substrains) is acceptable in publications after the first mention of the full proper designation, for example, B10 for C57BL/10, C3 for C3H; C for BALB/c; B6J for C57BL/6J; B6N for C57BL/6N; and D2 for DBA/2. Examples in rats are F344 for Fischer 344, WKY for Wistar-Kyoto, and BN for Brown Norway.

New mouse strain names should be registered through the Mouse Genome Database (MGD) at http://www.informatics.jax.org/mgihome/submissions/amsp_submission.cgi. Rat strain names should be registered through the Rat Genome Database (RGD) at http://rgd.mcw.edu/tools/strains/strainRegistrationIndex.cgi.

Inbred strains can gradually diverge genetically over time, leading to the emergence of substrains (see Chap. 1). For example, when lineages are isolated for over 20 generations from a shared ancestor, it becomes probable that genetic disparities between the lineages have arisen due to genetic drift. Substrains are given the symbol of the original strain they are derived from, followed by a forward slash and a substrain designation. This substrain designation is, in most instances, the lab code of the individual or laboratory originating the strain. For example, A/He is the substrain of the A inbred mice developed by Dr. W. Heston (He). Similarly, 129S2/SvPasCrl mice is the designation for the 129 substrain that was originally developed by Dr. L.C. Stevens (Sv), then introduced to the *Institut Pas*teur of Paris (Pas) and finally transferred for breeding to Charles River Laboratories (Crl). Examples from rats include the WKY/NHsd (Inotiv) and WKY/NCrl (Charles River Laboratories) substrains of the Wistar-Kyoto (WKY) inbred strain, originally at the National Institutes of Health.

Some strains used for gene mapping are generated crossing multiple inbred strains and should be designated appropriately, according to the breeding scheme that produced them. Comprehensive guidelines for naming Recombinant Inbred (RI), Collaborative Cross (CC), Recombinant Congenic (RC), Consomic and Conplastic strains can be accessed in the Guidelines for Nomenclature of Mouse and Rat Strains on the MGI webpage.

3.3 Nomenclature for Hybrid F1 Strains

Offspring resulting from the crossbreeding of two inbred strains of mice or rats in the same direction are genetically identical and can be denoted by the name of the two parental strains in uppercase (with the maternal strain listed first), followed by F1. It is important to note that reciprocal F1 hybrids are not genetically identical, leading to distinct designation. For example, the full designation for the offspring of a DBA/2 female and a C57BL/6J male is (DBA/2N x C57BL/6J)F1. The reciprocal cross designation is (C57BL/6J x DBA/2N)F1. Use of the approved abbreviations is acceptable, for example, B6C3F1 is the hybrid offspring from a cross between a female C57BL/6 and male C3H [11].

3.4 Genetically Altered Lines

Genetically altered (GA) lines, whether they harbor spontaneous, induced, targeted, or edited mutations, as well as randomly inserted transgenes, can exhibit a range of genetic backgrounds, spanning from purely inbred to partially mixed, and even encompassing complex genetic compositions. Lines derived from two standard inbred strains, one of which is the strain of origin (donor) of the mutant allele (see congenic strains below) at a backcross generation <*N5* may be designated using approved abbreviations from the recipient strain first, followed by the donor strain, separated by a *semicolon (;)* (Fig. 3.1). The abbreviation following the semicolon always represents the donor strain, for example, the strain of the ES cells in a KO/KI line. When more than two progenitor strains contribute to the genetic background or there is a genetic contribution from an unknown or non-inbred source, the line may be designated as "*STOCK*" (all capital letters), followed by a space (not a hyphen), and the symbol for the mutation or transgene [11] (Fig. 3.1). For example, STOCK $Abca4^{tm1Ght}$/J KO mice. For backcross generation >*N5* see congenic strains below.

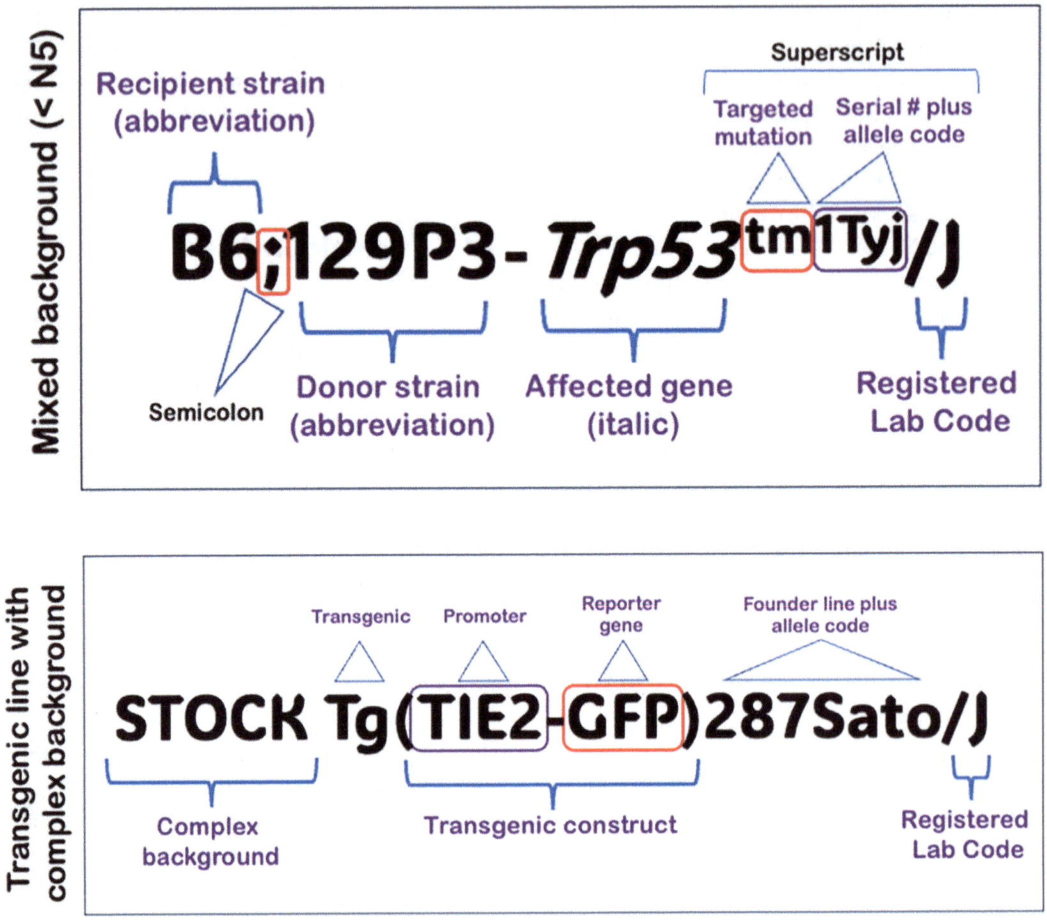

Fig. 3.1 *Nomenclature for lines with mixed, unknown, or non-inbred backgrounds.* Above: example of a KO (or KI) mouse line with mixed background. The donor strain is 129P3 (ES cells used for the targeting) and C57BL/6 (no information on substrain) is the background (recipient) strain. The semicolon separating the background and donor strain is used to show less than five backcross generations (<N5). Note that the symbol "tm" and the serial number plus allele code of this particular mutation are superscript. Below: example of a transgenic (Tg) line with unknown or non-inbred background. In this case, it is recommended to use "STOCK", all capitals and followed by a space (i.e., no hyphen), followed by the genetic modification (in this case a reporter transgene). Note that no component of a transgene symbol is italicized (or superscript), as these represent arbitrary insertions of foreign DNA material and are distinct from the native genome

3.5 Nomenclature for Co-isogenic and Congenic Strains (Including GA Lines)

Co-isogenic strains should be designated by the strain symbol (and where appropriate the substrain symbol) followed by a hyphen and the gene symbol (in italics) with the differential allele. For example, C57BL/10SnJ-Tyr^{c-11J}/J is a co-isogenic strain where a spontaneous mutation of the *Tyr* gene (Tyr^{c-11J} allele) appeared in a pure C57BL/10SnJ inbred background. An example in the rat is the co-isogenic ZUC-$Lepr^{fa}$ strain, where the spontaneous $Lepr^{fa}$ mutant allele arose on a pure ZUC inbred background. Spontaneous or chemically induced mutations are not the only genetic modifications that can be in co-isogenic state. For example, the FVB/NJ-Tg(Tyr)3412ARpw/EiJ line was generated by microinjecting the (Tyr)3412ARpw transgene into FVB/N fertilized eggs, generating a co-isogenic line on a pure FVB/N background (Fig. 3.2).

A strain is classified as *fully congenic* once it has undergone a minimum of ten backcross generations to the host background strain, with the initial hybrid or F1 generation counted as generation 1 (see Chap. 1). At this point, the unlinked donor genome is predicted to be less than 0.01%. However, when the strain reaches backcross generation N5, it is classified as an *incipient congenic*, with the donor genome contributing less than 3.2%. It is at this juncture that the congenic nomenclature can be applied. Alternatively, employing the marker-assisted "speed congenic" methodology allows achieving a fully congenic state in around five backcross generations (see Chap. 5).

Congenic strain (*N5 and above*) symbols (from left to right) contain: the official abbreviated symbol of the recipient strain followed by a *period (.)*; the abbreviation for the donor strain (in which the mutation originated); a hyphen; the introgressed allele(s) or congenic interval of interest; a forward slash; and the approved laboratory code (Fig. 3.3). For example, D2.HRS-*Hr*^{hr}/J is the designation for a congenic strain carrying the spontaneous "hairless" (*hr*) mutation developed by backcrossing the donor HRS inbred strain to DBA/2 and selecting for the *hr* allele (made and maintained at The Jackson Laboratory). If a chromosomal segment is transferred instead of a single mutation, the congenic line can be defined by listing the most proximal and the most distal markers demonstrated to be in the donor-derived segment (in parentheses), separated by a hyphen. For example, B6.D2-(*D3Mit25-D3Mit130*) is a congenic strain

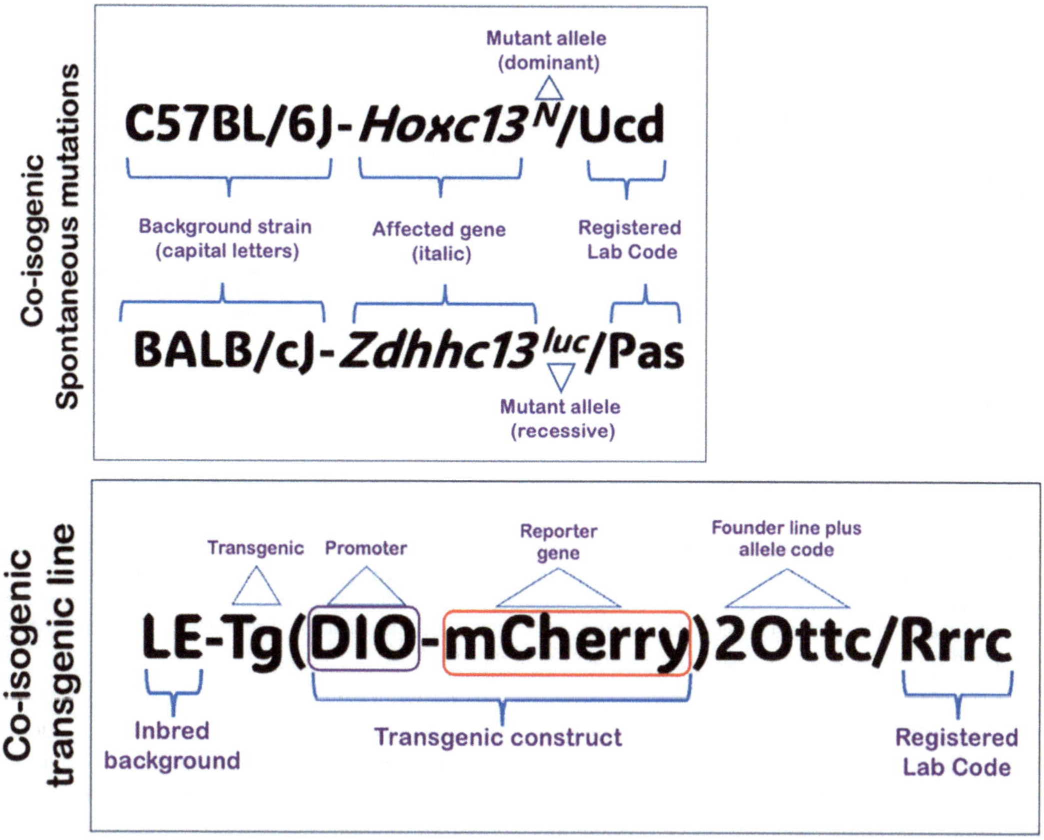

Fig. 3.2 *Nomenclature for co-isogenic lines with pure background.* Above: examples of two different spontaneous mouse mutations, one in C57BL/6 and the other in BALB/c background. Note that for co-isogenic strains, a hyphen separates the inbred background and the gene (italicized and first letter uppercase). The mutant allele is always superscript (first letter uppercase if dominant). Below: example of a rat transgenic (Tg) line with LE inbred background. Note that no component of a transgene symbol is italicized (or superscript), as these represent arbitrary insertions of foreign DNA material and are distinct from the native genome

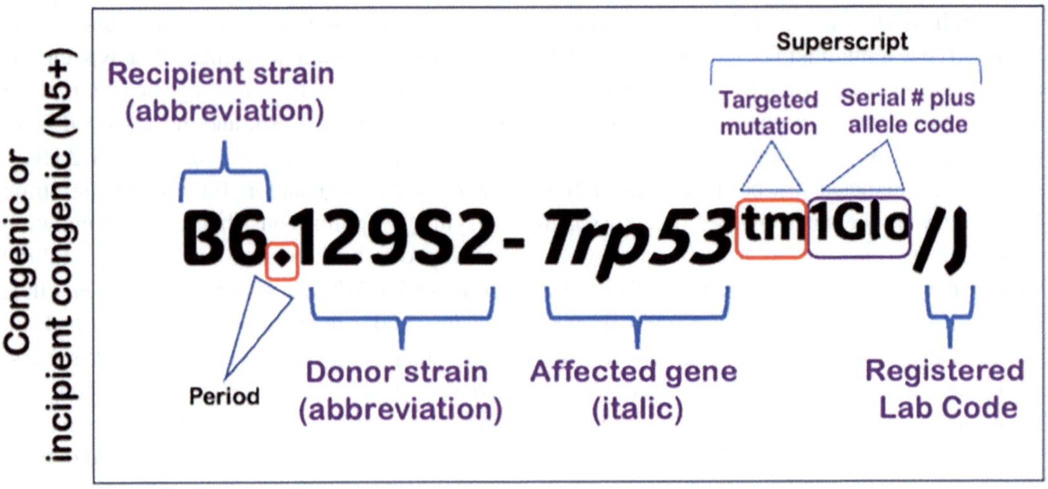

Fig. 3.3 *Nomenclature for congenic and incipient congenic lines.* Example of a congenic (or incipient congenic) line carrying a KO (or KI) targeted mutation with C57BL/6 as main background. Note that the official abbreviated symbol of the recipient strain is followed by a period, then the abbreviation for the donor strain (where the mutation was originated). Then, a hyphen separates the background information and the gene (italicized and first letter uppercase). The targeted mutant allele is always superscript. A strain is classified as an incipient congenic when reaches backcross generation N5

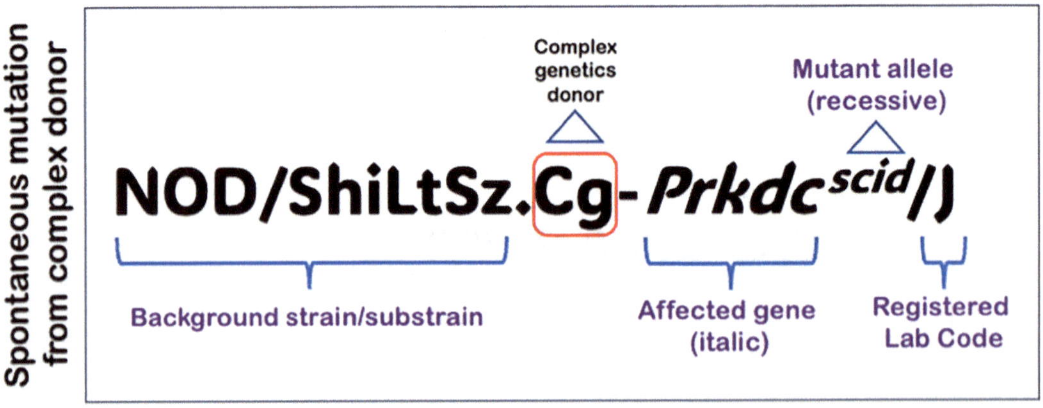

Fig. 3.4 *Nomenclature for lines with a complex donor component.* Example of a congenic line carrying the *scid* spontaneous mutation with NOD as main background. Note the use of the symbol Cg (complex genetics) as donor (where the mutation was originated). Cg should be used for strains carrying an allele (or alleles) coming from unknown or complex donor origins. A hyphen separates the background information and the gene (italicized and first letter uppercase). The mutant allele is always superscript

made by introducing into C57BL/6 J a segment from DBA/2 strain, extending between the two defined SSLP markers on chromosome 3. Incipient congenics may be given congenic nomenclature at N5, as long as the number of generations of backcrossing is clearly documented in information accompanying the strain.

In situations where the genetic background of the chromosomal segment containing the mutation is unknown or unclear, for example, due to the donor not being inbred, or have a *complex genetic* makeup, or involves an F1 hybrid with uncertainty about the parental segment in which the mutation emerged, the symbol "*Cg*" should be employed to signify this intricate genetic origin (Fig. 3.4). The utilization of the donor strain symbol or Cg is crucial to differentiate congenic strains from co-isogenic strains. The symbol Cg should also be used to designate a strain with multiple alleles that came from different donor origins [11].

3.6 Nomenclature for the Different Types of Mutant Alleles

Spontaneous mutations are typically named using phenotypical characteristics until the affected gene is identified (i.e., the gene and allele symbols are the same until the gene is known). In other words, where a gene is known only by a mutant phenotype, the gene is given the name and symbol of the first identified mutant, for example, nude, hairless, and obese mice, or fatty and tremor rats. Symbols of mutations conferring a recessive phenotype begin with a lowercase letter. Symbols for dominant or semidominant mutations begin with an uppercase letter (Fig. 3.2). Once the affected gene is identified, the symbol is placed above (superscript) the gene's name. For example, *hr* for the recessive allele hairless (Hr^{hr}), *Bbfc* for the dominant allele Babyface ($Sox9^{\mathrm{Bbfc}}$), and N for the semidominant allele *Naked* ($Hoxc13^{\mathrm{N}}$). New allelic mutations at the same locus, if they have the same phenotype, are given the same name with a Laboratory code preceded by a serial number. For example, $Tyr^{c\text{-}2J}$ is the second albino (c) mutation of the Tyrosinase gene identified at The Jackson Laboratory (J). All the different alleles of the same gene are represented by a superscript to the gene symbol, for example, $Ass1^{\mathrm{bar}}$, $Ass1^{\mathrm{fold}}$, $Ass1^{\mathrm{em1Gpt}}$, and $Ass1^{\mathrm{tm1Bay}}$ are all alleles (spontaneous and engineered) of the argininosuccinate synthetase 1 (*Ass1*) gene.

Random inserted transgenes, generated usually by pronuclear injection of DNA constructs (additive transgenesis), are given a symbol consisting of three parts: the symbol "*Tg*" denoting a transgene, the official gene symbol of the inserted DNA (in parentheses), the serial number of the line (founder), and the laboratory code of the originating lab. No component of a transgene symbol is italicized, as these represent arbitrary insertions of foreign DNA material and are distinct from the native genome (Figs. 3.1 and 3.2). For example, Tg(Ela1-Myc)160Bri is the symbol of a transgene that contains promoter sequences from the rat elastase 1 (*Ell a*) gene and exons 2–3 of the mouse *Myc* oncogene from Dr. Brinster laboratory (line ID 160 and lab code Bri). Note that the random insertion of a transgene into an endogenous gene (insertional mutagenesis) may produce a new allele of this gene. This new allele should be named as a mutant allele of the gene, with the symbol for the transgene as the superscript. For example, $awg^{Tg(GBtslenv)832Pkw}$ is the *abnormal wobbly gait* mutation (gene not identified yet) caused by the insertion of a transgene (from line 832), produced in the laboratory of Paul Wong. Regarding genotype notation, animals that are hemizygous for a transgene are designated Tg/0,[1] while homozygous are labeled as Tg/Tg.

Mutations that result from gene targeting in ES cells (classic KOs and KIs) are denoted by a three-part superscript symbol placed above the gene's name, incorporating "*tm*" to indicate a targeted mutation, a serial number for the line, and the laboratory code where the mutation was generated (Figs. 3.1 and 3.3). For example, $Trp53^{tm1Ty}$ is a KO allele of the tumor suppressor gene *Trp53* developed in the laboratory of Dr. Tyler Jacks (lab code Ty). However, this nomenclature does not differentiate KI from KO or conditional alleles, for example, the $Trp53^{tm1Glo}$ allele from Dr. Lozano laboratory (lab code Glo) is a humanized KI introducing an arginine to histidine substitution at residue R172 (R172H). Large-scale projects that systematically produce large numbers of targeted alleles may include a project abbreviation in parentheses as part of the allele designation. For example, $Gstm3^{tm1(KOMP)Vlcg}$ is a targeted allele created by Velocigene in the KOMP knockout project.

Similar to ES cell-derived targeted mutations, endonuclease-mediated mutations (ZFP,

[1]Transgenes are extra segments of DNA that have no corresponding "wild-type" sequence in the unmodified homologous chromosome in hemizygous animals, that is why it is recommended the use of "0" instead of "+" (typically used to denote wild-type alleles).

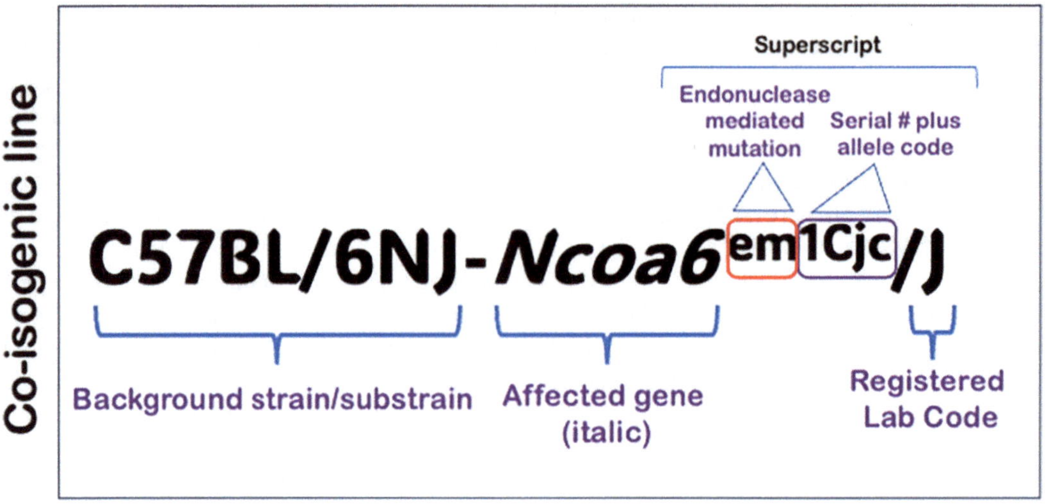

Fig. 3.5 *Nomenclature for lines with genome edited mutations.* Example of an endonuclease-mediated (em) mutation (e.g., CRISPR/Cas) in pure C57BL/6 background (co-isogenic line). A hyphen separates the inbred background and the gene (italicized and first letter uppercase). Note that the symbol "em" and the serial number plus allele code of this particular mutation are superscript

TALENs, CRISPR/Cas) are given the symbol of the mutated gene with a superscript consisting of three parts: the symbol "*em*" to denote an endonuclease-mediated mutation, a serial number from the laboratory of origin and the ILAR registered laboratory code of the laboratory where the mutation was produced (Fig. 3.5). For example, the $Trp53^{em1Yoli}$ allele is a single point mutation generated by the laboratory of Dr. Yong Li (lab code Yoli) using Zinc finger endonucleases; and the $Pten^{\mathrm{em1Miii}}$ KI allele was created by Dr. Miho Iijima (lab code Miii) using CRISPR/Cas9 technology.

For complete rules, recommendations and general principles of allele nomenclature and genotype documentation in GA animals refer to recent publications [2, 4, 9].

3.7 Nomenclature for Outbred Stocks

Outbred mice and rats are genetically uncharacterized; hence, no two individuals within an outbred stock share identical genetic makeup [3]. The intentional avoidance of inbreeding distinguishes outbred colonies, as their primary objective is to sustain maximal heterozygosity. The utilization of outbred stocks offers the advantage of cost-effectiveness due to their extended lifespan, resistance to diseases, and high reproductive capacity. Outbred mice and rats can be used in many experiments where the genetic component is not important (vasectomized males for inducing ovulation in female mice, recipient females for embryo transfer, etc.). However, a statement concerning outbred stocks that appears to be controversial nowadays is that they are a better choice for many experiments because they are genetically heterogeneous, like humans. An excellent review by Dr. Michael Festing is accessible via the Internet (http://www.isogenic.info/html/animal_models_in_research.html). We strongly recommend all scientists performing biological assays or other types of experiments to read Dr. Festing's comments [5]. Reading his recommendations will help to make the best possible choice of experimental animals and accordingly the better experimental design. Improved outbred stocks like J:ARC(S) and J:DO are now available and discussed in Chap. 1.

The first International Standardized Nomenclature for Outbred Stocks, including breeding systems, was completed and approved in 1971, published in 1972, and very recently reprinted by the ILAR Journal [6]. Unlike inbred strains, outbred stocks are identified by appending the Laboratory Code *before* the common stock root. For example, Crl:WI is a Wistar rat stock maintained by Charles River Laboratories, J:DO is the Diversity Outbred mouse stock maintained by The Jackson Laboratory, and RjOrl:SWISS is the albino "Swiss" mouse stock maintained by Janvier Labs (Lab code Rj, previously from Orleans, France).

3.8 Laboratory Codes

A crucial aspect of mouse and rat nomenclature involves the Laboratory Registration Code, often referred to as the Laboratory code [2]. This code, typically comprising three to four letters, with the initial letter in uppercase and the rest in lowercase, serves to distinguish a specific institution, laboratory, or researcher responsible for generating and potentially maintaining a populations of a given mouse or rat strain. Substrains, along with congenic and other strains that possess multiple indistinguishable variations, should be differentiated using Laboratory codes. These codes, coordinated by the Board on Animal Health Sciences, Conservation, and Research (BAHSCR) (National Academies of Sciences, Engineering, and Medicine), are utilized to discern specific institutions, laboratories, or researchers accountable for the development and preservation of distinct mouse or rat strains. (https://www.nationalacademies.org/ilar/lab-code-database). Some examples of Laboratory codes from institutions or vendors are "J" for The Jackson Laboratory, "N" for the National Institutes of Health of the United States, "Kyo" for Kyoto University, "Tac" for Taconic Biosciences, "Crl" for Charles River Laboratories, and "Hsd" for Harlan Sprague Dawley/Envigo/Inotiv.

3.9 Nomenclature for Mouse and Rat Genes

Mouse and rat gene symbols are *italicized* and begin with an uppercase letter followed by all lowercase letters, for example, *Lepr* for leptin receptor or *Foxn1* for forkhead box N1. Human gene symbols are also italicized but are written in all uppercase letters. Always remember that a gene may have multiple *synonyms*. These synonyms refer to the various names that have been used for the gene across different periods and have been linked to the gene in databases and publications. For further information regarding genes and alleles, we suggest referring to the Guidelines for Nomenclature of Genes, Genetic Markers, Alleles, and Mutations in Mouse and Rat (revised in September 2021) on the MGI webpage. For basic rules and guidelines to assist in the symbol and naming process, the Quick Guide to Nomenclature for Genes and the Quick Guide to Nomenclature for Alleles and Mutations on the MGI webpage can provide valuable insights.

References

1. Benavides F, Rulicke T, Prins JB, Bussell J, Scavizzi F, Cinelli P, Herault Y, Wedekind D. Genetic quality assurance and genetic monitoring of laboratory mice and rats: FELASA Working Group Report. Lab Anim. 2020;54:135–48.
2. Brayton CF. Laboratory codes in nomenclature and scientific communication (advancing organism nomenclature in scientific communication to improve research reporting and reproducibility). ILAR J. 2021;62:295–309.
3. Chia R, Achilli F, Festing MF, Fisher EM. The origins and uses of mouse outbred stocks. Nat Genet. 2005;37:1181–6.
4. Dobrowolski P, Buch T, Nagel-Riedasch S. Nomenclature for standardized designation of diploid genotypes in genetically modified laboratory animals. Lab Anim. 2023;57:371–80.
5. Festing MF. Inbred strains should replace outbred stocks in toxicology, safety testing, and drug development. Toxicol Pathol. 2010;38:681–90.
6. Festing M, Kondo K, Loosli R, Poiley SM, Spiegel A. International standardized nomenclature for outbred stocks of laboratory animals. ILAR J. 2021;62:362–6.

7. Guenet JL, Benavides F. Genetic monitoring of laboratory rodents. In: Patrinos GP, Ansorge W, editors. Molecular diagnostics. 2nd ed. Oxford: Oxford Academic Press; 2010.
8. Guenet JL, Benavides F, Panthier J, Montagutelli X. Genetics of the mouse. Berlin: Springer; 2015.
9. Perry MN, Smith CL. Murine allele and transgene symbols: ensuring unique, concise, and informative nomenclature. Mamm Genome. 2022;33: 108–19.
10. Sundberg JP, Schofield PN. A mouse by any other name. J Invest Dermatol. 2009;129:1599–601.
11. Valenzuela A. Nomenclature: naming your gene-modified mouse. Methods Mol Biol. 2020;2066:149–62.

4 Genetic Quality Controls for Inbred Strains and Outbred Stocks

Fernando Benavides and Thomas Rülicke

Abstract

Genetic quality control measures are imperative in maintaining the integrity and reliability of inbred strains and outbred stocks within laboratory animal facilities. In addition to health and microbiology monitoring, genetic quality control plays a crucial role in ensuring the accuracy of animal models. Therefore, it is essential that research studies utilizing laboratory rodents, particularly mice and rats, employ genetically defined animals. This chapter aims to outline the objectives and available methods for implementing genetic quality assurance programs in rodent facilities, highlighting two primary goals for inbred strains: (i) verifying the authenticity and uniformity of inbred strains and their substrains to ensure consistent genetic backgrounds and reliable colony maintenance, and (ii) detecting potential genetic contamination. When it comes to outbred rodents, the implementation of genetic quality controls presents some unique challenges since outbred stocks are bred for genetic heterogeneity. However, there are still important genetic quality control measures that can be applied to ensure the reliability and consistency of outbred populations.

Keywords

Inbred Strains · Quality Control · Genetic Monitoring · Microsatellites · Single Nucleotide Polymorphism · Background Characteriation

4.1 Introduction

In the past, several accidental outcrosses have occurred in different mouse and rat inbred colonies, leading to genetic contaminations [3, 65]. Most commercial breeders (vendors) and international repositories or resources are highly conscious of the risks associated with breeding mice and rats, and they actively engage in regular monitoring of their stocks and strains [19, 24, 37, 53, 67]. In addition, many of them take measures to protect their strains by archiving them as cryopreserved embryo and/or sperm banks (see Chap. 6). This proactive approach ensures prompt recovery of the strain whenever it becomes necessary. Regrettably, not all breeding colonies managers follow this diligent practice. Besides, archiving of outbred rodents by cryopreservation

F. Benavides (✉)
Department of Epigenetics and Molecular Carcinogenesis, The University of Texas MD Anderson Cancer Center, Houston, TX, USA
e-mail: fbenavid@mdanderson.org

T. Rülicke
Department of Biomedical Sciences, University of Veterinary Medicine Vienna, Vienna, Austria
e-mail: Thomas.Ruelicke@vetmeduni.ac.at

F. Benavides, A. K. Hansen (eds.), *Rodent Quality Control: Genes and Bugs*, Laboratory Animal Science and Medicine 2, https://doi.org/10.1007/978-3-031-59103-7_4

is unrealistic, as a very large number of animals would have to be included to avoid a bottleneck effect.

The objective for inbred strains and outbred stocks of laboratory rodents is to maintain their genetic quality and phenotypic characteristics consistently over generations. This stability is a prerequisite to obtain reproducible and comparable results in animal experiments. Genetic quality controls are indispensable in evaluating the uniformity and authenticity of a strain, as well as detecting genetic contamination and preventing genetic drift, which can lead to genetic variation within a strain over time. The implementation of these controls is vital to ensure the reliability and reproducibility of scientific research that relies on inbred strains [66]. In addition, these controls play a crucial role in monitoring breeding programs and selecting future breeders. In the case of outbred colonies, where direct testing of authenticity is not feasible, the level of genetic heterogeneity and the genetic equilibrium within the colony serves as an alternative indicator. Currently, genetic monitoring of mouse and rat strains primarily depends on the use of DNA markers with known locations on chromosomes [6, 30, 55, 58, 69, 76].

4.2 Genetic Monitoring of Inbred Strains

It is essential to adopt a comprehensive approach to genetic monitoring that encompasses not only molecular techniques but also considers phenotypic parameters like coat color, behavior, and breeding performance, which can be changed by spontaneous mutations in protein-coding sequences. While commercial breeders monitor their strains for genetic contamination using various sets of Single Nucleotide Polymorphisms (SNPs), they may not always account for genetic drift. As an exemplary measure to ensure the crucial genetic stability in their mouse strains, The Jackson Laboratory (JAX) has implemented a distinctive and patented Genetic Stability Program [67]. This program was designed to effectively limit cumulative genetic drift by rebuilding their foundation stocks from pedigreed, cryopreserved embryos every five generations. For example, starting in 2005, they began selling only C57BL/6 J mice derived from two chosen mice (Adam and Eve mice) through hundreds of frozen embryos of the duo's grandchildren, enough to last for 25–30 years [57]. The complete genome of the "Eve" female mouse was recently published [59]. It is important to highlight that thorough genetic monitoring should be conducted when retrieving strains from frozen stocks to ensure the absence of genetic contamination or incorrect genotypes prior to freezing. Many vendors maintain backup archives in an embryo bank, enabling the prompt establishment of a new nucleus colony if needed. The International Council for Laboratory Animal Science (ICLAS) has been actively promoting genetic monitoring programs and assisting academic institutions in their development and implementation to enhance overall quality assurance for mouse and rat models. The current recommendations from ICLAS have been reviewed in the work of Fahey et al. [25]. The JAX's Genetic Quality Control (GQC) Program is an example of an effective program that prevents genetic contamination by detecting and eliminating breeding errors early, before mice are given the chance to breed. In order to verify the genetic background, mice from expansion and production colonies from each inbred strain are randomly selected and SNP-genotyped at least once annually (Fig. 4.1).

The use of standard genetic nomenclature plays a crucial role in facilitating effective communication among researchers working with mice and rats. It establishes a common language that ensures accurate and unambiguous identification and description of the specific model being used. Consistency in naming strains across different laboratories and publications enhances reproducibility, enabling researchers to precisely identify and reproduce experimental findings. Furthermore, standardized nomenclature enables seamless integration and sharing of strain information across databases and repositories.

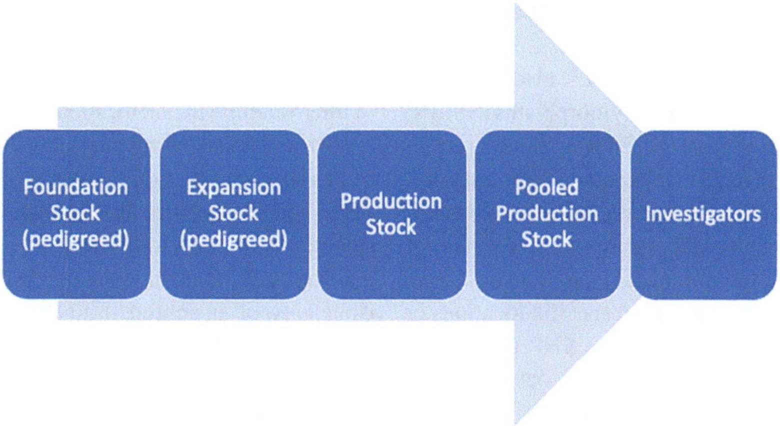

Fig. 4.1 JAX Genetic Quality Control Program. This is an example of a rigorous mouse colony management with proven husbandry practices and physically isolated foundation, expansion, and production stocks. Both foundation and expansion stocks maintain detailed pedigree records and are restricted to fewer than 10 generations from the main pedigree line. To ensure genetic integrity, production stocks are methodically replenished with mice sourced from the foundation stock. This type of program prevents genetic contamination by detecting and eliminating breeding errors before mice are given the chance to breed. In addition, animal technicians are constantly on the lookout for any phenotype deviations, such as alterations in coat color or physical abnormalities

Additionally, it simplifies the organization and maintenance of breeding records and pedigrees. The importance of standardized genetic nomenclature will be extensively addressed in Chap. 3, providing comprehensive insights into this critical aspect of research.

4.3 Current Practices for Genetic Monitoring

Inbred strains have well-defined sets of genetic markers with known locations on chromosomes. These variable (polymorphic) markers act as genetic flags, allowing researchers to compare the genetic profiles of individual inbred mice or rats. By analyzing the presence or absence of specific alleles from these markers, we can determine the strain's identity and detect any genetic contamination. Historically, genetic monitoring heavily relied on biochemical markers, such as enzyme polymorphisms, and immunological markers like the major histocompatibility complex (in fact, the mouse H2 haplotype may be the only tool available for typing congenic strains that differ only at their H2 loci). In addition, histocompatibility was often tested by skin grafting, which was a stressful procedure for the animals. However, these methods were often cumbersome and time-consuming. In recent years, there has been a shift toward more streamlined and scalable DNA genotyping techniques, which have largely replaced the older methods. Currently, SNPs have become the standard genetic marker used in quality assurance programs for mice and rats. They are widely employed to determine identity and assess genetic variation. Following SNPs, microsatellites are the second most commonly utilized DNA marker in genetic monitoring. Besides the genome-wide DNA markers used to determine identity, gene-specific genotyping (e.g., mutations or targeted alleles) are also important. This can be done using various methods such as polymerase chain reaction (PCR), DNA sequencing, or microarray analysis. Gene-specific genotyping allows researchers to distinguish between substrains, to assess zygosity among individuals within a strain and to identify and detect potential mislabeling. Chapter 5 has a section dedicated to explaining the quality controls needed for genetically modified (altered) strains.

4.3.1 Microsatellite Genotyping

In the mid-1990s, microsatellites, also known as Simple Sequence Length Polymorphisms (SSLPs) or Short Tandem Repeats (STRs), gained prominence as the preferred DNA markers for authenticating inbred strains. However, with the advent of readily accessible SNP genotyping, the use of microsatellites has diminished. Nonetheless, microsatellites still find utility in specific locations or species, particularly rats, where SNP genotyping may not be as readily available or accessible. SSLPs are regions of DNA characterized by the repetition of short nucleotide sequences located mostly in non-coding regions of the genome. These markers consist of a core sequence typically composed of 1–6 base pairs (bp), which is repeated multiple times in tandem along the DNA molecule (Fig. 4.2).

SSLP genotyping is an affordable and straightforward typing method that allows for easy detection of genetic variations within and between inbred strains [5, 31, 44, 54]. Animals are genotyped by analyzing PCR products amplified from the short, tandemly arranged DNA sequences. Genomic DNA primers are designed to unique sequences flanking the tandem repeats. The PCR products, typically around 100–300 bp in size, are analyzed using agarose or polyacrylamide gel electrophoresis. There are an enormous number of microsatellite loci in the mouse and rat genomes (~10^5), and it is generally possible to identify a set of markers whose amplification products define a strain-specific pattern. Routine analysis of DNA samples with carefully selected microsatellite markers will confirm both the authenticity and isogeneity of an inbred strain. Since microsatellites are multiallelic markers, each locus in a species can exhibit multiple alleles that can be differentiated by their individual amplification product sizes. For example, a single mouse or rat microsatellite can present up to eight different allele sizes among different strains. The use of fluorescently labeled primers to amplify microsatellite loci combined with capillary electrophoresis provide an even faster and more automated means for genetic monitoring [44]. With this method, the resulting PCR products can be distinguished from one another by both their size and the fluorescent dye associated with them, which aids in multiplexing the PCR reactions [12].

The Mouse Genome Informatics (MGI) webpage has comprehensive SSLP information, including primer sequences and size variations in

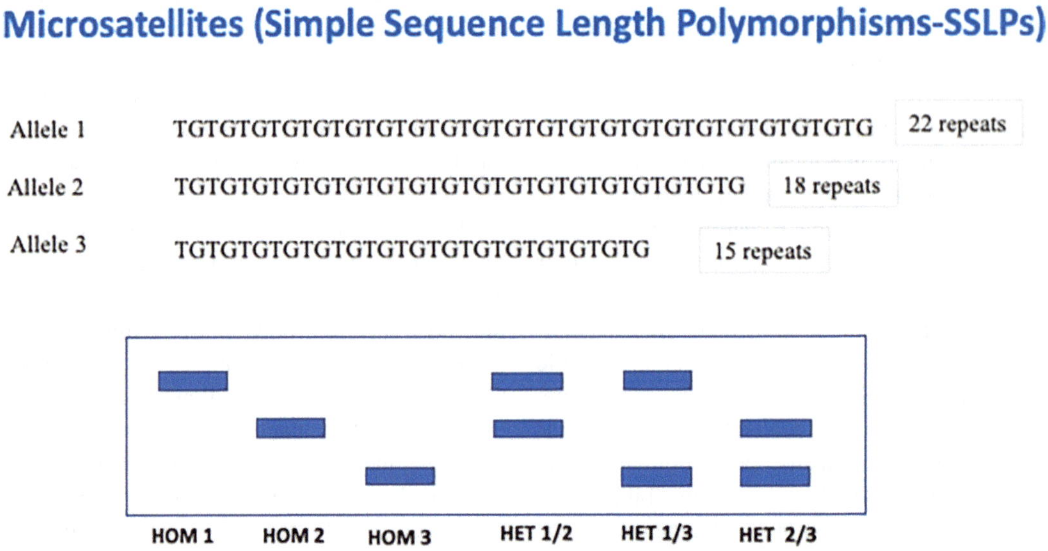

Fig. 4.2 Microsatellite markers. The figure shows three different alleles at a theoretical microsatellite locus composed of TG repeats (motif). Note that the number of repeats is variable, and this is the base of the polymorphism. Using specific primers flanking these repeats we can amplify and detect the various possible allele combinations in an electrophoresis gel, as shown in the illustration [4]. (Printed with permission from Springer Nature)

bp for several inbred mouse strains (http://www.informatics.jax.org/marker). Same type of information is presented for several rat strains in the marker section of the Rat Genome Database (RGD) (https://rgd.mcw.edu/rgdweb/search/markers.html?100). A collection of 357 mapped, highly polymorphic, SSLP markers for 179 inbred laboratory rat strains is available at The National BioResource Project Rat database (https://www.anim.med.kyoto-u.ac.jp/nbr/pedigree/sb.aspx).

The nomenclature system for microsatellites is based on the chromosome number, followed by a marker identifier or name, which is often a combination of letters and numbers. For instance, "D2Mit123" indicates the 123rd marker from the MIT group on mouse chromosome 2. An example in the rat is D5Rat81.

4.3.2 Single Nucleotide Polymorphism (SNP) Genotyping

SNP genotyping is a widely adopted approach for genetic monitoring, aimed at identifying single base-pair variations in the DNA sequence. This technique has gained popularity due to the ability to be outsourced and its affordability. SNPs, being the most prevalent form of genetic variation, occur in both coding and non-coding regions of the genome. Most SNPs are bi-allelic, meaning they exhibit two possible nucleotides at a given position. This results in three possible genotypes for an individual: homozygous for one nucleotide (e.g., C/C or T/T) or heterozygous with both nucleotides (e.g., C/T) (Fig. 4.3). SNP genotyping allows researchers to identify these specific alleles, providing insights into the identity of the sample analyzed. The abundance of SNPs (several millions in most species) and their bi-allelic nature make SNP genotyping an efficient method for genetic monitoring. Interestingly, human ancestry studies typically utilize SNP data from large-scale genotyping or sequencing projects such as the International HapMap Project, the 1000 Genomes Project, and commercial genotyping services like 23andMe and AncestryDNA.

SNP genotyping assays are currently based on allele-specific PCR (including KASPar™ fluorescent technology[1]) [52], real-time PCR (TaqMan®), direct sequencing, single-primer enrichment technology (SPET), and DNA arrays [1, 17, 49, 50]. Petkov and co-workers from JAX have described the allelic distribution of 235 SNPs in 48 mouse strains and selected a panel of 27 SNPs sufficient to characterize the majority of the ~300 inbred, wild-derived, congenic, consomic, and recombinant inbred strains maintained at The Jackson Laboratory [56]. This set of markers, encompassing all mouse chromosomes, is an excellent tool for detecting genetic contamination in mouse facilities. The same laboratory developed a larger set of informative SNPs selected from publicly available databases using Amplifluor genotyping [51]. The 2000 plus markers in this panel are informative and used to control 103 mouse strains maintained by JAX, including all the common inbred strains.

High-density microarrays developed for gene mapping purposes could also be used to create a complete SNP profile characterization for a new or poorly described inbred strain. For example, the Mouse Diversity Genotyping Array [50, 74] is a high-density mouse genotyping microarray (Affymetrix) that simultaneously assays over 600,000 SNPs. The Mouse Universal Genotyping Array (MUGA) in its MegaMUGA format has ~78,000 SNPs and the GigaMUGA provides more than 143,000 SNPs [50]. All of them were built on the Illumina Infinium platform. The GigaMUGA array also includes non-SNP probes intended for exploring copy number variations (CNV) of specific DNA segments. The MiniMUGA, a smaller and more affordable version of the original MUGA, was designed specifically for genetic monitoring of inbred mouse strains. The MiniMUGA array includes approximately 10,000 SNPs, which is sufficient to distinguish between most inbred mouse strains and detect genetic contamination [61]. This publication includes consensus genotypes for 241 inbred strains including classical, wild-derived, and recombinant inbred lines.

[1]KBioscience's competitive allele-specific PCR amplification of target sequences and endpoint fluorescence genotyping (KASPar™).

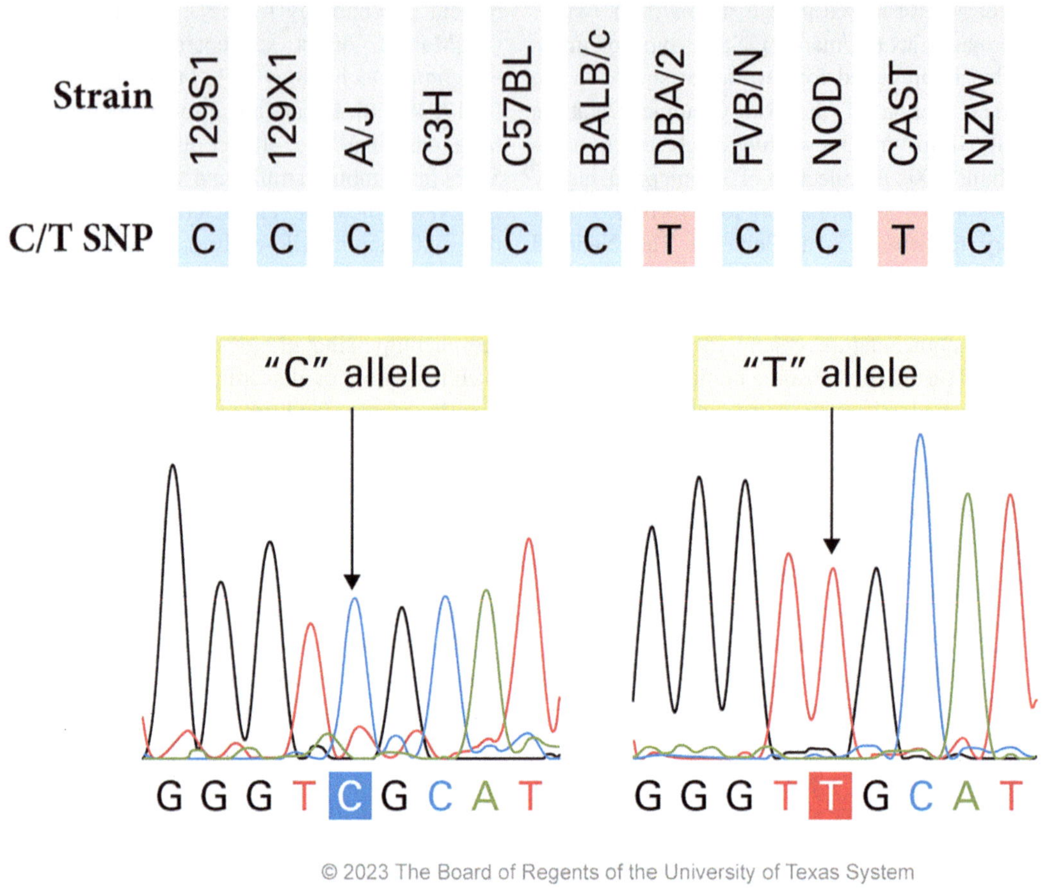

Fig. 4.3 Single Nucleotide Polymorphisms. SNPs are discrete DNA sequence variations occurring when a single nucleotide in the genome differs between members of the same species. SNPs are common, and they are scattered throughout the genome of all species. This allows the determination of a SNP pattern, which is unique to a given strain and accordingly can be used for assessing strain purity. The upper panel represents a C/T SNP that is polymorphic between strains DBA/2 and CAST (homozygous for the "T" allele) and other common inbred strains (homozygous for the "C" allele). The lower panel presents DNA sequencing electropherograms showing the SNP alleles (arrow)

The MiniMUGA includes a set of over 450 diagnostic alleles that are informative for differentiating between most inbred mouse strains. Additionally, these diagnostic SNPs can be utilized to identify substrains. Besides the SNPs, the array includes probes for the detection of constructs used in genetically engineered mice, like SV40 large T antigen, Cre recombinase, tetracycline-controlled transactivator (tTA), reverse improved tetracycline-controlled transactivator (rtTA), firefly luciferase, and CRISPR associated protein 9. The MiniMUGA array is a valuable tool for mouse background characterization allowing researchers to confirm the genetic identity and integrity of inbred strains, detect genetic contaminations or outcrossing events. This array has also been utilized in quantitative trait locus (QTL) mapping, aiming to identify genomic regions associated with specific traits or phenotypes [20]. In summary, the MiniMUGA provides a comprehensive overview of the genetic background of a mouse, offering valuable insights into various aspects such as genotyping quality, chromosomal sex, inbreeding, detected constructs, primary and secondary backgrounds, Y chromosome, and mitochondrial DNA analysis. Examples of MiniMUGA reports are provided in Fig. 4.4.

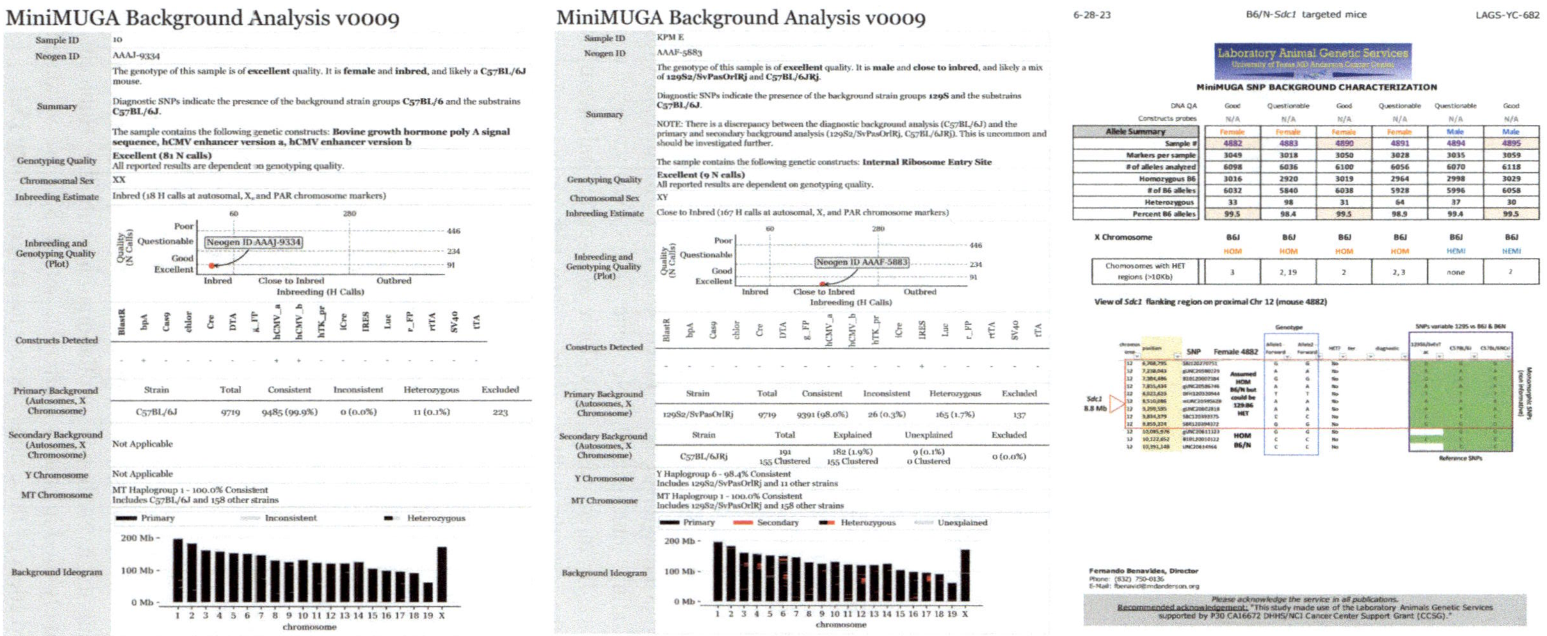

Fig. 4.4 The MiniMUGA SNP array. *Left panel:* Examples of the MiniMUGA (~9.6 K SNPS) analysis (first page) provided by Neogene (https://www.neogen.com/) via Transnetyx (https://www.transnetyx.com/monitoring). The summary on the left is from a female mouse that is almost pure C57BL/6 J (0.1% heterozygosity). The ideogram on the bottom shows a few heterozygous regions, but with inconsistent SNP results (gray color). This genetically engineered mouse is positive for some genetic construct probes (e.g., hCMV enhancer). The summary on the right is from a male mouse with a mixed background: mostly C57BL/6 J with 129S2 as secondary background (1.7% heterozygosity). The ideogram shows a few heterozygous regions (black is C57BL/6 J and red is 129S2). This genetically engineered mouse is also positive for one construct probe. *Right panel:* Example of in-house (MD Anderson Cancer Center) extensive analysis of the MiniMUGA SNP raw data. The analysis is based on SNPs that are polymorphic between B6 and 129S (~3000 SNPs). SNPs analyzed vary a little between the samples due to N/A calls (and quality of DNA). The estimated percentage of C57BL/6 background in theses samples ranges from 98.4% to 99.5%. The chromosomes showing residual heterozygosity (>10 Kb segments) are listed for each mouse. The screen capture below shows the targeted gene flanking region on proximal chromosome 12

Several publications have reported useful SNPs for the rat. Zimdahl and colleagues described a map with >12,000 gene-based SNPs from transcribed regions [77]. In another study, 485 SNPs were identified in 36 commonly used inbred rat strains [64]. More recently, the STAR (RATS backward) consortium reported a set of 20,000 SNPs across 167 inbred rat strains [16]. Another rat library was developed at Hubrecht institute, Utrecht, The Netherlands. This is a genome-wide set of 820 SNPs designed for the KASPar™ genotyping platform [52]. These SNPs were selected to be polymorphic between the BN and Wistar or SS strains and are evenly spread across the rat chromosomes.

4.3.3 Exome Sequencing as Potential Tool for Genetic Monitoring

Whole exome sequencing (WES) is a genomic technique that involves sequencing and analyzing the protein-coding regions of the genome. It is important to note that the term "exon" is sometimes mistakenly used interchangeably with "protein-coding" sequence. Thus, it is crucial to understand that WES only covers a fraction of the exons in the genome, excluding untranslated regions (UTRs), for example. In most cases, WES specifically targets the protein-coding regions of exons, although variants in the adjacent sequences may also be detected [2]. Currently, WES provides a robust method for discovering hereditary factors controlling rare Mendelian disorders in humans, as well as new mutations previously mapped through positional cloning in mice and rats [26]. WES can identify various genetic variations, including single nucleotide variants (SNVs), insertions, deletions, and structural variations within the exonic regions of the genome. By comparing the exome sequences of individuals within an inbred strain, researchers can detect genetic differences and assess the strain's genetic integrity. Even though this technique is not yet routinely used in animal facilities or by commercial breeders of mice and rats, the increasing availability of cost-effective benchtop sequencing equipment suggests that it may become more common soon, especially for the characterization of substrains. However, the biostatistical analysis and interpretation of the data will continue to pose challenges.

4.4 Outsourcing Genetic Monitoring for Mice and Rats

Microarray-based SNP genotyping is not typically used for small-scale, in-house genetic monitoring, but it is a good option for large breeders of inbred mice and rats. Most major laboratory animal companies offer mouse SNP panels for background characterization or "speed congenic" projects (see Chap. 2). When requesting SNP genotyping services, please note that, in a pairwise comparison, only a portion of the SNPs will exhibit polymorphism between the analyzed strains (e.g., ~30–40% for most classical inbred strain combinations). As mentioned, the MiniMUGA array [61] is a great option for routine genetic monitoring of mouse strains. It could also be utilized to generate a comprehensive SNP profile for newly created or poorly characterized inbred strains. The MiniMUGA testing is run by Neogene (https://www.neogen.com/) but it is available as a service only from Transnetyx, Inc. (https://www.transnetyx.com/monitoring). The MiniMUGA v2 Background Analysis Report tool, available at Transnetyx, aims to provide users with essential sample information derived from the MiniMUGA genotyping array and is provided with every sample genotyped. The myMUGA tool available at Transnetyx allows additional, customizable analysis of data generated from the MiniMUGA array.

In the case of rat inbred strains, besides some vendors, SNP assays are provided by LGC, Biosearch Technologies (https://www.biosearchtech.com/) as part of their genotyping service. The rat library used was developed in collaboration with the Hubrecht institute and offers a set of 820 pre-validated (KASPar™) assays for rat genotyping [52].

4.5 In-House Genetic Monitoring of Inbred Mice and Rats

In addition to implementing a robust genetic monitoring program for inbred colonies, it is highly recommended to procure animals from reputable vendors and refresh the breeding stock with animals from the same vendor within a maximum of ten generations (5 years). This approach is preferable to maintaining separate independent colonies of classical inbred strains. The separate breeding of an inbred colony over a significant period would result in a substrain due to the accumulation of undetected spontaneous mutations (see Chap. 1). By periodical replacement of the breeding nucleus by animals from the same reputable vendor, the formation of substrains carrying potential mutations can be effectively prevented, offering the additional advantage of better reproducibility and comparability of experimental results.

4.5.1 Phenotype Quality Control

When monitoring breeding colonies, it is vital to consider animal phenotypes. The vast majority of phenotypic characteristics of an animal are quantitative traits which are influenced by multiple genes, some of which may have pleiotropic effects, impacting multiple traits simultaneously. Moreover, quantitative traits such as body size and body weight are modified by environmental factors, such as the rearing and housing conditions of the animals, including different microbiomes. Consequently, an effective monitoring should encompass the measurement of an adequate number of animals. As previously mentioned, comprehensive monitoring should encompass the observation of innate behaviors, which can exhibit variations across different inbred strains. It is crucial to assess maternal care and rearing behaviors as part of routine breeding work. Additionally, it is important to consider obvious phenotypes of qualitative traits, such as coat color, as they can easily provide valuable insights during the monitoring process [40, 62].

Comprehensive genetic monitoring programs include the diligent monitoring of reproductive parameters, which involves evaluating the overall fertility of a given strain. A crucial indicator of genetic changes can be the routine calculation of the colony breeding index, which indicates the average number of born and weaned offspring per female and week, particularly under standardized environmental conditions. This parameter provides valuable insights into the reproductive performance and genetic stability of the strain being monitored. It is important to note that significant genetic alterations may be revealed through these parameters. To ensure the integrity of the results, it is imperative to conduct monitoring on a representative and randomly selected sample of breeding pairs. By avoiding selective bias, the monitoring process becomes more reliable and offers a comprehensive understanding of the strain's breeding capabilities. Notably, sudden increases in litter sizes or a rise in the breeding index [39] within an inbred strain strongly suggest the possibility of genetic contamination. Controlling for strain-specific pathologies is also important for discovering possible genetic contamination and genetic drift. Some examples of strain-specific pathologies in mice are teratocarcinomas in 129, thymic lymphomas in AKR, and Type 1 diabetes in the Non-obese Diabetic (NOD) (a model for Type 1 IDDM in man) [41]. Examples in the rat are Type 2 diabetes in the Zucker diabetic fatty rat (ZDF) and hypertension in the spontaneously hypertensive rat (SHR) [38].

The Jackson Laboratory has developed the Mouse Phenome Database (MPD) [8, 9] to collect baseline phenotypic data on the most popular inbred strains of mice through a coordinated international effort (Table 4.1). Also, the International Mouse Phenotyping Consortium (IMPC) was established in 2011 to test mutant lines through a broad-based primary phenotyping pipeline [29, 36]. The IMPC uses the International Mouse Phenotyping Resource of Standardized Screens (IMPReSS) phenotyping protocols, which are essential for the characterization of mouse phenotypes (see https://www.mousephenotype.org/impress) [46].

Table 4.1 This table was extracted from the Mouse Phenome Database (MPD) and represents red blood cell counts (measured in n/μL) performed on both female (left) and male (right) mice (8 weeks of age) from 11 different inbred strains maintained at The Jackson Laboratory

Females	Mean	SD	N	Males	Mean	SD	N
B6.129P2-*Apoe*^{tm1Unc}/J	10.4	± 0.272	10 mice	B6.129P2-*Apoe*^{tm1Unc}/J	10.4	± 0.281	10 mice
B6D2F1/J	11.1	± 0.280	10 mice	B6D2F1/J	11.3	± 0.464	12 mice
BALB/cByJ	10.5	± 0.253	21 mice	BALB/cByJ	10.8	± 0.377	20 mice
BALB/cJ	10.6	± 0.316	22 mice	BALB/cJ	10.7	± 0.613	21 mice
C3H/HeJ	9.48	± 0.244	13 mice	C3H/HeJ	10.1	± 0.254	11 mice
C57BL/6 J	10.8	± 0.477	42 mice	C57BL/6 J	10.6	± 0.383	34 mice
CBA/J	9.79	± 0.525	21 mice	CBA/J	10.3	± 0.391	23 mice
DBA/2 J	11.6	± 0.526	23 mice	DBA/2 J	12.0	± 0.516	20 mice
FVB/NJ	9.80	± 0.310	11 mice	FVB/NJ	10.8	± 0.350	14 mice
NOD.CB17-*Prkdc*^{scid}/J	9.50	± 0.290	11 mice	NOD.CB17-*Prkdc*^{scid}/J	9.40	± 0.259	12 mice
NOD/ShiLtJ	9.73	± 0.219	21 mice	NOD/ShiLtJ	10.3	± 0.293	26 mice

From MPD at http://phenome.jax.org/db/q?rtn=docs/home

The Phenotypes and Models Portal within the RGD [34] is a valuable resource for researchers working with rat strains and phenotypes. The portal offers a wealth of information on rat strains, including their genetic background, breeding details, and phenotypic characteristics. Researchers can access phenotypic data related to a wide range of traits, such as behavior, metabolism, cardiovascular function, and more [68].

4.5.2 Genotyping for a Small Set of Microsatellites Markers

It is recommended to conduct routine testing on in-house colonies using a small set of informative microsatellites to confirm their genetic integrity. Microsatellites are valuable for verifying the purity of animals within an inbred colony and ensuring the absence of genetic contamination. This verification becomes particularly significant in facilities where strains with identical coat colors are housed in the same room, as it poses an increased risk, especially in the absence of individually ventilated cage systems. By genotyping for a strategic microsatellite panel, potential genetic mix-ups or inadvertent contamination can be identified and addressed, thereby safeguarding the genetic integrity of the colony.

Performing SSLP testing in-house is feasible in most animal facilities. However, there is no standardized guideline for the number or ID of the markers to be used, as it varies depending on the specific situation and facility, as well as the strains being maintained. Nevertheless, a panel consisting of approximately 40–50 polymorphic SSLPs (or at least two per chromosome), which are evenly distributed across the autosomal chromosomes, can effectively detect, and rule out recent genetic contamination, given that the selected markers possess the capability to distinguish among the strains being analyzed. As shown in the example on Fig. 4.5, in cases of recent genetic contaminations, even five well-selected SSLPs are enough to clearly detect accidental outcrosses.

Typically, SSLP genotyping is carried out through standard endpoint PCR, and the necessary primers can be obtained from any reputable source. However, before initiating the PCR reaction, it is crucial to assess the quality and quantity of the DNA template acquired, preferentially from ear punches. This preliminary evaluation ensures that the DNA template is suitable for accurate and reliable genotyping analysis [10]. In order to provide a precise size standard for the expected alleles, it is essential to include control PCR reactions using DNA templates from authenticated samples representing each strain. These control reactions serve as a reliable reference for the expected band sizes. When analyzing the genotyping results, it is important to compare the observed band sizes to this internal standard rather than solely relying on estimations based on

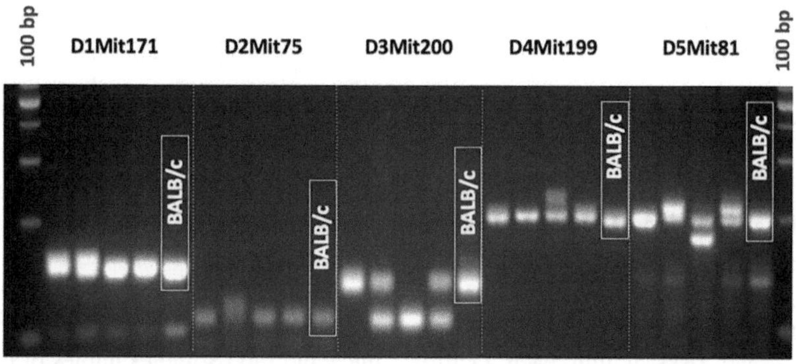

Fig. 4.5 Detecting genetic contamination with only five SSLPs. Example of genetic contamination detected by SSLP PCR. The picture shows an ethidium bromide-stained 4% agarose gel with the characteristic bands obtained after PCR amplification using genomic DNA from four mice supposedly belonging to the BALB/c strain (first four lanes), plus a standard DNA control for BALB/c (last lane). In this example, only five SSLP loci are shown, located on chromosomes 1–5. Note the presence of heterozygosity (two bands) and homozygosity for bands that do not match the standard for BALB/c. This is a clear case of loss of authenticity due to genetic contamination. The PCR products are compared with a 100 bp DNA ladder. The standard nomenclature for microsatellites is as follows: D [# of chromosome] [lab code] [ID of marker [4]. (Printed with permission from Springer Nature)

the DNA ladder. This visual comparison ensures accurate determination of the allele sizes and enhances the reliability of the genotyping analysis. After the amplification step, the PCR products are subsequently analyzed on a 4% agarose gel prepared using a wide range/standard 3:1 agarose mixture. This specific formulation provides strong and flexible gels suitable for effectively separating small PCR products (< 1 kb). To distinguish among SSLP alleles, amplification products should differ at least by 4 bp. The cost of SSLP genotyping is very low for facilities with access to thermocyclers, gel electrophoresis equipment, and gel imaging systems.

Recently, a panel consisting of 400 SSLPs was subjected to PCR amplification and agarose gel electrophoresis to assess their utility across ten classical mouse inbred strains. One of the advantages of these validated SSLPs is their widespread applicability in various research laboratories. Furthermore, the published study includes gel images, providing visual evidence of the amplified products. Additionally, the primer sequences used for amplification are disclosed, allowing for reproducibility and facilitating the implementation of these SSLPs [31].

Interpreting SSLP data is typically straightforward. In the case of inbred animals that are isogenic and homozygous, and assuming the animals are authentic, genotyping for a specific SSLP marker should result in the presence of only one band in the gel. This band, of the expected size listed on databases (e.g., MGI and RGD), represents homozygosity for a single allele. However, the presence of heterozygosity, indicated by two bands, or the presence of bands that do not align with those of the strain control DNA, should be considered as potential indications of strain contamination (Fig. 4.5). While mutations in SSLP loci are exceedingly rare, it is important to acknowledge that they are still possible and could potentially affect the genotyping results. The frequency for evaluating the colony depends on various factors such as the colony size, generation interval, and other considerations. As a general guideline, conducting strain identity testing once every 2 years is reasonable for facilities maintaining a limited number of colonies that are well-separated based on coat color and have minimal importations.

4.5.3 Genotyping for a Small Set of SNPs

As previously mentioned, SNPs have become the preferred choice for most commercial suppliers of laboratory mice and rats. When it comes to in-house authentication purposes, a suitable choice would be to use approximately 40 polymorphic SNPs evenly distributed across the chromosomes. The specific number should be adjusted based on the conditions and risks within each facility. However, it is important to note that while this approach can effectively detect recent genetic contamination, it does not provide any advantages over SSLP genotyping. Nevertheless, this technique has the advantage to distinguish between substrains when informative SNPs are employed. Small-scale SNP genotyping is now accessible through a range of platforms that vary in cost and automation capabilities. Among these options, allele-specific PCR is often considered the most convenient and cost-effective choice for minimally equipped laboratories [28]. One such method is Kompetitive Allele-Specific PCR (KASP), which is a variation of allele-specific PCR utilizing allele-specific oligo extension and fluorescence [51]. The KASP assay contains two different, allele-specific, competing forward primers with unique tail sequences and one reverse primer. The advantage of KASP is its compatibility with automation using 96- or 384-well plates and pipetting robots for standard PCR reactions. Plate readers equipped with suitable software are employed for the analysis of results, making the process more efficient and streamlined (Fig. 4.6).

In addition, hundreds of individual SNP KASPar™ assays for mice and rats are commercially available (e.g., LGC in the UK). This allows customizing the numbers and identities of SNPs for each situation and avoids using a fixed set of markers that may include non-informative SNPs, which cannot be avoided when using SNPs arrays. Another option is real-time PCR (TaqMan®) technology, which utilizes specific primers paired with a sequence specific fluorescent TaqMan probe. This method is effective and easy to automate; however, the cost per individual assay is comparatively high when compared to KASP assays. Additionally, it necessitates a more expensive real-time thermocycler.

If needed, information regarding the expected alleles (C, G, A, or T) for specific SNP/strain combinations, as well as their genomic locations, can be readily acces*sed through various databases and genome browsers. These resources provide comprehensive data for hundreds of thousands of SNPs and common inbred strains of mice and rats. Some examples of easily accessible databases and genome browsers include the MPD (Fig. 4.7), MGI, and RGD, as well as the Sanger Institute's Mouse Genomes Project (https://www.sanger.ac.uk/data/mouse-genomes-project/) and the mouse (http://useast.ensembl.org/Mus_musculus/Info/Index) and rat (http://useast.ensembl.org/Rattus_norvegicus/Info/Index) *Ensembl* genome browser. These platforms offer valuable information to assist in the interpretation and analysis of SNP data in the context of strain-specific genetic variations.

4.6 Discrimination of Substrains

The consensus is that if an inbred colony has been genetically isolated for more than 20 generations, it should be considered and designated a substrain, regardless of whether the strain has been confirmed to be genetically different from the parental strain. For instance, if the parent and new colony were separated, and they have each been bred for ten generations, the new colony and the parent colony are in fact 20 generations apart. This isolation ensures that the substrain develops unique genetic characteristics that distinguish it from the parent strain. For example, a pairwise comparison of substrains using the MegaMUGA array revealed 134 polymorphic SNPs between BALB/cJ and BALB/cByJ substrains, and 827 between C3H/HeJ and C3H/HeN substrains. [21]. Similar levels of SNP variability were described for several substrains of rats, like F344, SHR, and WKY [33].

If the objective is to determine the substrain to which a group of mice belongs, a limited set of SNPs can be selected based on the available data

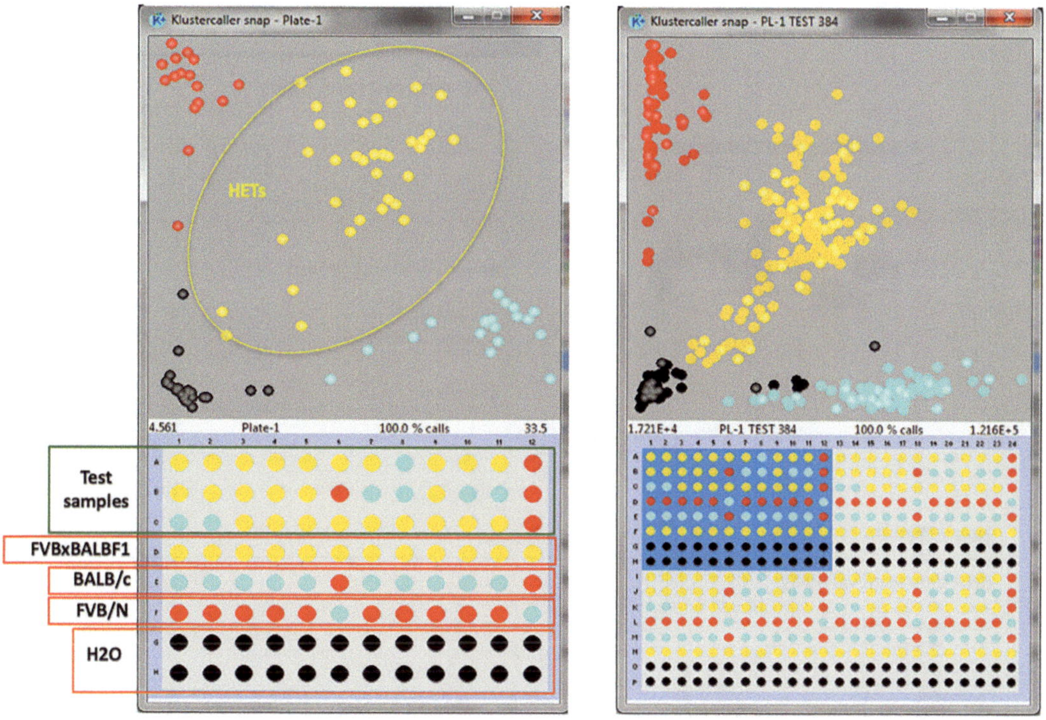

Fig. 4.6 KASP PCR for SNP genotyping. The figures show results obtained with the KlusterCaller™ software (LGC) after PCR amplification using KASP assays (LGC). When dual emission genotyping data from a fluorescent reader is imported into the KlusterCaller, proprietary algorithms built into the software automatically discern if the assay results are homozygous for allele 1 (red), homozygous for allele 2 (light blue) or heterozygous (yellow). In this example, 12 SNPs were analyzed (columns) in a 96-well plate (left), and the same layout (four replicates) in a 384-well plate (right). Both plates were loaded with an Eppendorf pipetting station. In the 96-well plate, rows A to C are test samples; row D is an (FVBxBALB/c)F1 control; row E is a BALB/c control, and row F is an FVB/N control. Yellow dots represent heterozygous SNP genotypes. Red or light blue dots represent homozygous genotypes (but not necessary a particular strain). Black dots are negative (water) controls. For example, the DNA sample in row A is heterozygous for SNPs 1–7 and 9–11, and homozygous for SNPs 8 (light blue) and 12 (red) (both BALB/c-like alleles). Keep in mind that all the SNPs used in this type of analysis are bi-allelic, so the genotypes can only be homozygous for allele 1, homozygous for allele 2, or heterozygous [6]

in SNP databases. As an example, different small SNP panels have been proposed to differentiate C57BL/6 J from C57BL/6 N [47, 48, 78]. As mentioned, the MiniMUGA SNP array can be useful for differentiating between substrains of inbred mice through the diagnostic SNPs [61]. Utilizing the MiniMUGA panel, approximately 250 SNPs display variability between the C57BL/6 J and C57BL/6 N substrains. On the other hand, microsatellites are not suitable for identifying substrains due to the insufficient availability of informative markers for comparing the majority of common substrains.

Testing for known spontaneous mutations can also aid in distinguishing between substrains. For instance, the presence of the *Crb1*^{rd8} mutation [45] has been observed in C57BL/6 N mice but not in the C57BL/6 J substrain. Conversely, the *Nnt* deletion [27] has been identified in C57BL/6 J but not in other "C57BL" strains. These examples underscore the significance of genetic characterization and the need to be aware of specific mutations associated with mouse substrains. Relevant genetic differences were found not only between substrains such as C57BL/6 J and C57BL/6 N, which have been bred separately since 1951. Genetic differences with phenotypic

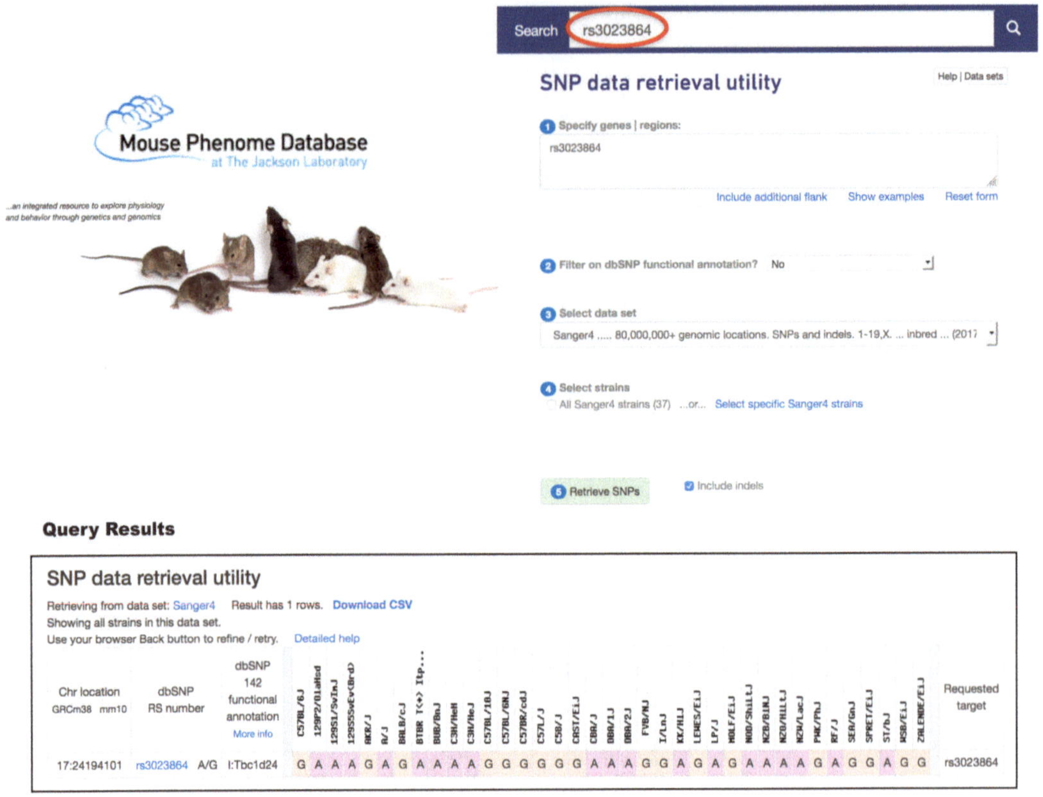

Fig. 4.7 SNP databases. This figure shows the results of a search using the Mouse Phenome Database (MPD) seeking for polymorphism between inbred mouse strains for a SNP (ID rs3023864) located on chromosome 17. In this case, the Sanger4 set of 37 inbred strains was selected as data set. The bottom of the figure is a screen capture of the results of the query showing the alleles (G or A) present in each of the strains (A = A/A; G = G/G). There are several other options for searching SNP data on-line (see text) [4]. (Printed with permission from Springer Nature)

effects were also found for animals of one of these substrains but provided by different commercial breeders [22, 35, 60, 72, 78]. By having knowledge of these mutations, researchers can make well-informed choices regarding substrain (vendor) selection (see Chap. 1).

4.7 Quality Controls for Breeding Procedures

Genetic quality control is routinely performed on all breeding animals in the nucleus or foundation colony, which is a self-reproducing system for maintaining the strain. However, commercial breeders may also selectively test the genetic integrity of animals in the expansion and production colonies if genetic contamination is suspected in individual animals. Foundation colonies consist of only a small number of breeding pairs, which are always full siblings, all descended from a common ancestral pair. The ancestral pair may be determined in each generation or at intervals of several generations based on genetic testing of their offspring. Once the ancestral pair has been determined, all animals not directly descended from that pair are removed from the foundation colony. Monogamous mating should be preferred in foundation colonies in order to unambiguously assign offspring to their parents. Where possible, strains with similar phenotypes, such as the same coat color, should be housed separately to avoid confusion.

The reproductive period of breeding females begins at 7–8 weeks of age and is limited to approximately 24 weeks or an average of 6 litters. Permanent mating takes advantage of the postpartum estrus and results in a greater number of litters per breeding female. Future breeding animals should not be selected from the first or last litters of a breeding pair. To extend the generation interval and thereby minimize genetic drift, late litters would be preferable for strain reproduction. However, due to the increasing frequency of spontaneous mutations in the germline of older individuals, the compromise is to select the next generation of breeding animals from the third and fourth litters.

One recommended scheme for breeding inbred strains is the pyramid-mating scheme. In this system, the foundation colony serves as the genetic and health standard and provides breeders for the top level of the pyramid in every barrier room. This top level, the nucleus colony, is composed of a relatively small number of pedigreed brother-sister mating pairs that produce breeders for the next level of the pyramid, in addition to replenishing itself. In larger colonies, the next level is called the expansion colony, and it provides breeders to the production colony, which in turn produces the animals for experiments (Fig. 4.1). The unidirectional flow of breed stock in this system helps to ensure that any genetic changes or mutations, which would be more likely to occur in the larger expansion or production colonies than in the smaller nucleus colony, will wash out within a single generation [6, 32].

Daily routine monitoring of breeding pairs includes the detection and documentation of pregnancies. The number of newborns should be counted on the day of birth and later compared with the number of weanlings to calculate the breeding success. The average litter interval can be calculated from the birth dates of successive litters. Individual tagging should be done at weaning to prevent mixing of juveniles. Separation of the sexes must also take place at weaning, since sexual maturity of females occurs at about 35 days of age, and that of males even slightly earlier. Breeding a nucleus colony of laboratory rodents involves three main aspects: (i) genetic and phenotypic evaluation of potential breeding animals from the current generation of offspring, (ii) selection of new breeding animals based on the results of genetic and phenotypic analysis, and (iii) mating of suitable animals. The overall breeding objective is the long-term genotypic and phenotypic stability of the nucleus colony across the generation line. The basis of evaluating potential breeding animals is the described methods for genetic and phenotypic quality control to prove their integrity. Error-free genetic quality control requires several measures, one of which is the comprehensive and individual recording of each animal's data using an appropriate recording system. The animal's pedigree, specific phenotypic characteristics of the strain and any welfare concerns are part of the basic data collection. In addition to the recording of animals in an electronic database, all cages shall be equipped with cage cards. The cards should contain all relevant information to identify the animals in the cage and provide information on the breeding success of mated animals. Cage cards are very important for routine daily inspection and for quickly locating animals in the room. In addition, they are an indispensable backup in case the IT system with the database fails, which can happen for a short time due to a power failure, but also due to hacker attacks on the facility's server.

Unique naming and individual identification of animals is a prerequisite for quickly locating each animal in the animal room and in the database, and for assigning quality control results to the correct individuals. This includes the accurate naming of the animals according to the Rules and Guidelines of Mouse and Rat Nomenclature (see Chap. 3). Not only is the correct designation of the strain important, but also information about any substrains and the place of origin of the animals by using the laboratory codes of the lab or vendor. An example is C57BL/6NJ, the NIH substrain separated from C57BL/6 J in 1951 that is now also bred and offered at JAX.

Several invasive and non-invasive methods are available for individual marking of laboratory mice and rats [18]. The most used and recommended method is ear punching according to a fixed numbering scheme [10]. The advantage of this method is that it is permanent and easy to read. On the other hand, the punched-out tissue pieces can be used to obtain genomic DNA for genetic quality control. Punching the ears is minimally stressful for the animals, technically easy to perform, and avoids a second procedure for tissue biopsy [15]. If a strain is to be permanently bred, a clear breeding strategy is required. This is especially true if, in addition to reproducing the foundation colony, animals are to be produced for experimental purposes. In order to avoid surplus animals, computer programs for breeding calculation using the established fertility parameters of the strain have proven to be valuable.

For outbred stocks, the intent is to minimize inbreeding, maintain heterozygosity and manage genetic drift that would otherwise lead to colony divergence. Ideally, outbred colonies should be maintained with ≥25 breeding pairs, all of which have to contribute to the next generation, in order to avoid an increase of the inbreeding coefficient per generation of more than 1%. Smaller colonies drift fast toward homozygosity because breeders are closely related [7].

4.8 Genetic Control for Outbred Stocks (the Value of Genetic Diversity)

Genetic monitoring of outbred stocks is much more complex because these animals are not genetically uniform. Outbred colonies are essentially a group of closely related animals, with shared ancestors and group identity, but that exhibit some level of genetic heterozygosity [13] (see Chap. 1). Most internationally available outbred stocks of the mouse, such as SWISS, NMRI, ICR(CD1), and MF1, are descended from seven females and two males supplied in 1926 from a breeding colony in Lausanne, Switzerland, to Clara Lynch, Rockefeller Institute in New York, USA [73]. Since outbred colonies form a population rather than a strain, it is difficult to establish a standard genetic monitoring program with only a few genetic markers (and a few individuals). However, with an adequate number of SNPs or SSLPs, allele frequencies within the population could give clues to the identity of the stock and of the success of the applied breeding strategy to stabilize its genetic constitution. For outbred stocks, genetic monitoring helps preserve the genetic heterogeneity and the allele pool of a colony. This complex process requires analyzing many animals and comparing this data with historical data documenting the alleles present, their frequency, and the level of heterozygosity in that colony. An example of outbred genotyping is the use of the MUGA SNP array for genetic characterization of the Diversity Outbred (DO) mouse population, an extremely heterogeneous stock recently developed from eight founder inbred strains [14] (see Chap. 1).

One of the main concerns associated with in-house outbred stocks is the limited number of animals typically present in the breeding colony. This restricted population size can result in a reduced representation of alleles within the stock, leading to increased genetic drift and higher inbreeding coefficients. Furthermore, unintended directional selection in the reproduction of the breeding nucleus can additionally result in shifting of allele frequencies. As a result, these colonies are neither truly outbred nor inbred and their genetic constitution will most likely change over time. While SSLPs or SNPs can be utilized to estimate the level of heterozygosity and genetic stability within the colony, it is preferable to obtain outbred rodents from vendors that maintain large colonies with several hundreds of nucleus breeding pairs and strictly follow recommended breeding schemes to minimize inbreeding through mating of unrelated individuals. Since it is usually not feasible to maintain an adequate number of breeders (e.g., 50 pairs using a rotational breeding strategy), purchasing outbred rodents from reputable vendors is generally recommended when using outbred rodents.

Besides the use of genetic markers, the characterization of outbred stocks should involve assessing phenotypic traits and calculating mean values and standard deviations of measurable characteristics. Clear deviations in the expression of phenotypic traits may indicate changes in the gene pool of the population. The degree of genetic heterogeneity in outbred colonies is influenced by their breeding history. Factors such as poor selection of future breeding stock, deviation from rotational breeding systems, or the bottleneck effect, due to a limited number of breeding animals to establish the imported stock or rederivation of a contaminated colony can result in reduced genetic heterogeneity and changes in phenotypic characteristics. Conversely, outcrossing events in outbred stocks can increase genetic heterogeneity. However, the so-called blood replacement will not only change the gene pool of the population, but also its specific phenotypic characteristics. For this reason, outbred stocks are consistently bred as a closed population without crossbreeding.

Genetic control of outbred stocks primarily focuses on avoiding inbreeding and maintaining stable genetic diversity across multiple generations. The aim is to prevent the loss of genetic variation and promote the genetic and phenotypic stability of the population. This objective is essentially achieved by managing the outbred stock by means of correct selection of an appropriate number of breeding animals in each generation and their mating according to an established breeding scheme. Furthermore, unintended directional selection in the reproduction of the breeding nucleus can also result in shifting of allele frequencies. To avoid this, the breeding pairs of the stock have to contribute equally to the formation of the next generation. The monitoring of genetic diversity within the stock can additionally ensure that inbreeding is minimized, and genetic heterogeneity is maintained over time.

4.9 Background Characterization for Genetically Altered Lines

This topic will be extensively discussed in Chap. 5. However, we would like to provide a brief comment here, as the proliferation of genetically altered (GA) lines is intensifying the issue of undefined "mixed backgrounds" in experimental rodents. This is of particular concern for inducible and conditional models that necessitate the crossing of two separate lines, such as Cre-expressing lines with "floxed" gene lines. The influence of genetic background on phenotype, particularly through modifier genes, makes this issue very significant. Mutations, whether spontaneous or induced, introduced into a new genetic background may not manifest the anticipated phenotype as expected from the donor strain [23, 42]. The phenotype of a mutant is not only the result of the considered genetic modification but also arises in interaction with other hereditary factors. Due to these considerations, it is crucial on the one hand to characterize the genetic background of each GA line. On the other hand, transitioning to a different genetic background may provide insights into the impact of genetic modifiers influencing the expression of the mutant phenotype. In any case, selecting a defined and reproducible genetic background is essential for characterizing the phenotypic impact of a genetic modification. This approach is also advantageous for ensuring the availability of appropriate control animals in the investigation [6, 11, 43, 71, 75].

Genetic markers that are uniformly distributed throughout the genome can be utilized to estimate the proportions of the genome originating from different inbred backgrounds in a GA line. This background characterization involves using polymorphic markers to differentiate between the suspected inbred strains. Among mice, commonly employed strains in genetic targeting typically included C57BL/6 J, C57BL/6 N, and various 129 substrains [63]. Historically, ES cells derived exclusively from 129 substrains have been used for the development of knockout and knock-in mice through homologous recombination. Nowadays, CRISPR/Cas technology is revolutionizing the field of genetic engineering and opening new possibilities for modifying the genomes of any mouse or rat inbred strain or stock, as well as various other species. By doing background characterizations, researchers can obtain valuable information about the genetic composition and contributions from different inbred strains within a given GA line. This knowledge is essential for understanding and interpreting phenotypic traits and ensuring appropriate selection of control animals. The accessibility of line-specific historical information is crucial, and we highly recommend utilizing a rodent passport that includes details about GA animals [70].

References

1. Andrews KR, Hunter SS, Torrevillas BK, Cespedes N, Garrison SM, Strickland J, Wagers D, Hansten G, New DD, Fagnan MW, Luckhart S. A new mouse SNP genotyping assay for speed congenics: combining flexibility, affordability, and power. BMC Genomics. 2021;22:378.
2. Aspden JL, Wallace EWJ, Whiffin N. Not all exons are protein coding: addressing a common misconception. Cell Genom. 2023;3:100296.
3. Benavides FJ. Genetic contamination of an SJL/J Mouse Colony: rapid detection by PCR-based microsatellite analysis. Contemp Top Lab Anim Sci. 1999;38:54–5.
4. Benavides F, Guénet J-L. Rodent Genetics. In: Experimental design and reproducibility in preclinical animal studies. Germany: Springer Nature; 2021.
5. Benavides F, Glasscock E, Coghlan LG, Stern MC, Weiss DA, Conti CJ. PCR-based microsatellite analysis for differentiation and genetic monitoring of nine inbred SENCAR mouse strains. Lab Anim. 2001;35:157–62.
6. Benavides F, Rulicke T, Prins JB, Bussell J, Scavizzi F, Cinelli P, Herault Y, Wedekind D. Genetic quality assurance and genetic monitoring of laboratory mice and rats: FELASA working group report. Lab Anim. 2020;54:135–48.
7. Berry, M. M., Linder, C.C. 2007. Breeding systems: considerations, genetic fundamentals, genetic background and strain types. Fox Jg, B. S., Davisson Mt, Newcomer CE, Quimby FE, And Smith AL (ed.) The mouse in biomedical research: volume 1: history, wild mice and genetics Academic.
8. Bogue MA, Grubb SC, Walton DO, Philip VM, Kolishovski G, Stearns T, Dunn MH, Skelly DA, Kadakkuzha B, Tehennepe G, Kunde-Ramamoorthy G, Chesler EJ. Mouse phenome database: an integrative database and analysis suite for curated empirical phenotype data from laboratory mice. Nucleic Acids Res. 2018;46:D843–50.
9. Bogue MA, Ball RL, Walton DO, Dunn MH, Kolishovski G, Berger A, Lamoureux A, Grubb SC, Gerring M, Kim M, Liang H, Emerson J, Stearns T, He H, Mukherjee G, Bluis J, Davis S, Desai S, Sundberg B, Kadakkuzha B, Kunde-Ramamoorthy G, Philip VM, Chesler EJ. Mouse phenome database: curated data repository with interactive multi-population and multi-trait analyses. Mamm Genome. 2023;34:509–19.
10. Bonaparte D, Cinelli P, Douni E, Herault Y, Maas M, Pakarinen P, Poutanen M, Lafuente MS, Scavizzi F, Federation Of European Laboratory Animal Science Associations Working, G. FELASA guidelines for the refinement of methods for genotyping genetically-modified rodents: a report of the Federation of European Laboratory Animal Science Associations Working Group. Lab Anim. 2013;47:134–45.
11. Bourdi M, Davies JS, Pohl LR. Mispairing C57BL/6 substrains of genetically engineered mice and wild-type controls can lead to confounding results as it did in studies of JNK2 in acetaminophen and concanavalin A liver injury. Chem Res Toxicol. 2011;24:794–6.
12. Bryda EC, Riley LK. Multiplex microsatellite marker panels for genetic monitoring of common rat strains. J Am Assoc Lab Anim Sci. 2008;47:37–41.
13. Chia R, Achilli F, Festing MF, Fisher EM. The origins and uses of mouse outbred stocks. Nat Genet. 2005;37:1181–6.
14. Churchill GA, Gatti DM, Munger SC, Svenson KL. The diversity outbred mouse population. Mamm Genome. 2012;23:713–8.
15. Cinelli P, Rettich A, Seifert B, Burki K, Arras M. Comparative analysis and physiological impact of different tissue biopsy methodologies used for the genotyping of laboratory mice. Lab Anim. 2007;41:174–84.
16. Consortium S, Saar K, Beck A, Bihoreau MT, Birney E, Brocklebank D, Chen Y, Cuppen E, Demonchy S, Dopazo J, Flicek P, Foglio M, Fujiyama A, Gut IG, Gauguier D, Guigo R, Guryev V, Heinig M, Hummel O, Jahn N, Klages S, Kren V, Kube M, Kuhl H, Kuramoto T, Kuroki Y, Lechner D, Lee YA, Lopez-Bigas N, Lathrop GM, Mashimo T, Medina I, Mott R, Patone G, Perrier-Cornet JA, Platzer M, Pravenec M, Reinhardt R, Sakaki Y, Schilhabel M, Schulz H, Serikawa T, Shikhagaie M, Tatsumoto S, Taudien S, Toyoda A, Voigt B, Zelenika D, Zimdahl H, Hubner N. SNP and haplotype mapping for genetic analysis in the rat. Nat Genet. 2008;40:560–6.
17. Cui SF, Zhou Q, Qu XH. Snp genotyping for the genetic monitoring of laboratory mice by using a microarray-based method with dualcolour fluorescence hybridisation. Altern Lab Anim. 2012;40:155–63.
18. Dahlborn K, Bugnon P, Nevalainen T, Raspa M, Verbost P, Spangenberg E. Report of the Federation of European Laboratory Animal Science Associations Working Group on animal identification. Lab Anim. 2013;47:2–11.
19. Davisson MT, Taft RA. Strategies for managing an ever increasing mutant mouse repository. Brain Res. 2006;1091:255–7.
20. Delpero M, Arends D, Freiberg A, Brockmann GA, Hesse D. QTL-mapping in the obese Berlin fat mouse identifies additional candidate genes for obesity and fatty liver disease. Sci Rep. 2022;12:10471.
21. Didion JP, Buus RJ, Naghashfar Z, Threadgill DW, Morse HC 3rd, De Villena FP. SNP array profiling of mouse cell lines identifies their strains of origin and reveals cross-contamination and widespread aneuploidy. BMC Genomics. 2014;15:847.
22. Dobrowolski P, Fischer M, Naumann R. Novel insights into the genetic background of genetically modified mice. Transgenic Res. 2018;27:265–75.
23. Doetschman T. Influence of genetic background on genetically engineered mouse phenotypes. Methods Mol Biol. 2009;530:423–33.

24. Eppig JT, Motenko H, Richardson JE, Richards-Smith B, Smith CL. The international mouse strain resource (IMSR): cataloging worldwide mouse and ES cell line resources. Mamm Genome. 2015;26:448–55.
25. Fahey JR, Katoh H, Malcolm R, Perez AV. The case for genetic monitoring of mice and rats used in biomedical research. Mamm Genome. 2013;24:89–94.
26. Fairfield H, Srivastava A, Ananda G, Liu R, Kircher M, Lakshminarayana A, Harris BS, Karst SY, Dionne LA, Kane CC, Curtain M, Berry ML, Ward-Bailey PF, Greenstein I, Byers C, Czechanski A, Sharp J, Palmer K, Gudis P, Martin W, Tadenev A, Bogdanik L, Pratt CH, Chang B, Schroeder DG, Cox GA, Cliften P, Milbrandt J, Murray S, Burgess R, Bergstrom DE, Donahue LR, Hamamy H, Masri A, Santoni FA, Makrythanasis P, Antonarakis SE, Shendure J, Reinholdt LG. Exome sequencing reveals pathogenic mutations in 91 strains of mice with Mendelian disorders. Genome Res. 2015;25:948–57.
27. Freeman H, Shimomura K, Cox RD, Ashcroft FM. Nicotinamide nucleotide transhydrogenase: a link between insulin secretion, glucose metabolism and oxidative stress. Biochem Soc Trans. 2006;34:806–10.
28. Gaudet M, Fara AG, Beritognolo I, Sabatti M. Allele-specific Pcr in SNP genotyping. Methods Mol Biol. 2009;578:415–24.
29. Groza T, Gomez FL, Mashhadi HH, Munoz-Fuentes V, Gunes O, Wilson R, Cacheiro P, Frost A, Keskivali-Bond P, Vardal B, McCoy A, Cheng TK, Santos L, Wells S, Smedley D, Mallon AM, Parkinson H. The international mouse phenotyping consortium: comprehensive knockout phenotyping underpinning the study of human disease. Nucleic Acids Res. 2023;51:D1038–45.
30. Guenet JL, Benavides F. Genetic monitoring of laboratory rodents. In: Patrinos GP, Ansorge W, editors. Molecular diagnostics. 2nd ed. Elsevier/Academic Press, Amsterdam, 2010.
31. Gurumurthy CB, Joshi PS, Kurz SG, Ohtsuka M, Quadros RM, Harms DW, Lloyd KC. Validation of simple sequence length polymorphism regions of commonly used mouse strains for marker assisted speed congenics screening. Int J Genomics. 2015;2015:735845.
32. Hedrich HJ. The laboratory mouse. Elsevier; 2012.
33. Hermsen R, De Ligt J, Spee W, Blokzijl F, Schafer S, Adami E, Boymans S, Flink S, Van Boxtel R, Van Der Weide RH, Aitman T, Hubner N, Simonis M, Tabakoff B, Guryev V, Cuppen E. Genomic landscape of rat strain and substrain variation. BMC Genomics. 2015;16:357.
34. Kaldunski ML, Smith JR, Hayman GT, Brodie K, De Pons JL, Demos WM, Gibson AC, Hill ML, Hoffman MJ, Lamers L, Laulederkind SJF, Nalabolu HS, Thorat K, Thota J, Tutaj M, Tutaj MA, Vedi M, Wang SJ, Zacher S, Dwinell MR, Kwitek AE. The rat genome database (RGD) facilitates genomic and phenotypic data integration across multiple species for biomedical research. Mamm Genome. 2022;33:66–80.
35. Kang MJ, Gong JE, Kim JE, Choi HJ, Bae SJ, Choi YJ, Lee SJ, Seo MS, Kim KS, Jung YS, Cho JY, Lim Y, Hwang DY. Influence of three Balb/c substrain backgrounds on the skin tumor induction efficacy to Dmba and TPA cotreatment. Lab Anim Res. 2020;36:30.
36. Koscielny G, Yaikhom G, Iyer V, Meehan TF, Morgan H, Atienza-Herrero J, Blake A, Chen CK, Easty R, Di Fenza A, Fiegel T, Grifiths M, Horne A, Karp NA, Kurbatova N, Mason JC, Matthews P, Oakley DJ, Qazi A, Regnart J, Retha A, Santos LA, Sneddon DJ, Warren J, Westerberg H, Wilson RJ, Melvin DG, Smedley D, Brown SD, Flicek P, Skarnes WC, Mallon AM, Parkinson H. The international mouse phenotyping consortium web portal, a unified point of access for knockout mice and related phenotyping data. Nucleic Acids Res. 2014;42:D802–9.
37. Kuramoto T, Satoshi N, Yamasaki K-I, Kumafuji K, Sakakibara Y, Neoda Y, Takizawa A, Kaneko T, Otsuki M, Hashimoto R, Voigt B, Mashimo T, Serikawa T. Genetic quality control of the rat strains at the National Bio Resource Project–Rat. Interdiscip Bio Central. 2010;2(12):1–7.
38. Kurtz TW, Montano M, Chan L, Kabra P. Molecular evidence of genetic heterogeneity in Wistar-Kyoto rats: implications for research with the spontaneously hypertensive rat. Hypertension. 1989;13:188–92.
39. Lambert R. Breeding strategies for maintaining colonies of laboratory mice: a Jackson Laboratory resource manual. Bar Harbor: The Jackson Laboratory; 2009.
40. Lamoreux ML, Delams V, Larue L, Bennet DC. The colors of mice. A model genetic network. Wiley-Blackwell; Chichester, West Sussex, UK 2010.
41. Leiter EH, Prochazka M, Coleman DL. The non-obese diabetic (NOD) mouse. Am J Pathol. 1987;128:380–3.
42. Linder CC. The influence of genetic background on spontaneous and genetically engineered mouse models of complex diseases. Lab Anim (NY). 2001;30:34–9.
43. Linder CC. Genetic variables that influence phenotype. ILAR J. 2006;47:132–40.
44. Mashimo T, Voigt B, Tsurumi T, Naoi K, Nakanishi S, Yamasaki K, Kuramoto T, Serikawa T. A set of highly informative rat simple sequence length polymorphism (SSLP) markers and genetically defined rat strains. BMC Genet. 2006;7:19.
45. Mattapallil MJ, Wawrousek EF, Chan CC, Zhao H, Roychoudhury J, Ferguson TA, Caspi RR. The Rd8 mutation of the Crb1 gene is present in vendor lines of C57BL/6N mice and embryonic stem cells, and confounds ocular induced mutant phenotypes. Invest Ophthalmol Vis Sci. 2012;53:2921–7.
46. Meehan TF, Conte N, West DB, Jacobsen JO, Mason J, Warren J, Chen CK, Tudose I, Relac M, Matthews P, Karp N, Santos L, Fiegel T, Ring N, Westerberg H, Greenaway S, Sneddon D, Morgan H, Codner GF, Stewart ME, Brown J, Horner N, International Mouse Phenotyping, C, Haendel M, Washington N, Mungall CJ, Reynolds CL, Gallegos J, Gailus-Durner V, Sorg T, Pavlovic G, Bower LR, Moore M, Morse I, Gao

X, Tocchini-Valentini GP, Obata Y, Cho SY, Seong JK, Seavitt J, Beaudet AL, Dickinson ME, Herault Y, Wurst W, De Angelis MH, Lloyd KCK, Flenniken AM, Nutter LMJ, Newbigging S, Mckerlie C, Justice MJ, Murray SA, Svenson KL, Braun RE, White JK, Bradley A, Flicek P, Wells S, Skarnes WC, Adams DJ, Parkinson H, Mallon AM, Brown SDM, Smedley D. Disease model discovery from 3,328 gene knockouts by the international mouse phenotyping consortium. Nat Genet. 2017;49:1231–8.
47. Mekada K, Abe K, Murakami A, Nakamura S, Nakata H, Moriwaki K, Obata Y, Yoshiki A. Genetic differences among C57BL/6 substrains. Exp Anim. 2009;58:141–9.
48. Mekada K, Hirose M, Murakami A, Yoshiki A. Development of SNP markers for C57BL/6N-derived mouse inbred strains. Exp Anim. 2015;64:91–100.
49. Moran JL, Bolton AD, Tran PV, Brown A, Dwyer ND, Manning DK, Bjork BC, Li C, Montgomery K, Siepka SM, Vitaterna MH, Takahashi JS, Wiltshire T, Kwiatkowski DJ, Kucherlapati R, Beier DR. Utilization of a whole genome SNP panel for efficient genetic mapping in the mouse. Genome Res. 2006;16:436–40.
50. Morgan AP, Fu CP, Kao CY, Welsh CE, Didion JP, Yadgary L, Hyacinth L, Ferris MT, Bell TA, Miller DR, Giusti-Rodriguez P, Nonneman RJ, Cook KD, Whitmire JK, Gralinski LE, Keller M, Attie AD, Churchill GA, Petkov P, Sullivan PF, Brennan JR, McMillan L, Pardo-Manuel De Villena F. The mouse universal genotyping array: from substrains to subspecies. G3 (Bethesda). 2015;6:263–79.
51. Myakishev MV, Khripin Y, Hu S, Hamer DH. High-throughput SNP genotyping by allele-specific PCR with universal energy-transfer-labeled primers. Genome Res. 2001;11:163–9.
52. Nijman IJ, Kuipers S, Verheul M, Guryev V, Cuppen E. A genome-wide SNP panel for mapping and association studies in the rat. BMC Genomics. 2008;9:95.
53. Ostermeier GC, Wiles MV, Farley JS, Taft RA. Conserving, distributing and managing genetically modified mouse lines by sperm cryopreservation. PLoS One. 2008;3:e2792.
54. Otsen M, Den Bieman M, Winer ES, Jacob HJ, Szpirer J, Szpirer C, Bender K, Van Zutphen LF. Use of simple sequence length polymorphisms for genetic characterization of rat inbred strains. Mamm Genome. 1995;6:595–601.
55. Petkov PM, Cassell MA, Sargent EE, Donnelly CJ, Robinson P, Crew V, Asquith S, Haar RV, Wiles MV. Development of a SNP genotyping panel for genetic monitoring of the laboratory mouse. Genomics. 2004a;83:902–11.
56. Petkov PM, Ding Y, Cassell MA, Zhang W, Wagner G, Sargent EE, Asquith S, Crew V, Johnson KA, Robinson P, Scott VE, Wiles MV. An efficient SNP system for mouse genome scanning and elucidating strain relationships. Genome Res. 2004b;14:1806–11.
57. Reardon S. Lab mice's ancestral 'Eve' gets her genome sequenced. Nature. 2017;551:281.
58. Reifenberg K, Hedrich H, Wedekind D, Howells N. Objective and methods of genetic monitoring of isogenic mouse and rat strains. GV-SOLAS Specialist Information; Society for Laboratory Animal Science GV-SOLAS, Germany 2014.
59. Sarsani VK, Raghupathy N, Fiddes IT, Armstrong J, Thibaud-Nissen F, Zinder O, Bolisetty M, Howe K, Hinerfeld D, Ruan X, Rowe L, Barter M, Ananda G, Paten B, Weinstock GM, Churchill GA, Wiles MV, Schneider VA, Srivastava A, Reinholdt LG. The genome of C57BL/6J "eve", the mother of the laboratory mouse genome reference strain. G3 (Bethesda). 2019;9:1795–805.
60. Siersbaek MS, Ditzel N, Hejbol EK, Praestholm SM, Markussen LK, Avolio F, Li L, Lehtonen L, Hansen AK, Schroder HD, Krych L, Mandrup S, Langhorn L, Bollen P, Grontved L. C57BL/6J substrain differences in response to high-fat diet intervention. Sci Rep. 2020;10:14052.
61. Sigmon JS, Blanchard MW, Baric RS, Bell TA, Brennan J, Brockmann GA, Burks AW, Calabrese JM, Caron KM, Cheney RE, Ciavatta D, Conlon F, Darr DB, Faber J, Franklin C, Gershon TR, Gralinski L, Gu B, Gaines CH, Hagan RS, Heimsath EG, Heise MT, Hock P, Ideraabdullah F, Jennette JC, Kafri T, Kashfeen A, Kulis M, Kumar V, Linnertz C, Livraghi-Butrico A, Lloyd KCK, Lutz C, Lynch RM, Magnuson T, Matsushima GK, McMullan R, Miller DR, Mohlke KL, Moy SS, Murphy CEY, Najarian M, O'brien L, Palmer AA, Philpot BD, Randell SH, Reinholdt L, Ren Y, Rockwood S, Rogala AR, Saraswatula A, Sassetti CM, Schisler JC, Schoenrock SA, Shaw GD, Shorter JR, Smith CM, St Pierre CL, Tarantino LM, Threadgill DW, Valdar W, Vilen BJ, Wardwell K, Whitmire JK, Williams L, Zylka MJ, Ferris MT, McMillan L, Manuel De Villena FP. Content and performance of the MiniMUGA genotyping array: a New tool to improve rigor and reproducibility in mouse research. Genetics. 2020;216:905–30.
62. Silvers WK. The coat colors of mice. A model for mammalian gene action and interaction. Springer; 1979.
63. Simpson EM, Linder CC, Sargent EE, Davisson MT, Mobraaten LE, Sharp JJ. Genetic variation among 129 substrains and its importance for targeted mutagenesis in mice. Nat Genet. 1997;16:19–27.
64. Smits BM, Guryev V, Zeegers D, Wedekind D, Hedrich HJ, Cuppen E. Efficient single nucleotide polymorphism discovery in laboratory rat strains using wild rat-derived SNP candidates. BMC Genomics. 2005;6:170.
65. St Lezin EM, Pravenec M, Wong A, Wang JM, Merriouns T, Newton S, Stec DE, Roman RJ, Lau D, Morris RC Jr, et al. Genetic contamination of Dahl SS/Jr rats. Impact on studies of salt-sensitive hypertension. Hypertension. 1994;23:786–90.

66. Strobel MC, Reinholdt LG, Malcolm RD, Pritchett-Corning K. Genetic monitoring of laboratory mice and rats. 3rd ed. Laboratory Animal Medicine: Elsevier Inc; 2015.
67. Taft RA, Davisson M, Wiles MV. Know thy mouse. Trends Genet. 2006;22:649–53.
68. Vedi M, Smith JR, Thomas Hayman G, Tutaj M, Brodie KC, De Pons JL, Demos WM, Gibson AC, Kaldunski ML, Lamers L, Laulederkind SJF, Thota J, Thorat K, Tutaj MA, Wang SJ, Zacher S, Dwinell MR, Kwitek AE. 2022 updates to the rat genome database: a findable, accessible, interoperable, and reusable (FAIR) resource. Genetics. 2023;224:iyad042.
69. Wedekind D, Reifenberg K, Hedrich HJ. Genetic monitoring of inbred strains. In: Hedrich HJ, editor. The laboratory mouse. Elsevier; 2012.
70. Wells DJ, Playle LC, Enser WE, Flecknell PA, Gardiner MA, Holland J, Howard BR, Hubrecht R, Humphreys KR, Jackson IJ, Lane N, Maconochie M, Mason G, Morton DB, Raymond R, Robinson V, Smith JA, Watt N. Assessing the welfare of genetically altered mice. Lab Anim. 2006;40:111–4.
71. Wolfer DP, Crusio WE, Lipp HP. Knockout mice: simple solutions to the problems of genetic background and flanking genes. Trends Neurosci. 2002;25:336–40.
72. Wotjak CT. C57BLack/BOX? The importance of exact mouse strain nomenclature. Trends Genet. 2003;19:183–4.
73. Yalcin B, Nicod J, Bhomra A, Davidson S, Cleak J, Farinelli L, Osteras M, Whitley A, Yuan W, Gan X, Goodson M, Klenerman P, Satpathy A, Mathis D, Benoist C, Adams DJ, Mott R, Flint J. Commercially available outbred mice for genome-wide association studies. PLoS Genet. 2010;6:e1001085.
74. Yang H, Ding Y, Hutchins LN, Szatkiewicz J, Bell TA, Paigen BJ, Graber JH, De Villena FP, Churchill GA. A customized and versatile high-density genotyping array for the mouse. Nat Methods. 2009;6:663–6.
75. Yoshiki A, Moriwaki K. Mouse phenome research: implications of genetic background. ILAR J. 2006;47:94–102.
76. Yoshiki A, Ballard G, Perez AV. Genetic quality: a complex issue for experimental study reproducibility. Transgenic Res. 2022;31:413–30.
77. Zimdahl H, Nyakatura G, Brandt P, Schulz H, Hummel O, Fartmann B, Brett D, Droege M, Monti J, Lee YA, Sun Y, Zhao S, Winter EE, Ponting CP, Chen Y, Kasprzyk A, Birney E, Ganten D, Hubner N. A SNP map of the rat genome generated from cDNA sequences. Science. 2004;303:807.
78. Zurita E, Chagoyen M, Cantero M, Alonso R, Gonzalez-Neira A, Lopez-Jimenez A, Lopez-Moreno JA, Landel CP, Benitez J, Pazos F, Montoliu L. Genetic polymorphisms among C57BL/6 mouse inbred strains. Transgenic Res. 2011;20:481–9.

Genetic Quality Controls for Genetically Altered Rodents

5

Fabien Delerue and Fernando Benavides

Abstract

Genetic quality controls for genetically altered (GA) rodents are imperative in any laboratory or animal facility engaged in breeding or utilizing mutant models. Ensuring the validity of the animal model requires robust genetic quality control measures. For studies involving GA rodents, especially mice and rats, the use of genetically controlled and defined animals should be enforced. Rigorous quality controls ensure that the desired genetic alterations are successfully incorporated into the animals, validating the model's relevance to the study, contributing to the reproducibility of experiments, allowing other researchers to validate and build upon the findings. These controls also play a crucial role in promoting the ethical use of animals in research. This chapter outlines current methods employed to accurately define the genetic composition of GA lines.

F. Delerue
Department of Genetics, The University of Texas MD Anderson Cancer Center, Houston, TX, USA
e-mail: fdelerue@mdanderson.org

F. Benavides (✉)
Department of Epigenetics and Molecular Carcinogenesis, The University of Texas MD Anderson Cancer Center, Houston, TX, USA
e-mail: fbenavid@mdanderson.org

Keywords

Quality control · Offtarget mutations · Genetic background · Speed congenics · Backcrossing

5.1 Introduction

Downstream of a genetic alternation in animals, two types of controls can be implemented. Depending on the size of the intended genetic modification, and the depth of the quality control, researchers can use quantitative (e.g., PCR) or qualitative (i.e., sequencing) techniques. Note that quantitative assessment is usually not sufficient for thorough quality control and should be complemented with a quality assessment before establishing a colony and using rodents for research purposes. As previously mentioned in Chap. 2, off-target mutations are rare in the context of generating GA rodents [38]. However, on-target mutations that are detrimental or otherwise unintended may occur with high frequency when performing gene editing. For instance, concatemers (array of more than one copy of a transgene) or bystander mutations in the vicinity of the CRISPR-induced double stranded break are often reported [47]. These undesired mutations are often undetected using only one type of quality control, and a combination of techniques to ensure thorough analysis of the newly generated

F. Benavides, A. K. Hansen (eds.), *Rodent Quality Control: Genes and Bugs*, Laboratory Animal Science and Medicine 2, https://doi.org/10.1007/978-3-031-59103-7_5

GA mice is highly recommended. Understanding the genome-wide background (e.g., inbred, or mixed background) is also crucial for researchers using GA lines in experiments, as genetic variations, even subtle ones, can impact experimental outcomes. Characterizing the genetic background helps ensure the reliability and reproducibility of research results. Although beyond the scope of this chapter, it is also critical that GA rodents be ultimately assessed at the cellular and physiological level before being used in biomedical research. For instance, if a transgenic mouse carrying a transgene coding for a given protein in a tissue specific manner is to be generated, it is imperative (following genotyping and sequencing) that the intended overexpression be checked (for instance using specific antibodies against the protein) in the tissue of interest. Finally, genetic quality controls alone are not sufficient to ensure that a mouse model is suitable to address a research question.

5.2 Characterization of GA Lines with Large Genetic Modifications

Deletions or insertions involving a relatively small number of base pairs (bp), typically less than a hundred or a few hundred, might be considered small-scale modifications. Deletions or insertions spanning thousands or more bp are generally considered large-scale modifications. These changes can have substantial effects on the structure and function of genes and regulatory elements.

5.2.1 Large Insertions

Mice or rats carrying large genomic alternations are usually easy to detect using Polymerase Chain Reaction (PCR) genotyping. Indeed, when a large transgene is inserted in the genome of a mouse, its DNA sequence is known, and often not found in the murine genome (hence, this transgene is sometimes referred to as "exogenous DNA"). As such, researchers can easily design one (or more) pair of PCR primers that can specifically bind the transgene, using online tools such as Primer-Blast (https://www.ncbi.nlm.nih.gov/tools/primer-blast/index.cgi). Such primers are then used to detect the presence or absence of the transgene in a given animal. For this purpose, PCR amplification is able to amplify even a single copy of the transgene to a yield (i.e., several millions of copies) that can be visualized on an agarose gel. Note that PCR genotyping is only quantitative and only indicates whether a genetic modification is present or absent in a DNA sample. However, it is one of the most commonly used quality control for GA animals, and biochemists Kary Mullis and Michael Smith were awarded the 1993 Nobel prize in chemistry for inventing the PCR. Given that large transgenes can be inserted into the mouse genome either randomly or by gene targeting (see Chap. 2), it is imperative that PCR positive animals be further analyzed and sequenced.

When a transgene is randomly inserted, technologies such as the Targeted Locus Amplification (TLA) can be used to determine the number of copies and location of the integrated transgene [15]. Recently, the Oxford Nanopore Technologies (ONT©) long read sequencing (LRS) has been developed as a third generation type of sequencing able to sequence DNA molecules of several kilobases and even megabases. LRS has been used as an alternative to TLA to perform the quality control of randomly inserted transgenes in mice [8, 63]. Following gene targeting, LRS can also be used to perform quality control at the integration site [29], allowing for an assessment of the whole targeted region following homologous recombination.

5.2.2 Large Deletions

Large structural variations (e.g., translocations, inversions, deletions, etc.) are relatively more difficult to detect using conventional PCR genotyping. Although it is in principle possible to design primers to detect and analyze deletions of any size, the break points (even after gene editing) are not always known, and the cell repair mechanism that stick two DNA strands back together are not fully understood. In mice, the most common techniques to analyze large deletions are the quantitative PCR (qPCR) and the droplet digital PCR (ddPCR) [27]. As previously mentioned, these methods should be complemented with sequencing for complete quality control. Protocols specifics to mouse studies have been developed [27] and successfully applied to newly generated mouse and rats lines [41]. Alternative or complementary techniques are continuously developed and tested, and new technologies such as the Optical Genome Mapping [48] may be added to the workflow of quality controls of the GA rodents.

5.3 Characterization of Subtle Genomic Changes

Subtle genomic changes are usually more difficult to detect than their large counterparts, yet the quality control is critical to ensure that a link between a subtle genomic change and a phenotypic expression is warranted. Indeed, most human genetic disorders are linked to discreet mutations such as point mutations (also referred to as single nucleotide variants—SNV). The presence or absence (quantitative analysis) of such point mutations in the mouse genome can be assessed by conventional techniques such as tetra primer-ARMS-PCR [61] or Sanger sequencing. However, these first generation technologies are usually not ideal to accurately detect subtle genomic changes [44]. Alternative methods based on enzymatic reactions (e.g., Restriction Fragment Length Polymorphism—RFLP) [23] and fluorimetric analysis of the composition of the DNA (e.g., High Resolution Melting Analysis—HRMA) [56] have been developed and usually more accurate, although only quantitative.

Several qualitative (i.e., sequencing) methods are available to analyze newly generated GA rodents. Recently, a survey of validations strategies following CRISPR/Cas9 editing in mice showed that genotyping based on Sanger sequencing analysis can miscall allele or miss discreet mutations [42]. Sanger sequencing being a first generation type of sequencing, the analysis of gene editing outcomes using second generation (i.e., Next Generation Sequencing—NGS) is far more accurate and highly recommended to detect subtle genomic changes in newly generated GA rodents [34].

5.4 Influence of Genetic Background on Phenotypes

The proliferation of GA lines in the last decades has intensified the issue of ambiguous "mixed backgrounds" in experimental rodents. This concern is intensified especially in the case of inducible and conditional models which necessitate the interbreeding of two distinct lines (such as Cre-expressing lines combined with "floxed" gene lines) that might possess different genetic backgrounds. Considering that genetic background significantly impacts phenotype, primarily due to the involvement of modifier genes, the introduction of mutations (spontaneous, induced or genetically engineered) or transgenes into a novel background might not manifest the anticipated phenotype [16, 17, 26, 31, 46]. An early instance highlighting the importance of genetic background in mutant mice was observed with the diabetes (*Lepr^{db}*) mutation. While it led to transient diabetes in a C57BL/6 background, it resulted in full-fledged diabetes in the closely related C57BLKS strain [24]. The impacts of genetic background effects have been documented across various domains of biomedical research, influencing mouse models pertaining to cancer [18, 21], immunology [10, 19], and human diseases [9, 13]. Some examples include background effects on survival rate in *Egfr* (epidermal growth factor receptor) KO mice [53], effects on tumor incidence and spectrum in *Trp53* KO mice [25] and *Pten* KO mice [21], and effects on acetaminophen-induced liver injury in *Jnk2* KO mice [7], to name only a few. There are also examples in rats, like the influence of genetic background on prostate tumorigenesis in Pb-SV40 transgenic rats [3] and in a model of Hirschsprung disease [14]. For this reason, every

GA rodent line should be characterized in terms of their genetic background. Moreover, the knowledge of the genetic background of a GA line is also important for the selection of the corresponding control animals [7].

5.5 Background Characterization of GA Lines

In the case of classical transgenesis (random insertion), knowing the strain utilized for pronuclear injection, along with the breeding procedures employed to maintain the line (including founder identification), is vital. An issue frequently encountered with transgenic mice pertains to the utilization of hybrid F1 embryos for the transgene injection, followed by crosses of founders with randomly chosen wild-type mice. This results in a mixed background line where each mouse has a different combination of the parental genomes present in the injected hybrid F1 embryo (e.g., D2B6F1). For targeted mutagenesis using ES cells, it is essential to know the origin of the ES cell line used for the targeting, plus the preliminary crosses done for the germline transmission test and the establishment of the line (if known, number of backcrosses, etc.). If the ES cell line belongs to a different strain than the one selected to cross the founders, the resulting KO/KI line may have a mixed background.

Bigenic lines carrying two independent transgenes, two targeted genes, or a combination of both, are also a concern regarding mixed backgrounds. As an illustration, the Tet-on (or Tet-off) inducible system requires the creation of two separate transgenic lines, which are subsequently crossed to produce animals carrying both the responder and transactivator constructs within the same organism [55]. Likewise, conditional systems, such as Cre-LoxP, Dre-Rox, or Flp-Frt recombination methods, need the presence of two distinct genetic modifications in the experimental animals. For example, mice that carry both the Cre transgene (or Cre KI locus) and the floxed targeted gene in their genomes. In the examples provided, it is frequently observed that mouse cohorts exhibit mixed genetic backgrounds. This is primarily due to the initial single-gene lines (such as the Cre-expressing line and the floxed line) originating from diverse backgrounds or outbred stocks.

By utilizing evenly distributed genetic markers across the entire genome, it is feasible to calculate the proportional contributions arising from different inbred sources that make up the genetic background. Currently, single nucleotide polymorphisms (SNPs) have become the standard genetic marker used in genome scans for mouse and rat GA lines (see Chap. 4). Following SNPs, microsatellites (SSLPs) are the second most commonly DNA marker utilized for background characterizations. Commercial companies and institutional core facilities offer the process of background characterization for this purpose. In a typical background characterization procedure, polymorphic markers are utilized to differentiate between suspected inbred strains. In the context of mice, this often involves discerning between C57BL/6 and 129 strains due to the historical fact that ES cells used for the development of KO and KI mice via homologous recombination (see Chap. 2) were originally exclusively derived from 129 substrains [45]. Instead, wild-type C57BL/6 females were typically used to prove germline transmission from the chimeras. Without subsequent backcrosses, this scheme resulted in a B6;129 mixed background. Nonetheless, the landscape is evolving fast due to the presence of ES cell lines originating from alternate strains, notably including C57BL/6N, and particularly the emergence of genome editing methods (e.g., CRISPR/Cas) that facilitate the direct creation of specific modifications in a broader spectrum of mouse or rat strains [20].

The amount and type of genetic markers for a background characterization have not been standardized, however, a set of 100 SNPs, polymorphic between the suspected strains, should suffice to have a reasonable estimation (Fig. 5.1). A very important part of marker selection is determining how the GA line was developed and subsequently crossed and maintained. Access to detailed historical information concerning the analyzed GA line holds greatest significance, underscoring the impor-

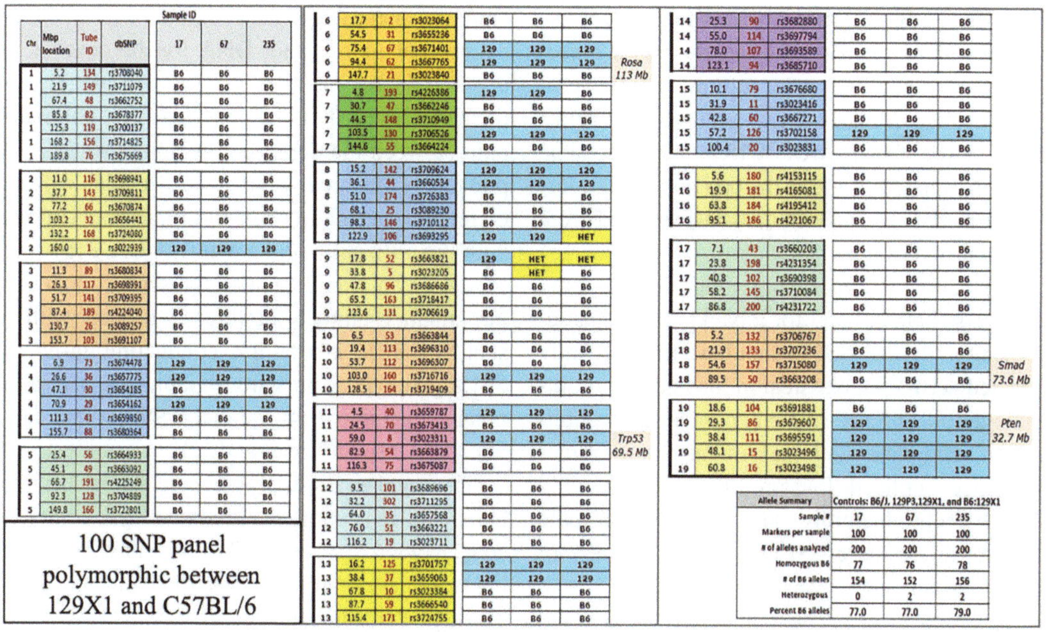

QC	Excelent	Excelent	Excelent	Excelent	Excelent	Excelent	Excelent	Excelent
other markers	hCMV	hCMV	hCMV	hCMV	hCMV	hCMV	hCMV	Cre+
Allele Summary	Male	Female	Male	Female	Female	Male	Male	Male
Sample #	**2C**	**3B**	**4A**	**4C**	**5B**	**5C**	**6A**	**7A**
Markers per sample	**3230**	**3221**	**3231**	**3228**	**3234**	**3117**	**3118**	**3234**
# of alleles analyzed	**6460**	**6442**	**6462**	**6456**	**6468**	**6234**	**6236**	**6468**
Homozygous 129	**2692**	**2764**	**2497**	**2408**	**2826**	**2479**	**2435**	**2625**
# of 129 alleles	**5384**	**5528**	**4994**	**4816**	**5652**	**4958**	**4870**	**5250**
Heterozygous	**538**	**457**	**734**	**820**	**408**	**638**	**683**	**609**
Percent 129S1 alleles	**91.7**	**92.9**	**88.6**	**87.3**	**93.7**	**89.8**	**89.0**	**90.6**
X Chromosome	**129S1**	**129S1**	**129S1**	**129S1 BALB/c**	**129S1**	**129S1**	**129S1**	**129S1**
	HEMI	**HOM**	**HEMI**	**RECOM**	**HOM**	**HEMI**	**HEMI**	**HEMI**

Fig. 5.1 Background characterization of GA lines. The *upper panel* shows an example of a background characterization for three quadruple KO/KI mice (*Rosa26*, *Trp53*, *Smad*, and *Pten*) using a panel of 100 SNPs. The SNPs were selected to be polymorphic between 129X1 (ES cells used for the targeting) and C57BL/6 (background strain), and evenly distributed along all the autosomal chromosomes (first column on each panel). The estimated percentage of C57BL/6 in these three mice ranges from 77% to 79%. The *lower panel* shows an example of a background characterization for eight mice (backcrossed from BALB/c to 129S1) using the MiniMUGA SNP panel (~3,200 SNPs were selected for this analysis for being plymorphic between BALB/c and 129S1). The estimated percentage of 129S1 in these N3 mice ranges from 88.6% to 93.7%. The MiniMUGA also reports the sex of the mice and the presence of other sequences, like Cre and hCMV in this example

tance of employing a rodent-passport that encompasses comprehensive data on GA animals [58]. Once this information is known, a set of polymorphic markers can be selected to estimate the composition of the genetic background. Comprising approximately 9600 SNPs, the MiniMUGA array [43] serves as an excellent alternative for characterizing the genetic background of a GA line (Fig. 5.1) (see Chap. 4). The MiniMUGA testing is run by Neogene (https://www.neogen.com/) but it is available as a service only from Transnetyx, Inc. (https://www.transnetyx.com/monitoring). The

MiniMUGA Background Analysis Report tool furnished with every genotyped sample has the objective of providing users with information extracted from the MiniMUGA genotyping array. Furthermore, the myMUGA tool, also accessible through Transnetyx, offers the option for additional and personalized analysis of data generated from the MiniMUGA array.

When SSLP genotyping represent the only marker system at hand for a background characterization, the 400 SSLPs that have been recently verified across ten classical mouse inbred strains emerges as a highly valuable choice. This publication, complete with gel images and primer sequences utilized for amplification, not only ensures reproducibility but also simplifies the integration of these SSLPs into the process of characterizing genetic backgrounds [22].

In any case, the problem of mixed background can be circumvented altogether by (i) injecting transgenes or Cas9-sgRNA into inbred embryos from the strain of choice; (ii) modifying the gene of interest in ES cells from the preferred background strain (e.g., using C57BL/6N ES cells); and (iii) crossing KO/KI founders with mice of the same strain as the ES cells used for the targeting. Finally, if the GA has already been developed or acquired from a collaborator or repository, a background characterization should be performed, and if needed, a fully congenic strain should be developed, either by classical or marker-assisted backcrossing. Periodic backcrossing of a congenic strain to the background strain also minimizes divergence and keeps the congenic strain genetically close to the strain background of control animals.

5.6 Marker-Assisted Backcrossing as Refinement

The use of DNA markers has allowed for a much more rapid and rigorous process of congenic strain development called *marker-assisted* backcrossing or *speed congenics* [2, 30, 54, 57]. This process relies on using polymorphic genetic markers covering the whole genome to determine the percentage of donor genome present in the animals, then selecting the animals with the lowest percentage of donor DNA for the next backcross to the recipient strain. Common practice is the use of at least 100 genetic markers, polymorphic between the donor and recipient strains. As previously detailed, SNP panels available from commercial vendors or institutional cores, along with the utilization of the MiniMUGA array [43] serves as an excellent tool for speed congenic. For smaller animal facilities equipped with a standard PCR laboratory, conducting in-house SSLP genotyping presents a viable alternative. A notable example is the validated SSLP panel outlined by Gurumurthy et al. [22].

This marker-assisted procedure diminishes the number of generations required to achieve complete congenicity (e.g., from N10 to N5), thereby reducing strain development time by approximately 50%. Through the utilization of marker-assisted backcrosses and the appropriate quantity of animals, it becomes feasible to attain at least 80% recipient background at N2, approximately 94% at N3, and roughly 99% at N4; in contrast to the conventional mean values of 75.0%, 87.5%, and 93.7%, respectively, without marker-assisted backcrosses (see also Chap. 1). A graphic depicting the speed congenic process is shown in Fig. 5.2.

Ideally, the backcross procedure is started with a donor female and a recipient male. Then, F1 mutant males will carry the correct Y-chromosome and after mating to a recipient female, males of the N2 generation will carry the correct X- and Y-chromosome of the recipient strain (avoiding the use of genetic markers on these chromosomes) [16]. Also, the speed congenic strategy can be further accelerated by different methods of assisted reproduction [35]. Notably, although a large number of molecular markers are necessary to perform efficient and reliable genotyping during the first backcross generation (N2), this number decreases rapidly because once a marker is typed "homozygous" for the allelic form of the background strain, it is no longer necessary to genotype the offspring of the future generations for this marker because it is permanently fixed. A flowchart depicting the standard speed congenic timeline is shown in Fig. 5.3.

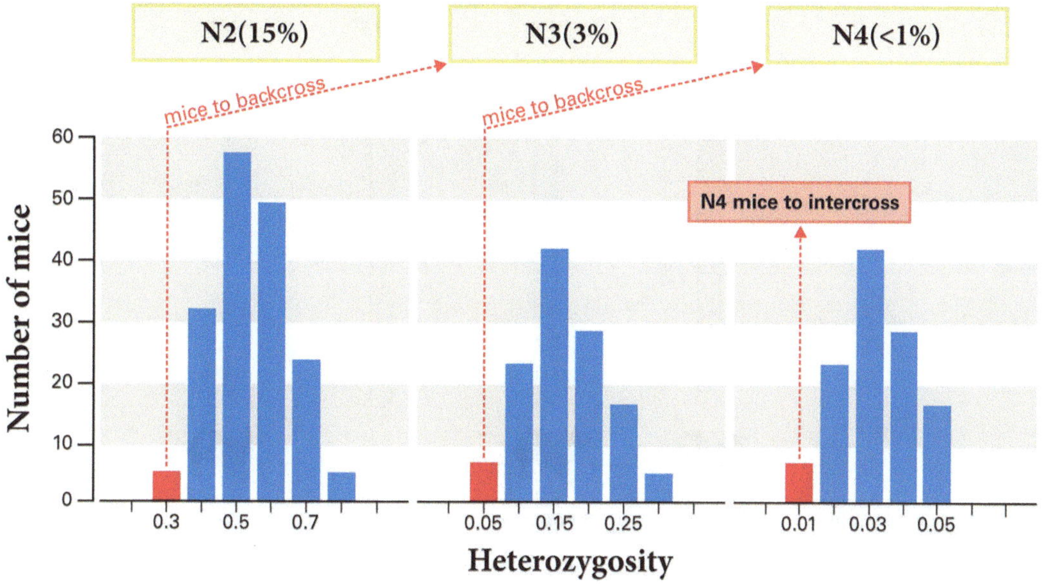

From: Wakeland et al. *Immunology Today* 1997, 18, 472-477.

Fig. 5.2 Speed Congenics. As shown in the graphics, the amount of donor versus background (recipient) chromosomal segments in each backcross (N) generation (shown as percentage of heterozygosity in the polymorphic genetic markers analyzed) follows a normal (Gaussian) distribution, with an average of 75% background genome in generation N2, 87.5% in N3, and 93.8% in N4. What does the trick is the possibility of choosing those "extreme" animals with very low percentage of heterozygous markers in each generation. The N2 mouse selected as breeder on the example shows 15% heterozygous markers, which means a higher background genome (~85%) than the average 75%. By repeating this breeder selection process at each generation, the animals are globally >99% recipient background after five generations of crosses instead of the ten generations needed in the normal process. (Reprinted from Wakeland et al. [57], with permission from Elsevier)

- **Start crossing donor (carrier) female with recipient strain male to generate F1 carrier males (PI).**
- **Backcross F1 males to generate ~10 N2 carrier males (~25%) (PI)**
- **Scan N2 carriers with SNPs and select the best breeders (Service)**
- **Cross best N2 males with several recipient strain females (PI)**
- **Generate ~10 N3 carrier males (~25%) (PI)**
- **Genotype N3 mice for heterozygous SNPs in N2 analysis (Service)**
- **Cross best N3 males with several recipient strain females (PI)**
- **Repeat same scheme at N4 (and if necessary, at N5)**

Fig. 5.3 Speed Congenic timeline. Selecting at each backcross generation the breeder with the lowest percentage of introgressed (donor) DNA greatly accelerates the establishment of a congenic strain. It is important to note that genotyping requires many polymorphic DNA markers only for the first backcross progeny (N2). Once a marker is characterized homozygous it is no longer necessary to type it in the forthcoming generations. Although carrier males (heterozygous for the gene of interest) are typically recommended as "best breeders," females can also be used, as long as they have high percentages of the recipient genome. The prediction of ~99% recipient genome at N5 is based on the use of 20 best breeders (carriers) at each generation [30], however, this number is not always available and fewer breeders can be used, with disparate results, depending also on chance. PI: principal investigator laboratory. Service: facility core or outsourced service

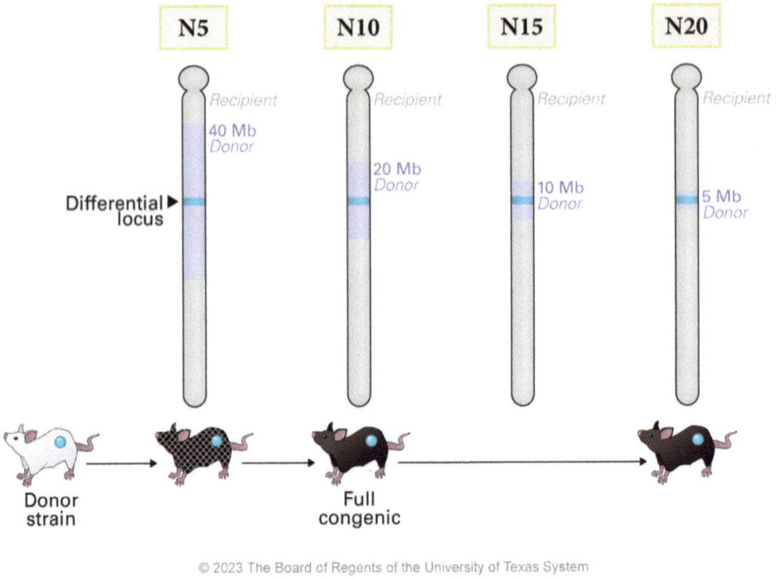

Fig. 5.4 The flanking region in congenic strains. The region flanking the gene of interest in congenic strains can pose a challenge as it may impact the phenotype of the strain under investigation. This is due to the potential presence of modifier genes within the donor strain's chromosomal region, which could interact with genes from the receipient strain, resulting in unintended genetic effects that might influence the interpretation of experimental outcomes. Consequently, minimizing the size of flanking regions in congenic mouse strains is crucial in order to mitigate the likelihood of confounding effects. As shown in the figure, even at backcross generation N10 the size of the flanking (donor) region (lavender color) could be as big as 20 megabases (Mb), potentially carrying hundreds of genes

For the optimal development of speed congenic strains, it is recommended to incorporate an additional group of polymorphic markers situated adjacent to the spontaneous mutation, transgene, or targeted gene of interest. This enables the identification of breeders possessing the minimal amount of flanking (donor) DNA, originating from the ES cell line, the genome where a transgene or Cas9-sgRNA was introduced, or the location where a spontaneous mutation emerged. This strategic inclusion helps mitigate potential complications arising from a modifier gene situated near the genetic alteration [11, 60], a situation also known as the flanking genes issue (Fig. 5.4). This effort may require breeding many mice to obtain breeders with small introgressed segments.

5.7 Genetic Stability Programs and Cryobanking

In the case of co-isogenic and congenic GA lines, strategic breeding methods and genetic stability initiatives serve to mitigate substrain divergence arising from genetic drift, as well as to avoid genetic contamination due to inadvertent crosses with other strains. To restrict genetic drift, it is advised to minimize the number of generations for in-house breeding to 10 (~5 years), while also submitting the GA lines to repositories. This submission facilitates archiving as frozen embryos and/or sperm, ensuring the line's security and offering a mechanism for replenishing the breeding stock. This recommendation aligns with The Jackson Laboratory Genetic Stability Program

(GSP) and serves to prevent the cumulative effects of genetic drift [5, 50]. A key to implementing a successful genetic stability program is to invest in education and training of the personnel. Good colony management helps prevent genetic contamination and genetic drift. Caretakers must recognize strain characteristics and animals that deviate from strain norms. Phenotypically indistinguishable GA lines should not be housed in the same room (or rack), and procedures should be in place to avoid line mix-ups.

To maintain a GA line on a pure inbred background, backcrossing heterozygous carriers with pure inbred mice is recommended. Maintaining and archiving heterozygous mice (e.g., +/−) on inbred backgrounds is ideal even when homozygous or multiple mutants are needed for research, because crossing heterozygotes to produce homozygotes (25% −/−) will also produce proper littermate controls (50% +/−; 25% +/+). Employing accurate standard nomenclature to delineate the animal model and maintaining a comprehensive account of the genetic background are of paramount importance (see Chap. 3). When designing experiments involving multiple lines, opting for a shared genetic background is advisable. This choice facilitates experiment reproducibility across different instances.

As previously mentioned, in the event that a new GA line is generated on an unspecified or inaccurate genetic background, a recommended course of action is to subject it to backcrossing to the appropriate genetic background. Following this, it is advisable to cryopreserve the line (see Chap. 6). Verification of the viability of recovered sperm (by in vitro fertilization) or embryos after thawing, along with validation of the anticipated genotypes, should then be conducted. Historically, embryo freezing has been the favored method for preserving mouse lines. However, sperm freezing is gaining traction as a more convenient option, thanks to the utilization of advanced assisted reproductive technologies [36, 49, 51, 52]. Exchanging GA lines between research facilities currently requires only transporting frozen sperm [32, 39, 40].

Cryopreservation strategies have been adopted for long-term storage of embryos and gametes in several large, centralized repositories including the Mutant Mouse Resource & Research Center (MMRRC) [1], the EMMA/INFRAFRONTIER (European Mutant Mouse Archive) [12, 59], the Knock Out Mouse Project (KOMP) Repository [28], The Jackson Laboratory Repository [36], The Center for Animal Resources and Development (CARD) [33] and the Riken Bio Resource Center [62], which can provide cryopreserved material or live mice to laboratories (see Chap. 6). These repositories facilitate the availability of these strains to the worldwide scientific community and provide a backup for a potential loss of a strain. The International Mouse Strain Resource (IMSR) is a searchable online database of mouse strains, stocks, and mutant ES cell lines available worldwide, including inbred, mutant, and genetically engineered strains (http://www.findmice.org/) (see Chaps. 1 and 2).

5.8 Reproducibility of Studies

In recent years, there has been a significant rise in concerns regarding the credibility of scientific results, as highlighted by a recent survey conducted by the Nature publishing group. The survey revealed that an overwhelming majority of the researchers interviewed were convinced that there is currently a reproducibility crisis [4]. For studies using mice and rats, and important part is to start with a comprehensive documentation of the imported GA lines to be used [6]. The ARRIVE guidelines, which stands for "Animal Research: Reporting of In Vivo Experiments," provide a set of recommendations to improve the reporting of research involving animals. These guidelines are designed to enhance the transparency and reproducibility of animal studies, thereby contributing to the overall quality of scientific research. The ARRIVE guidelines 1.0 were first introduced in 2010 and updated recently with guidelines 2.0 [37]. For animal experimentation, some key components of the ARRIVE guidelines include genetic modification status, genotype, strain and substrain, information about the source of animals (e.g., vendor or in-house breeding), and use of standard nomenclature. Adherence to these guidelines is encouraged by scientific journals to ensure the quality and transparency of published animal research.

References

1. Amos-Landgraf J, Franklin C, Godfrey V, Grieder F, Grimsrud K, Korf I, Lutz C, Magnuson T, Mirochnitchenko O, Patel S, Reinholdt L, Lloyd KCK. The Mutant Mouse Resource and Research Center (MMRRC): the NIH-supported National Public Repository and distribution archive of mutant mouse models in the USA. Mamm Genome. 2022;33:203–12.
2. Andrews KR, Hunter SS, Torrevillas BK, Cespedes N, Garrison SM, Strickland J, Wagers D, Hansten G, New DD, Fagnan MW, Luckhart S. A new mouse SNP genotyping assay for speed congenics: combining flexibility, affordability, and power. BMC Genomics. 2021;22:378.
3. Asamoto M, Hokaiwado N, Cho YM, Shirai T. Effects of genetic background on prostate and taste bud carcinogenesis due to SV40 T antigen expression under probasin gene promoter control. Carcinogenesis. 2002;23:463–7.
4. Baker M. 1,500 scientists lift the lid on reproducibility. Nature. 2016;533:452–4.
5. Benavides F, Rulicke T, Prins JB, Bussell J, Scavizzi F, Cinelli P, Herault Y, Wedekind D. Genetic quality assurance and genetic monitoring of laboratory mice and rats: FELASA Working Group Report. Lab Anim. 2020;54:135–48.
6. Birling MC, Fray MD, Kasparek P, Kopkanova J, Massimi M, Matteoni R, Montoliu L, Nutter LMJ, Raspa M, Rozman J, Ryder EJ, Scavizzi F, Voikar V, Wells S, Pavlovic G, Teboul L. Importing genetically altered animals: ensuring quality. Mamm Genome. 2022;33:100–7.
7. Bourdi M, Davies JS, Pohl LR. Mispairing C57BL/6 substrains of genetically engineered mice and wild-type controls can lead to confounding results as it did in studies of JNK2 in acetaminophen and concanavalin A liver injury. Chem Res Toxicol. 2011;24:794–6.
8. Bryant WB, Yang A, Griffin SH, Zhang W, Rafiq AM, Han W, Deak F, Mills MK, Long X, Miano JM. CRISPR-Cas9 Long-read sequencing for mapping transgenes in the mouse genome. CRISPR J. 2023;6:163–75.
9. Calyjur PC, Almeida Cde F, Ayub-Guerrieri D, Ribeiro AF Jr, Fernandes Sde A, Ishiba R, Santos AL, Onofre-Oliveira P, Vainzof M. The mdx mutation in the 129/Sv background results in a milder phenotype: transcriptome comparative analysis searching for the protective factors. PLoS One. 2016;11:e0150748.
10. Carlucci F, Cortes-Hernandez J, Fossati-Jimack L, Bygrave AE, Walport MJ, Vyse TJ, Cook HT, Botto M. Genetic dissection of spontaneous autoimmunity driven by 129-derived chromosome 1 Loci when expressed on C57BL/6 mice. J Immunol. 2007;178:2352–60.
11. Chen S, Kadomatsu K, Kondo M, Toyama Y, Toshimori K, Ueno S, Miyake Y, Muramatsu T. Effects of flanking genes on the phenotypes of mice deficient in basigin/CD147. Biochem Biophys Res Commun. 2004;324:147–53.
12. Consortium, I. INFRAFRONTIER – providing mutant mouse resources as research tools for the international scientific community. Nucleic Acids Res. 2015;43:D1171–5.
13. Cowin RM, Bui N, Graham D, Green JR, Yuva-Paylor LA, Weiss A, Paylor R. Genetic background modulates behavioral impairments in R6/2 mice and suggests a role for dominant genetic modifiers in Huntington's disease pathogenesis. Mamm Genome. 2012;23:367–77.
14. Dang R, Torigoe D, Suzuki S, Kikkawa Y, Moritoh K, Sasaki N, Agui T. Genetic background strongly modifies the severity of symptoms of Hirschsprung disease, but not hearing loss in rats carrying Ednrb(sl) mutations. PLoS One. 2011;6:e24086.
15. De Vree PJ, De Wit E, Yilmaz M, Van De Heijning M, Klous P, Verstegen MJ, Wan Y, Teunissen H, Krijger PH, Geeven G, Eijk PP, Sie D, Ylstra B, Hulsman LO, Van Dooren MF, Van Zutven LJ, Van Den Ouweland A, Verbeek S, Van Dijk KW, Cornelissen M, Das AT, Berkhout B, Sikkema-Raddatz B, Van Den Berg E, Van Der Vlies P, Weening D, Den Dunnen JT, Matusiak M, Lamkanfi M, Ligtenberg MJ, Ter Brugge P, Jonkers J, Foekens JA, Martens JW, Van Der Luijt R, Van Amstel HK, Van Min M, Splinter E, De Laat W. Targeted sequencing by proximity ligation for comprehensive variant detection and local haplotyping. Nat Biotechnol. 2014;32:1019–25.
16. Dobrowolski P, Fischer M, Naumann R. Novel insights into the genetic background of genetically modified mice. Transgenic Res. 2018;27:265–75.
17. Doetschman T. Influence of genetic background on genetically engineered mouse phenotypes. Methods Mol Biol. 2009;530:423–33.
18. Dorman A, Baer D, Tomlinson I, Mott R, Iraqi FA. Genetic analysis of intestinal polyp development in Collaborative Cross mice carrying the Apc (Min/+) mutation. BMC Genet. 2016;17:46.
19. Farkas B, Boldizsar F, Tarjanyi O, Laszlo A, Lin SM, Hutas G, Tryniszewska B, Mangold A, Nagyeri G, Rosenzweig HL, Finnegan A, Mikecz K, Glant TT. BALB/c mice genetically susceptible to proteoglycan-induced arthritis and spondylitis show colony-dependent differences in disease penetrance. Arthritis Res Ther. 2009;11:R21.
20. Fernandez A, Josa S, Montoliu L. A history of genome editing in mammals. Mamm Genome. 2017;28:237–46.

21. Freeman D, Lesche R, Kertesz N, Wang S, Li G, Gao J, Groszer M, Martinez-Diaz H, Rozengurt N, Thomas G, Liu X, Wu H. Genetic background controls tumor development in PTEN-deficient mice. Cancer Res. 2006;66:6492–6.
22. Gurumurthy CB, Joshi PS, Kurz SG, Ohtsuka M, Quadros RM, Harms DW, Lloyd KC. Validation of simple sequence length polymorphism regions of commonly used mouse strains for marker assisted speed congenics screening. Int J Genomics. 2015;2015:735845.
23. Hirasawa T, Ohara T, Makino S. Genetic typing of the mouse Ob mutation by PCR and restriction enzyme analysis. Exp Anim. 1997;46:75–8.
24. Hummel KP, Coleman DL, Lane PW. The influence of genetic background on expression of mutations at the diabetes locus in the mouse. I. C57BL-KsJ and C57BL-6J strains. Biochem Genet. 1972;7:1–13.
25. Kuperwasser C, Hurlbut GD, Kittrell FS, Dickinson ES, Laucirica R, Medina D, Naber SP, Jerry DJ. Development of spontaneous mammary tumors in BALB/c p53 heterozygous mice. A model for Li-Fraumeni syndrome. Am J Pathol. 2000;157:2151–9.
26. Linder CC. The influence of genetic background on spontaneous and genetically engineered mouse models of complex diseases. Lab Anim (NY). 2001;30:34–9.
27. Lindner L, Cayrou P, Jacquot S, Birling MC, Herault Y, Pavlovic G. Reliable and robust droplet digital PCR (ddPCR) and RT-ddPCR protocols for mouse studies. Methods. 2021;191:95–106.
28. Lloyd KC. A knockout mouse resource for the biomedical research community. Ann N Y Acad Sci. 2011;1245:24–6.
29. Low BE, Hosur V, Lesbirel S, Wiles MV. Efficient targeted transgenesis of large donor DNA into multiple mouse genetic backgrounds using bacteriophage Bxb1 integrase. Sci Rep. 2022;12:5424.
30. Markel P, Shu P, Ebeling C, Carlson GA, Nagle DL, Smutko JS, Moore KJ. Theoretical and empirical issues for marker-assisted breeding of congenic mouse strains. Nat Genet. 1997;17:280–4.
31. Montagutelli X. Effect of the genetic background on the phenotype of mouse mutations. J Am Soc Nephrol. 2000;11(Suppl 16):S101–5.
32. Moreno-Del Val G, Munoz-Robledano P. Revitalizing genetically-modified mouse strains using frozen-thawed sperm after up to 192 h of refrigerated epididymis transportation. Lab Anim. 2017;51:526–9.
33. Nakagata N, Yamamura K. Current activities of CARD as an international core center for mouse resources. Exp Anim. 2009;58:343–50.
34. Narina S, Connelly JP, Pruett-Miller SM. High-throughput analysis of CRISPR-Cas9 editing outcomes in cell and animal models using CRIS.py. Methods Mol Biol. 2023;2631:155–82.
35. Ogonuki N, Inoue K, Hirose M, Miura I, Mochida K, Sato T, Mise N, Mekada K, Yoshiki A, Abe K, Kurihara H, Wakana S, Ogura A. A high-speed congenic strategy using first-wave male germ cells. PLoS One. 2009;4:e4943.
36. Ostermeier GC, Wiles MV, Farley JS, Taft RA. Conserving, distributing and managing genetically modified mouse lines by sperm cryopreservation. PLoS One. 2008;3:e2792.
37. Percie Du Sert N, Hurst V, Ahluwalia A, Alam S, Avey MT, Baker M, Browne WJ, Clark A, Cuthill IC, Dirnagl U, Emerson M, Garner P, Holgate ST, Howells DW, Karp NA, Lazic SE, Lidster K, Maccallum CJ, Macleod M, Pearl EJ, Petersen OH, Rawle F, Reynolds P, Rooney K, Sena ES, Silberberg SD, Steckler T, Wurbel H. The ARRIVE guidelines 2.0: updated guidelines for reporting animal research. PLoS Biol. 2020;18:e3000410.
38. Peterson KA, Khalouei S, Hanafi N, Wood JA, Lanza DG, Lintott LG, Willis BJ, Seavitt JR, Braun RE, Dickinson ME, White JK, Lloyd KCK, Heaney JD, Murray SA, Ramani A, Nutter LMJ. Whole genome analysis for 163 gRNAs in Cas9-edited mice reveals minimal off-target activity. Commun Biol. 2023;6:626.
39. Raspa M, Fray M, Paoletti R, Montoliu L, Giuliani A, Collaborations-and-Projects/European-Mouse, E. M. I. T. W. G. E. A. H. W. I. E. I.-R.-I. I, Scavizzi F. A new, simple and efficient liquid nitrogen free method to cryopreserve mouse spermatozoa at −80 degrees C. Theriogenology. 2018a;119:52–9.
40. Raspa M, Fray M, Paoletti R, Montoliu L, Giuliani A, Group, E. M. I. T. W, Scavizzi F. Long term maintenance of frozen mouse spermatozoa at −80 degrees C. Theriogenology. 2018b;107:41–9.
41. Schaeffer L, Lindner L, Pavlovic G, Herault Y, Birling MC. CRISMERE chromosome engineering in mouse and rat. Methods Mol Biol. 2023;2631:277–97.
42. Sentmanat MF, Peters ST, Florian CP, Connelly JP, Pruett-Miller SM. A survey of validation strategies for CRISPR-Cas9 editing. Sci Rep. 2018;8:888.
43. Sigmon JS, Blanchard MW, Baric RS, Bell TA, Brennan J, Brockmann GA, Burks AW, Calabrese JM, Caron KM, Cheney RE, Ciavatta D, Conlon F, Darr DB, Faber J, Franklin C, Gershon TR, Gralinski L, Gu B, Gaines CH, Hagan RS, Heimsath EG, Heise MT, Hock P, Ideraabdullah F, Jennette JC, Kafri T, Kashfeen A, Kulis M, Kumar V, Linnertz C, Livraghi-Butrico A, Lloyd KCK, Lutz C, Lynch RM, Magnuson T, Matsushima GK, McMullan R, Miller DR, Mohlke KL, Moy SS, Murphy CEY, Najarian M, O'Brien L, Palmer AA, Philpot BD, Randell SH, Reinholdt L, Ren Y, Rockwood S, Rogala AR, Saraswatula A, Sassetti CM, Schisler JC, Schoenrock SA, Shaw GD, Shorter JR, Smith CM, St Pierre CL, Tarantino LM, Threadgill DW, Valdar W, Vilen BJ, Wardwell K, Whitmire JK, Williams L, Zylka MJ, Ferris MT, McMillan L, Manuel De Villena FP. Content and performance of the MiniMUGA genotyping array: a new tool to improve rigor and reproducibility in mouse research. Genetics. 2020;216:905–30.
44. Simkin D, Papakis V, Bustos BI, Ambrosi CM, Ryan SJ, Baru V, Williams LA, Dempsey GT, McManus

OB, Landers JE, Lubbe SJ, George AL Jr, Kiskinis E. Homozygous might be hemizygous: CRISPR/Cas9 editing in iPSCs results in detrimental on-target defects that escape standard quality controls. Stem Cell Rep. 2022;17:993–1008.
45. Simpson EM, Linder CC, Sargent EE, Davisson MT, Mobraaten LE, Sharp JJ. Genetic variation among 129 substrains and its importance for targeted mutagenesis in mice. Nat Genet. 1997;16:19–27.
46. Sittig LJ, Carbonetto P, Engel KA, Krauss KS, Barrios-Camacho CM, Palmer AA. Genetic background limits generalizability of genotype-phenotype relationships. Neuron. 2016;91:1253–9.
47. Skryabin BV, Kummerfeld DM, Gubar L, Seeger B, Kaiser H, Stegemann A, Roth J, Meuth SG, Pavenstadt H, Sherwood J, Pap T, Wedlich-Soldner R, Sunderkotter C, Schwartz YB, Brosius J, Rozhdestvensky TS. Pervasive head-to-tail insertions of DNA templates mask desired CRISPR-Cas9-mediated genome editing events. Sci Adv. 2020;6:eaax2941.
48. Smith AC, Neveling K, Kanagal-Shamanna R. Optical genome mapping for structural variation analysis in hematologic malignancies. Am J Hematol. 2022;97:975–82.
49. Sztein JM, Farley JS, Mobraaten LE. In vitro fertilization with cryopreserved inbred mouse sperm. Biol Reprod. 2000;63:1774–80.
50. Taft RA, Davisson M, Wiles MV. Know thy mouse. Trends Genet. 2006;22:649–53.
51. Takeo T, Nakagata N. Reduced glutathione enhances fertility of frozen/thawed C57BL/6 mouse sperm after exposure to methyl-beta-cyclodextrin. Biol Reprod. 2011;85:1066–72.
52. Takeo T, Sztein J, Nakagata N. The CARD method for mouse sperm cryopreservation and in vitro fertilization using frozen-thawed sperm. Methods Mol Biol. 2019;1874:243–56.
53. Threadgill DW, Dlugosz AA, Hansen LA, Tennenbaum T, Lichti U, Yee D, Lamantia C, Mourton T, Herrup K, Harris RC, et al. Targeted disruption of mouse EGF receptor: effect of genetic background on mutant phenotype. Science. 1995;269:230–4.
54. Van Laere AS, Tromme A, Delaval L, Farnir F, Blomet J, Desmecht D. A timely, user-friendly, and flexible marker-assisted speed congenics method. Transgenic Res. 2023;32:451–61.
55. Vitale-Cross L, Amornphimoltham P, Fisher G, Molinolo AA, Gutkind JS. Conditional expression of K-ras in an epithelial compartment that includes the stem cells is sufficient to promote squamous cell carcinogenesis. Cancer Res. 2004;64:8804–7.
56. Vossen RH. Genotyping DNA variants with high-resolution melting analysis. Methods Mol Biol. 2017;1492:17–28.
57. Wakeland E, Morel L, Achey K, Yui M, Longmate J. Speed congenics: a classic technique in the fast lane (relatively speaking). Immunol Today. 1997;18:472–7.
58. Wells DJ, Playle LC, Enser WE, Flecknell PA, Gardiner MA, Holland J, Howard BR, Hubrecht R, Humphreys KR, Jackson IJ, Lane N, Maconochie M, Mason G, Morton DB, Raymond R, Robinson V, Smith JA, Watt N. Assessing the welfare of genetically altered mice. Lab Anim. 2006;40:111–4.
59. Wilkinson P, Sengerova J, Matteoni R, Chen CK, Soulat G, Ureta-Vidal A, Fessele S, Hagn M, Massimi M, Pickford K, Butler RH, Marschall S, Mallon AM, Pickard A, Raspa M, Scavizzi F, Fray M, Larrigaldie V, Leyritz J, Birney E, Tocchini-Valentini GP, Brown S, Herault Y, Montoliu L, De Angelis MH, Smedley D. EMMA – mouse mutant resources for the international scientific community. Nucleic Acids Res. 2010;38:D570–6.
60. Wolfer DP, Crusio WE, Lipp HP. Knockout mice: simple solutions to the problems of genetic background and flanking genes. Trends Neurosci. 2002;25:336–40.
61. Ye S, Dhillon S, Ke X, Collins AR, Day IN. An efficient procedure for genotyping single nucleotide polymorphisms. Nucleic Acids Res. 2001;29:E88.
62. Yoshiki A, Ike F, Mekada K, Kitaura Y, Nakata H, Hiraiwa N, Mochida K, Ijuin M, Kadota M, Murakami A, Ogura A, Abe K, Moriwaki K, Obata Y. The mouse resources at the RIKEN BioResource center. Exp Anim. 2009;58:85–96.
63. Yu C, Caothien R, Pham A, Tam L, Alcantar T, Bacarro N, Reyes J Jr, Jackson M, Nakao B, Roose-Girma M. ASIS-Seq: transgene insertion site mapping by nanopore adaptive sampling. Methods Mol Biol. 2023;2631:135–53.

6 Cryopreservation of Valuable Mouse and Rat Lines

Martina Crispo, María Noel Meikle, and Thomas Rülicke

Abstract

The application of cryopreservation technologies to generate a cryobank of embryos and germ cells from genetically valuable mouse and rat lines is a standard procedure applied in many rodent facilities and large repositories around the world. Slow freezing was the first methodology developed for embryo cryopreservation and is still suitable and reliable. Despite this, the ultra-rapid vitrification technique is gaining more adepts since it is faster, consistent, and does not require expensive equipment. For sperm freezing, standardized protocols using commercial media are now available with excellent outcomes, reducing the number of donor animals required, the final cost, and ensuring the preservation for future in vitro fertilization. In this chapter, we will summarize the past, present, and future of the main techniques used nowadays and their benefits, with the aim to help readers understand the relevance to preserve the germplasm (i.e., spermatozoa, oocytes, and resulting embryos) of valuable lines.

M. Crispo · M. N. Meikle
Laboratory Animal Biotechnology Unit, Institut Pasteur de Montevideo, Montevideo, Uruguay
e-mail: crispo@pasteur.edu.uy; manomeikle@pasteur.edu.uy

T. Rülicke (✉)
Department of Biomedical Sciences, University of Veterinary Medicine Vienna, Vienna, Austria
e-mail: Thomas.Ruelicke@vetmeduni.ac.at

Keywords

Cryopreservation · Embryo freezing · Sperm freezing · Vitrification · In vitro fertilization · Archiving

6.1 Introduction

The number of laboratory rodents developed as models to study physiological and pathophysiological processes in mammalian species or used in pharmacological and toxicological studies has grown exponentially since the early 1980s. Fueled by advancements in genetic engineering techniques, the conventional inbred and outbred stocks of mice and rats have seen a significant expansion with the inclusion of an increasing array of genetically altered (GA)[1] models. In the realm of biomedical research, these novel models have played a pivotal role in advancing biomedical knowledge over the past three decades.

In the early 1980s, using retroviral vectors or microinjection into the pronucleus of zygotes, DNA constructs could be stably transferred into the genome of a mouse (the first transgenic

[1] There are numerous terms used to describe genetic changes in rodents. Genetically engineered mice (GEM) or genetically modified mice (GMM) are typically used to describe any type of genetic modification in the mouse. We use the term genetically altered (GA) here to also include those carrying spontaneous or chemically induced mutations.

F. Benavides, A. K. Hansen (eds.), *Rodent Quality Control: Genes and Bugs*, Laboratory Animal Science and Medicine 2, https://doi.org/10.1007/978-3-031-59103-7_6

animal species) [11, 17, 28]. By the late 1980s, researchers achieved another milestone by creating the first transgenic rats [53]. Subsequently, in the following years, scientists successfully isolated mouse embryonic stem (ES) cells and maintained their pluripotent developmental state in vitro, establishing a foundation for gene targeting [30]. However, the first successful derivation of rat ES cells was reported only in 2008 by researchers at Kyoto University [38].

These reverse genetics techniques have facilitated the analysis of the function and regulation of specific genetic factors within the intricate framework of a mammalian organism [5]. This involves modifying not only protein-coding genes but also non-coding regulatory RNAs, which act as important post-transcriptional regulators. The generation of GA laboratory rodents has been recently further accelerated by the development of endonuclease-based methods, in particular, CRISPR/Cas9 gene editing [23, 34, 67] (see Chap. 2). Besides mice and rats, other GA rodents are slowly being generated, for example, Syrian hamsters [87] and prairie voles [8]. The resulting ever-increasing number of newly generated mutants far exceeds the capacity of experimental animal facilities and can only be managed by archiving as many lines as possible and keeping them as living models only during the experimental phase [48].

Globally, cryopreserved lines and strains, particularly those of the laboratory mouse and rat, are gathered in central repositories and shared with the scientific community. This practice significantly contributes to the advancement of the Replacement, Reduction, and Refinement (3Rs) principles [6]. This not only ensures prompt accessibility without the need for repetitive generation of identical GA lines [44] but also eliminates the necessity of transporting live animals around the world. Archiving plays a crucial role in minimizing animal usage by eliminating the necessity for constant breeding, which often leads to a surplus of animals. Notably, cryopreserving a new GA line in its early stages prevents genetic drift, thereby maintaining the integrity of the original phenotype. Lastly, archiving in a central repository aid in ensuring adherence to international guidelines for mouse and rat strain nomenclature (see Chap. 3).

The first successful and reproducible method of mouse embryo cryopreservation was introduced in 1972 by the pioneer work of Witthingham, Leibo, and Mazur [94]. Approximately two decades later, Takeshima, Nakagata, and Ogawa achieved practical success in archiving mouse sperm [85]. Currently, cryopreservation techniques for embryos and sperm are well-established practices employed in archiving strains and GA lines as well as in genetic stability programs for standard inbred strains (see Chap. 4). Practically, various methods exist for archiving embryos, differing in cryoprotectant selection, process duration, and the utilization of diverse devices. In recent years, there has been significant improvement in the cryopreservation outcomes of mouse spermatozoa, aligning with the ongoing optimization of in vitro fertilization (IVF) conditions. Additionally, successful cryopreservation of rat sperm has been achieved. On the contrary, oocyte cryopreservation still encounters specific challenges that need to be addressed before it can be widely adopted in research facilities.

The following sections provide a discussion of the advantages and disadvantages associated with various methods of cryopreserving germplasm from laboratory mice and rats. It also explores the applications for each method and outlines the techniques for revitalizing lines and strains from frozen sperm and embryos.

6.2 Mouse and Rat Embryo Cryopreservation

Embryo cryopreservation is a crucial and widely used technique in various fields of research, including genetics, reproductive biology, as well as medical applications. In rodent facilities maintaining GA lines, embryo cryopreservation is the preferred choice for storing complex lines carrying more than one genetic modification. This method ensures the preservation of the complete diploid genome of the model, along with the mitochondrial DNA of the strain. *Slow freezing* and *vitrification* are the two primary approaches employed for preserving mouse and rat embryos in this context (Fig. 6.1).

Since both techniques have their advantages and are used in different settings, it is important to choose the appropriate cryopreservation method based on the specific needs and requirements of the species and the animal facility. After successfully recovering mouse strains cryopreserved for over 30 years, there is a growing belief that frozen mouse embryos can be archived for extended durations without compromising their developmental capacity. Notably, even a simulated storage period exceeding 2000 years did not reveal any discernible increase in the incidence of mutations attributable to normal cosmic rays [27]. Therefore, properly cryopreserved embryos can be a valuable resource for preserving GA lines.

6.2.1 Slow Freezing Method

The slow method of embryo cryopreservation relies on the use of moderate concentrations of cryoprotective agents (CPAs) such as ethylene glycol, propylene glycol, dimethyl sulfoxide or glycerol. These CPAs cause the necessary dehydration of the embryonic cells to avoid excessive intracellular ice formation. The cryopreservation procedure utilizes the cells' efforts to reach an osmotic equilibrium with the surrounding milieu of the hyperosmotic freezing solution. In this process, the free intracellular water is removed from the cells and replaced by the permeable CPAs. The process is characterized by a relatively rapid shrinkage of the cells after exposing to the freezing medium and a subsequently slow return to original size because the small water molecules cross the membrane during dehydration more quickly than the CPAs can enter the cells. To prevent osmotic shock, it is recom-

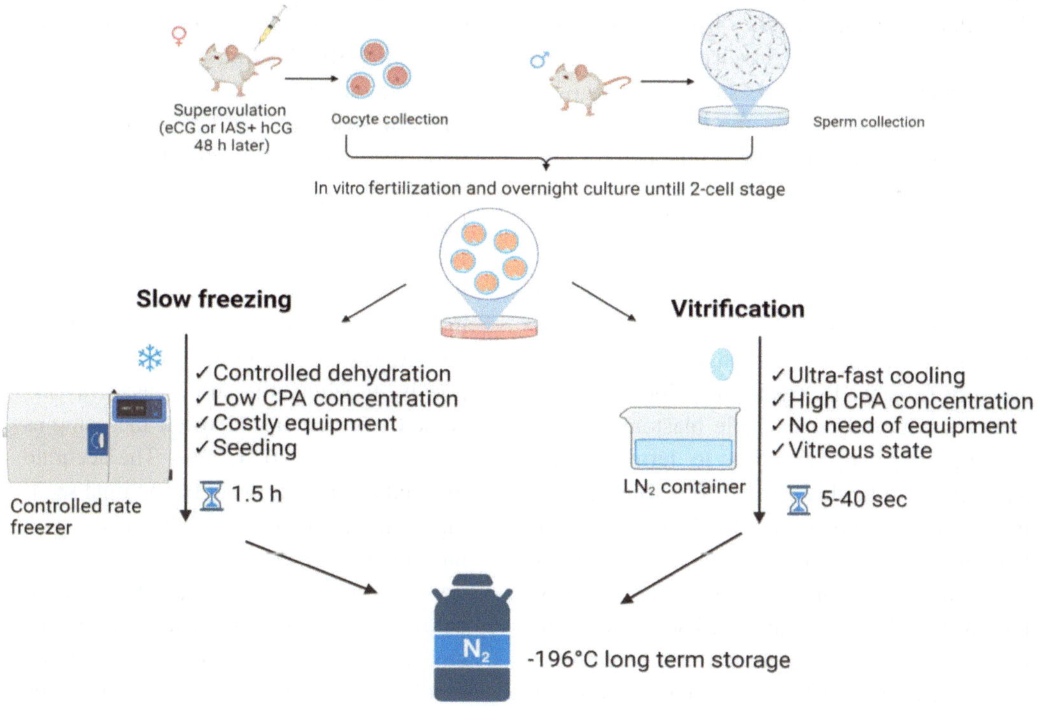

Fig. 6.1 Primary methodologies for embryo cryopreservation. In this schematic illustration, two-cell mouse embryos are obtained through in vitro fertilization following the superovulation of the donor female and the retrieval of fresh sperm from the donor male. Details of the slow freezing and vitrification procedures are elaborated in the accompanying text (see also Box 6.1 and Box 6.2). (Created with BioRender.com)

mended to incrementally increase the concentration of CPAs before cooling. Additionally, a gradual removal of CPAs from the medium after warming is advisable.

It is also important to note that CPAs will enable supercooling, a process of lowering the temperature of the medium below its freezing/melting point without crystallization. However, for successful cryopreservation, extracellular ice formation must occur before cells can spontaneously form intracellular ice. To avoid the risk of supercooling, all freezing protocols include active seeding to induce precisely timed ice formation of the surrounding medium before reaching a subzero temperature at which the cytoplasm can nucleate [68]. During thawing or warming, the removal of CPAs should be performed to return the embryo to a physiological environment and avoid possible toxicity of the CPA. The cooling-warming rate is key to protect the embryo from cryoinjury [31, 37].

As a rule of thumb it is recommended to cryopreserve at least 200 embryos per strain or line, depending on the genetic background [63]. However, the quantity of frozen embryos required is also contingent on factors such as the zygosity of the genetic modification, the anticipated recovery rate, and whether the line is intended for self-use or for archiving in a repository for broader access within the scientific community. Nowadays, it is possible to freeze all pre-implantation stages of mouse and rat embryos. In mice, most large repositories use 2-cell embryos for archiving (Fig. 6.1), although 8-cell embryos are very resistant to cryoinjury. Notably, 2-cell embryos with only one viable blastomere post-thawing frequently progress to term and are advised not to be disregarded [65]. Blastomeres in an 8-cell embryo are smaller than those in earlier developmental stages, containing less water. They exhibit continued development after thawing, even if 2–3 cells have died. For 2-cell and 8-cell stages, it is necessary to flush the embryos from the oviduct at 1.5 and 2.5 days post-coitum, respectively (this is technically more challenging than flushing later stages from uterine horns).

Currently, most mouse facilities are performing superovulation of donor females plus IVF to produce embryos for cryopreservation. Fertilized zygotes are left in overnight culture to obtain 2-cell embryos to be cryopreserved, with high survival rates [25, 86]. This approach has also been tested in rats with good outcomes [33]. A novel technique called ultra-superovulation relies on the use of an inhibin antiserum (IAS) combined with equine chorionic gonadotropin (eCG) promoting follicle growth and increasing 4–5 times the number of ovulated oocytes in mice [80]. A model is deemed "archived" when the thawed embryos are transferred to pseudopregnant females, and the desired genotype is successfully obtained in the live pups. A concise protocol for slow freezing is presented in Box 6.1.

Box 6.1: Slow Freezing Procedure

Slow freezing is based on the equilibrium established between the rate of water loss from the cell and the rate of which this water forms extracellular ice crystals [43]. Controlled freezing rates range between 0.3 and 1 °C/min to subzero temperatures of −30 to −80 °C before samples are stored in liquid nitrogen (LN_2) [10]. Generally, CPAs are used at 1 or 1.5 M, permeating into the embryos with low toxicity. Approximately, 15–20 embryos are usually loaded with freezing medium into a 0.25 mL plastic straw which is sealed tightly before being placed in the programmable controlled rate freezer. Usually seeding is performed to trigger crystallization by touching the straws using a LN_2 cooled tweezer or rod when the samples reached around −6 °C [29]. The accurate induced crystallization at the optimal seeding temperature of the media serves as a nucleus for further controlled and uniform crystal growth. If crystallization is induced below −7 °C, the survival rate of the embryos decreases drastically and without seeding the embryos will not survive the freezing procedure [93]. After crystallization of the medium, the seeding temperature should be maintained in the cooling bath for a few minutes to allow the embryos to adapt to the new osmotic conditions in the surrounding ice. Once the freezer has reached the set end temperature, the straws with the embryos will be transferred to a LN_2 tank (−196 °C) for longtime storage. Depending on the freezing rate and the end temperature the complete procedure takes a minimum of 1.5 h and needs the use of specific equipment and LN_2. Whether the thawing of the embryos is slow or fast depends on the temperature at which the straws with embryos are transferred from the freezing machine to the liquid nitrogen Dewar. Rapid thawing in a 37 °C warming bath (~1000 °C/min) is usually combined with a set end temperature of the freezer at −32 °C. In contrast, slow freezing to −80 °C requires controlled slow thawing (~8 °C/min) to achieve high survival rates [18]. Thawing procedure generally uses 1 M sucrose, a non-permeating CPA, which act by reducing the osmotic imbalance between the intra- and extracellular compartments and the velocity of water movement across the membranes. To control the risk of excessive swelling, stepwise decreasing concentrations of sucrose are used in thawing solutions.

6.2.2 Vitrification Method

Vitrification, a non-equilibrium method, is a cutting-edge technique for cryopreserving mouse embryos. Unlike slow freezing methods, vitrification involves rapid cooling, preventing the formation of ice crystals and promoting a glass-like, amorphous solid state. This approach has shown success in maintaining the viability of mouse embryos during cryopreservation, making it a valuable method. Vitrification methods started to be used in the mid 1980's with the seminal publication of Rall and Fahy [62], and have gained widespread use in mouse facilities (Fig. 6.1). Many reports have stated the benefits of vitrification over slow freezing [18, 68, 93]. Nowadays, it is also the preferred method of choice in human assisted reproductive techniques clinics [61]. The use of a device called Cryotop® (a plastic strip with small, flat wells that can hold individual embryos) has improved the efficiency of mouse embryo vitrification [1, 14, 41]. Other approaches such as minimum volume using a vitrification spatula (VS) device [47, 89] were reported with acceptable outcomes in mice. A concise protocol for vitrification is presented in Box 6.2.

Box 6.2: Vitrification Procedure

Vitrification induces ultra-rapid cooling of the samples, comprising a few steps in which embryos are fast cooled (up to 2500 °C/min) from room temperature to low subzero in the vapor phase of LN_2 (≤ −130 °C) or directly in LN_2 (−196 °C) in seconds. This, combined with high CPA concentrations (5–7 M) prevent ice formation and the supercooled water is converted into a highly viscous glass; thus, called vitrification [68]. This technique has the advantages of using cheap equipment and shorter processing time than the controlled rate of slow freezing [31]. There are many protocols for vitrification procedures and devices used, e.g. open-pulled straws, double straw system, electron microscope grids, Cryotop®, CryoLoop™, nylon and metal mesh, spatula, among others [4, 21, 49]. Those devices allow to place the embryos (usually 20–30) in a minimum drop volume to produce a higher and faster cooling rate. Warming rates should also be very fast, maintaining the vitreous state and avoiding the CPA toxicity and the possibly intracellular ice formation.

6.3 Sperm Cryopreservation in the Mouse

Following the discovery of the effect of glycerol as CPA in the 1950s, the cryopreservation of sperm has been used in farm animal species in connection with artificial insemination [60]. It became evident early on that the suitability of sperm for this procedure varied significantly among different mammalian species. Despite rigorous efforts, satisfactory results remained elusive for an extended period when using sperm from mice and rats. The first breakthrough with mouse sperm was achieved in the early 1990s by Takeshima, Nakagata, and Ogawa [74]. The key of the success in mouse sperm freezing was the use of a CPA consisting of a non-permeable combination of 3% skim milk and 18% raffinose in diH_2O [85]. This CPA gave the best and most reproducible results compared to several other protocols, mostly based on glycerol, and is still used as the method of choice for cryopreservation of mouse spermatozoa. However, further development was required after it became apparent that fertilization rates after IVF with frozen sperm from C57BL/6 mice were significantly worse than those from other inbred strains tested [57, 73]. Since C57BL/6 is in practice the standard inbred background for the maintenance of GA mouse lines (including international programs), intensive work has been done to further develop mouse cryopreservation of this inbred strain.

It is important to consider that when sperm is frozen, only half of the genome is preserved because sperm cells are haploid. This is a concern only if the intention is to preserve an inbred strain that is unique and not available elsewhere. Mitochondrial DNA is almost exclusively inherited through the oocyte [42], though it is also present in the midpiece of the sperm cell, albeit in significantly smaller quantities compared to nuclear DNA. Nevertheless, sperm freezing has several advantages over cryopreservation of preimplantation embryos, especially for lines with an inbred background where the number of embryos obtained after superovulation is low. First, usually only 2–3 sexually mature males are required as sperm donors. Thus, sperm freezing reduces the number of animals required to about one-tenth compared to embryo freezing. Second, sperm freezing is an efficient alternative or complement to embryo archiving, as it can significantly reduce the time, labor, and space required for animal maintenance. Finally, it does not require a controlled rate freezer, only liquid nitrogen. Despite efforts it has not yet been possible to develop a satisfactory method of artificial insemination in laboratory mice and rats, so efficient and safe IVF is required to revitalize mice archived by sperm freezing.

6.3.1 Advances in Mouse Sperm Cryopreservation and IVF

Initially, assisted reproduction techniques were used to improve fertilization rates of frozen C57BL/6 sperm. These included partial zona pellucida dissection (PZD), partial zona pellucida incision (ZIP), subzonal insemination (SUZI), and intracytoplasmic sperm injection (ICSI). However, these techniques are difficult to perform and require specialized equipment such as lasers, piezo elements, and micromanipulators. In addition, they damage the zona pellucida, which is thought to protect the oocytes and preimplantation embryos from infections. For these reasons, IVF without assistance has been chosen as the standard for revitalizing cryopreserved sperm [72].

The process of cryopreservation induces several changes in sperm quality that are directly related to their ability to fertilize. These include membrane and DNA damage, impaired capacitation, and oxidative stress due to overproduction of reactive oxygen species (ROS). Indeed, successful fertilization typically demands a more extended incubation period for thawed spermatozoa compared to fresh spermatozoa, before introducing them to oocytes in IVF. The addition of the ROS reducing agent alpha-monothioglycerol (MTG) to the freezing medium further increased the fertilization rate in IVF with frozen-thawed C57BL/6 sperm [59]. The addition of the amino acid L-glutamine to the freezing medium also enhanced the protective property of the CPA and resulted in a significant improvement in the IVF ability of this inbred strain. The observed cryoprotective effect of L-glutamine is thought to be due to the protection of the lipid layer membrane of spermatozoa [78].

Sperm capacitation is a prerequisite for fertilization. A fundamental process of mammalian sperm capacitation is cholesterol efflux from the sperm membrane, which is induced and supported in vitro by the addition of cholesterol acceptors such as albumin or β-cyclodextrins to the preincubation medium. Methyl-beta-cyclodextrin (MBCD) has previously been shown to be a particularly efficient cholesterol acceptor and to accelerate cholesterol efflux [15]. In practical application, MBCD in the preincubation medium significantly increased the fertilization rate in IVF with frozen-thawed C57BL/6 sperm, even after a shortened preincubation time [58, 77].

Continuous development of CPAs and preincubation media ensured that a sufficient proportion of preserved spermatozoa were fertilizable after thawing, even from the freeze sensitive C57BL/6 spermatozoa. To further improve fertilization rates, additional work was done to optimize the fertilization environment during IVF. An important step in this direction was the addition of reduced glutathione to the fertilization medium [7]. Since oxidized glutathione had no effect in the IVF medium, the effect of reduced glutathione as an antioxidant could be due to the protection of spermatozoa from oxidative stress, thus improving their motility and DNA integrity. During IVF with glutathione, a significant increase of the perivitelline space of the oocytes was also observed as a result of an expansion of the zona pellucida. The positive effect of glutathione on the fertilization rate could therefore also be due to a structural modification of the zona pellucida that facilitates sperm penetration [79].

The protocols for mouse sperm cryopreservation and its use in IVF for revitalization of archived lines have been modified several times according to new knowledge. Fortunately, the media required for mouse sperm cryopreservation, sperm preincubation, and IVF are now commercially available and of high quality. The optimized protocols, developed at The Jackson Laboratory [75, 76] and the Nakagata Laboratory at the Center for Animal Resources and Development (CARD) [84], have been published and are available as valuable tools for archiving and revitalizing mouse sperm (Fig. 6.2). Furthermore, validated protocols and training videos for mouse spermatozoa freezing techniques and IVF procedures are provided for download by the European Mouse Mutant Archive at https://www.infrafrontier.eu/emma/cryopreservation-protocols/.

Box 6.3: Sperm Freezing Procedures

Regardless of the protocol used for archiving mouse spermatozoa, some general requirements should be considered for a successful procedure. Young adults ideally in the age of 3–6 months are preferred as donors. One week before cryopreservation, they should be individually separated to avoid testosterone suppression, frequently induced by the dominant male in grouped cages. Euthanize a male mouse, cut the two cauda epididymides and place them in a filter paper to remove the fat and surrounded blood. Several incisions should be made to the epididymides in a CPA drop to release the sperm to be maintained in a hot plate for 3–5 min, spinning the dish. Straws should be prepared with the following sequence: 100 uL of mHTF, 15 mm of air bubble, 10 uL of sperm suspension, 15 mm of air bubble and seal at the end. Around ten sperm straws should be obtained from each male. Finally, the samples should be cooled down in a freezing canister floating on liquid nitrogen in a cryobiological container. Around 10 min later the straws should be immersed into nitrogen and immediately stored in a liquid nitrogen tank using a suitable canister (Fig. 6.2). Sperm cryopreservation can also be performed using cryovials. For references see Takeo et al. [83].

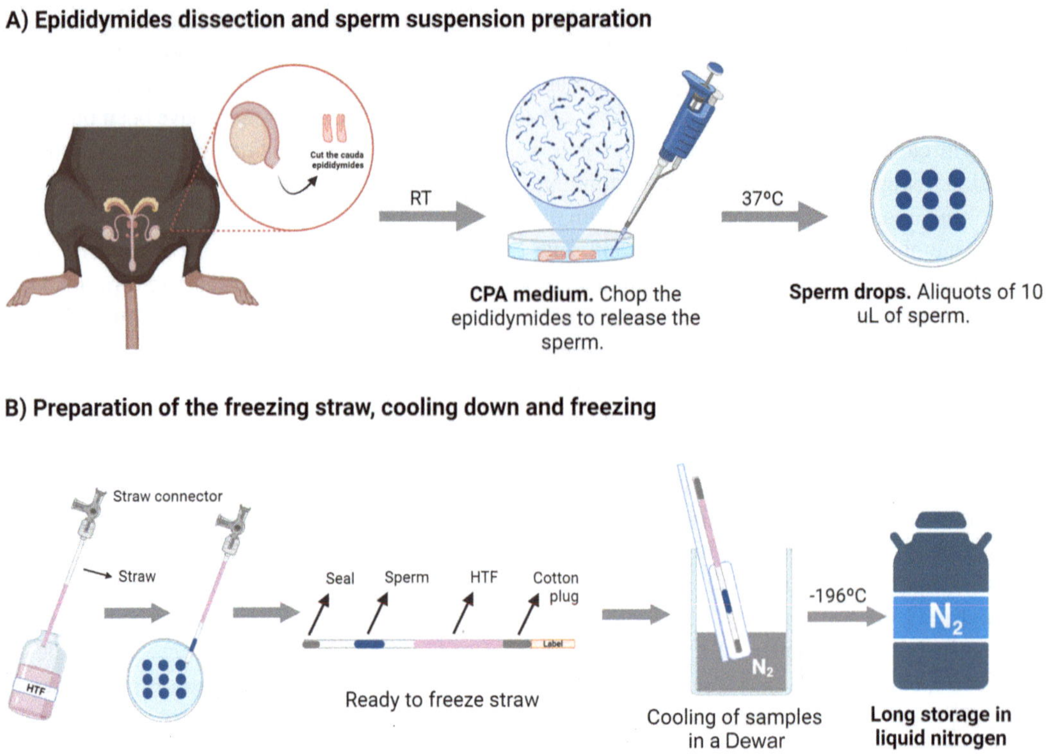

Fig. 6.2 Mouse sperm cryopreservation. With this procedure, 2–3 males (3–6 months old) are usually required as donors, providing around 20–30 straws with 10 uL of sperm each. (A): the first step is the dissection of the cauda epididymis and the preparation of the sperm suspension in drops. (B): the second part is the preparation of the straws followed by the cooling and storage of the sperm (note that cryovials instead of straws can also be used) (see Box 6.3). Different protocols can be found with these general steps [83]. (Created with BioRender.com)

6.3.2 Sperm Collection

For the successful cryopreservation of mouse spermatozoa, it is imperative not to overlook the accompanying techniques. The ultimate outcome, namely the birth of offspring, is influenced by every step, from the method of sperm collection to the successful transfer of embryos post-IVF. Sperm for storage is routinely collected from the cauda epididymis and vas deferens of freshly euthanized males. This method guarantees a very large amount of sperm, so usually only 2–3 males are needed to safely archive a strain or line (Fig. 6.2). Young adults ideally in the age of 3–6 months are preferred as donors. Preliminary fertility testing is not recommended since this would result in unnecessary surplus animals. However, sperm quality should be analyzed before and immediately after freezing. This ensures the assessment of the donor's sperm quality, the suitability of the sperm sample for cryopreservation and subsequent revitalization, and the successful completion of the cryopreservation process.

The main quality parameters are the number of viable spermatozoa in the sample, their general and progressive motility, and the percentage of spermatozoa with a normal morphology. The analysis can be performed under a microscope using simple counting chambers, or using a computer assisted sperm analyzer (CASA). The results of analysis of frozen/thawed sperm samples provide a reliable assessment of the quality of the archived spermatozoa, although there are no fixed, generally accepted limit values for the parameters mentioned. However, IVF and embryo culture to blastocyst hatching or even embryo transfer to take the resulting embryos to term may be essential to ensure their fertilization capacity.

Pathogenic microorganisms such as the highly infectious murine hepatitis virus and mouse parvovirus have been found in the testes and cauda epididymis of infected mice [2, 66, 70]. Therefore, the microbiological quality of donors used for sperm cryopreservation should be considered. Although infection of embryos during IVF with currently known mouse pathogens can be excluded as long as the zona pellucida is intact and the embryos are washed extensively before transfer to pseudopregnant surrogates, there is a risk of microbiological contamination of archived samples. Animals with a specified pathogen-free status should therefore generally be preferred as donors.

An alternative, which does not require the killing of the donor males, is epididymal sperm aspiration. However, the amount of sperm obtained by this method is highly variable and significantly lowers overall [9, 19, 50, 51]. In addition, sperm retrieved by this technique have been shown to be more likely to have morphological abnormalities and reduced motility, which limits their suitability for cryopreservation [95]. As further methods for nonterminal sperm collection, attempts were made at electroejaculation, the pharmacologic induction and flushing of the uterus of females previously mated. In particular, electrical stimulation and the ejaculation by the action of drugs had widely varying results and undesirable side effects, some of which were fatal to the animals. In contrast, ejaculated sperm obtained by uterine flushing could be successfully used for IVF and cryopreservation [69, 92]. The method can also be applied in such a way that neither males nor females need to be killed to obtain sperm [24]. Also, with the interruption of natural mating, ejaculated mouse sperm can be obtained in larger quantities without killing the donors. If the male was separated from the female only 1–2 s after onset of the ejaculatory shudder, ejaculation occurred immediately following and the vaginal plug could be recovered along with the ejaculate from the penile cup of the erected penis [46].

6.3.3 Storage and Recovery of Frozen Sperm

For short-term storage or transport, mouse spermatozoa can be kept refrigerated for a few days. A feasible method was developed using a preservation medium for epididymal storage containing 10% DMSO and 100 µg/mL quercetin (a plant-derived antioxidant). For example, C57BL/6 sperm can be stored at 4 °C for 7 days and still achieved 80% fertilization rate after IVF [96]. Long-term storage of cryopreserved sperm requires extremely low temperatures and is typically conducted in liquid nitrogen at −196 °C. This entails the use of specialized nitrogen tanks (often called Dewars after its inventor, Sir James Dewar) for storage and a continuous provision of liquid nitrogen. The maintenance of these low temperatures during transport represents an additional expense.

Freeze-drying of mouse spermatozoa has been successfully tested as an alternative option for sperm storage and transport [91]. Following freeze-drying, spermatozoa lose their motility and vitality, yet retain the capacity to activate the oocyte. Therefore, revitalization can only be achieved through ICSI, requiring specialized laboratory equipment and expertise. Freeze-dried (lyophilized) sperm can be stored and transported at room temperature for short durations. However, for extended storage, freeze-dried mouse sperm must also be preserved at temperatures below −80 °C. Despite continuous efforts to improve mouse and rat freeze-dried and vacuum-dried sperm, these methods are not yet routinely used in these rodent species [90].

The recovery of frozen mouse sperm is a critical aspect of line preservation. After cryopreservation, mouse sperm can be thawed and recovered for use in assisted reproductive techniques. The thawing process involves carefully warming the sperm to regain its motility and viability. Typically, a stepwise reduction of cryoprotectant agents is employed to avoid potential damage caused by osmotic stress. Following the thawing process, the sperm becomes suitable for use in IVF (very unusually for ICSI), facilitating the

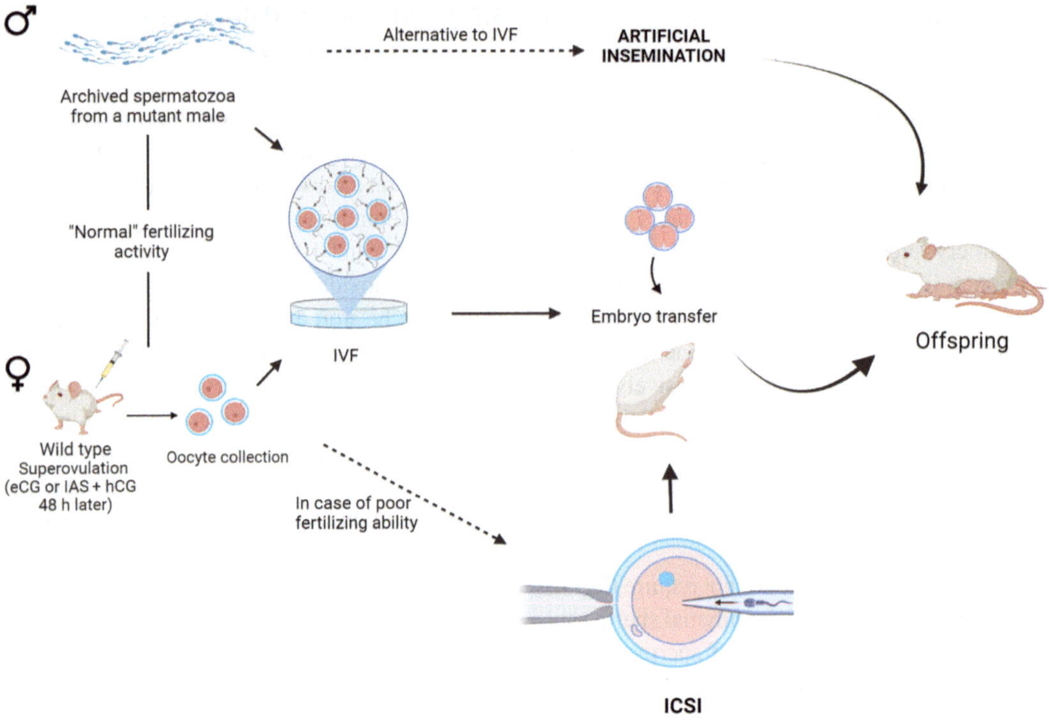

Fig. 6.3 Recovery of cryopreserved mouse sperm. This schematic illustration presents the available methodologies for the recovery of frozen sperm. Following cryopreservation, the thawed sperm undergoes a stepwise reduction of cryoprotectant agents to prevent osmotic stress. The recovered sperm is then utilized in IVF procedures after the superovulation of female mice. Although very rarely used, the ICSI option is depicted, providing alternative approaches for fertilization. For details, refer to the text. (Created with BioRender.com)

production of live mice with the anticipated genotype. The success of the recovery process is crucial for maintaining valuable mouse strains and ensuring the conservation of genetic resources (Fig. 6.3).

6.4 Sperm Cryopreservation in the Rat

The advancement of endonuclease techniques has streamlined the creation of genetically modified rats [64, 67]. As rat models are more suitable than mice for certain studies, in part due to their size, the numbers of rats as models in biomedical research is increasing and a safe and efficient method of archiving is needed for this species as well. Therefore, sperm cryopreservation is an important alternative to embryo cryopreservation also for laboratory rats. Numerous experiments have shown that rat spermatozoa are very sensitive to changes in their external environment and that the established freezing protocols (including CPAs) used for mouse spermatozoa could not be easily applied to rat spermatozoa. The pioneers of archiving mouse sperm with a CPA based on lactose and chicken egg yolk have recently published a successful protocol in EGFP transgenic male rats with Sprague Dawley outbred background [55]. Optimized protocols for rat sperm cryopreservation and IVF, including Brown Norway, Long Evans, F344, and Wistar inbred strains, have been recently published [81].

As in the laboratory mouse, the success of storage by sperm cryopreservation in the rat depends on the results of the revitalization, which is carried out by IVF and therefore also requires efficient oocyte retrieval. Superovulation of the rat, previously induced by injection of eCG and hCG, was significantly improved by additional treatment with IAS and luteinizing hormone-releasing hormone (LHRH) [33]. However, important criteria for improvement were also identified for the IVF itself, which is difficult to perform in laboratory rats and is characterized by poor results. The fertilization rate decreases significantly with the time that elapses between the euthanasia of the hormone-treated females and the removal of the fallopian tubes. This time should not exceed 1 min, so a rapid method, such as cervical dislocation, which is applicable to immature rats, is strongly preferable over euthanasia with CO_2 or an overdose of isoflurane. A positive effect has also been demonstrated for a short period of time, not exceeding 3 min, in the recovery of cumulus-oocyte complexes from the ampulla. With strict adherence to these guidelines, reproducible IVF rates of more than 90% have been achieved for various inbred strains and stocks of rats [32].

6.5 Oocyte Cryopreservation

While oocyte cryopreservation is a technique primarily developed in human-assisted reproductive clinics to preserve fertility [16, 35], its widespread implementation in rodents is yet to be fully realized. It is well known that the larger the cell to be cryopreserved, the greater the damage due to the higher water content, the low surface area to volume ratio, and the formation of ice crystals in the cytoplasm. This is the case of mature oocytes (MII stage), which in rodents can have a size of ~100 μm in diameter with a high quantity of water inside, making them a challenging cell to cryopreserve. Despite several efforts to overcome these technical difficulties such as using low concentrations of trehalose and DMSO [22] or the addition of N-acetyl cysteine after warming [52], oocyte survival and fertilization ability after thawing or warming are still not satisfactory in mice and rats. The use of enhanced vitrification methods can partially address this low efficiency rate, as the ultra-rapid freezing technique prevents structural damage if executed proficiently [14, 54]. Jadoon et al. compared the cryopreservation of mouse oocytes either by slow freezing or vitrification, reporting a greater survival rate of vitrified versus slow frozen oocytes (70.3% vs. 12.5%), and also higher fertilization rates (15.86% vs. 4.6%) [36].

The oocytes vitrification technique is similar to embryo vitrification and consists mainly of a two-step procedure, in which increasing concentrations of CPAs, such as a mixture of ethylene glycol, DMSO, and sucrose, are added, with equilibration steps that can take up to 15 min depending on the protocol. Subsequently, the samples are loaded into the corresponding device and immediately plunged into liquid nitrogen for long-term storage. DMSO has been reported as a crucial CPA in mouse oocytes, improving cryo-survival and developmental potential of embryos up to blastocyst without altering maternal imprinted gene expression [13]. Devices are the same as the ones used for embryo vitrification and are described above. A comprehensive review of cryodevices can be found in a recent article [45]. Although this technology is simpler and faster than slow freezing, some considerations such as cooling rate, viscosity, and volume of vitrification solution should be taken into account to obtain acceptable survival rates [4]. In rats this approach is still underdeveloped, with some advancements using immature oocytes [26, 39, 56, 81].

Nowadays, the need for an oocytes cryobank does not appear to be a requirement in standard rodent facilities. This is due in part to the fact that oocytes preserve only a haploid genome, as in the case of sperm. Additionally, working with oocytes poses more challenges compared to male gametes. Archiving oocytes from inbred strains commonly used for IVF, such as C57BL/6, might be beneficial in minimizing the number of females required when surplus animals are used as donors. However, it is essential to have proficient mastery of IVF techniques within the facility. Unless there are advancements in terms of user-friendly application, survival rates, fertilization success, pregnancy, and birth rates, the preferred option remains the cryopreservation of embryos or sperm as a backup.

6.6 Cryopreservation of Ovarian Tissue in the Mouse

A feasible alternative to archiving female germ cells is the cryopreservation of ovarian tissue. Although this method was established several years ago for use with mouse inbred strains, it is rarely employed today [12, 71]. This may be due to the more demanding surgical procedure of ovariectomy and subsequent ovarian tissue transplantation. One way to facilitate the procedure is to transfer pieces instead of whole ovaries. For this purpose, the ovaries can be divided into 2, 4, or even 8 pieces of equal size. In experiments with frozen ovary fragments, pregnancy rates of 90–100% were achieved in recipients after orthotopic transplantation [40]. Although the average litter size decreases somewhat as the amount of ovarian tissue transplanted decreases, archiving smaller pieces of ovarian tissue may be advantageous because the two ovaries from one donor can be transferred bilaterally to more than one recipient. A prerequisite for successful transplantation and subsequent ovarian function is histocompatibility between donor and recipient females. This limitation has been overcome by the availability of immunodeficient mutants like the NSG (NOD.Cg-*Prkdc*^{scid};*Il2rg*^{tm1Wjl}/SzJ) mice, allowing transplantation of ovarian tissue from mismatched or genetically undefined donors [40].

Several protocols for slow freezing and vitrification of ovary tissue from mice have been published [88]. These have tested different cryoprotectants alone or in combination and with cryotubes or straws for longtime storage. Although all published protocols have been successful in cryopreserving ovarian tissue, there are significant differences in latency to first pregnancy and litter size. Using straws, 1.5 M DMSO as cryoprotectant, seeding at −7 °C and slow freezing at 0.3 °C/min to −40 °C, results for syngeneic transplantation of C57BL/6 ovarian halves were similar to control animals of the same strain after transplantation of fresh, untreated tissue [40].

6.7 Archiving Mouse and Rat Models

Archiving repositories, primarily located in the northern hemisphere, provide cryopreservation services for mouse and rat embryos and sperm. This approach serves as a means to safeguard valuable GA lines from various global sources. Stringent protocols with standardized operating procedures are meticulously followed to ensure the delivery of a high-quality service that remains reproducible over time. Numerous repositories are accessible for receiving live animals, thereafter, cryopreserving either embryos or sperm (charges fees may apply to partially cover repository costs). This facilitates the archiving of genotypes and the establishment of a shared resource [20].

For mice, the Mutant Mouse Resource & Research Center (MMRRC, https://www.mmrrc.org/) funded by the NIH is one of the largest centers in the USA and holds more than 50,000 mutant alleles as live mice, cryopreserved germplasm, and/or mutant embryonic stem cells [3]. The Canadian Mouse Mutant Repository offer a similar approach, with more than 1000 mouse lines in the stock. Other repositories are located in Japan, like RIKEN (https://mus.brc.riken.jp) or CARD (https://cardmice.com) [82], Australia (Animal Resources Centre, ARC, https://www.wa.gov.au/organisation/animal-resources-centre), and Europe (European Mouse Mutant Archive, EMMA, https://www.infrafrontier.eu/emma/). The International Mouse Strain Resource (IMSR) (https://www.findmice.org/) is a searchable online database of mouse strains, stocks, and mutant ES cell lines available worldwide, including inbred, mutant, and genetically engineered strains. No repositories can be found in Latin America, although some facilities such as the Laboratory Animals Biotechnology Unit of Institut Pasteur de Montevideo in Uruguay are covering the regional demand for embryo and sperm archiving for local researchers.

In rats, with thousands of models developed for a variety of applications, two international repositories can be found, the Rat Resource and Research Center (RRRC, https://www.rrrc.us) in the USA and the National Bio Resource Project for the Rat (NBRP, https://www.anim.med.kyoto-u.ac.jp/nbr/) in Japan. The gene editing capabilities to specifically modify the rat genome using CRISPR/Cas9 means that more GA lines of laboratory rats will be generated in the future and archiving them will become important. Repositories such as the EMMA are already including cryopreservation of laboratory rats in their programs.

The option to archive mouse and rat lines in multiple repositories worldwide, coupled with inclusive access for researchers across various disciplines, renders these resources invaluable for advancing biomedical knowledge. In summary, cryopreservation aligns with the 3Rs principles by replacing the need for live animals for certain purposes, reducing the overall number of animals required, preventing the duplication of efforts, and refining experimental methods. In an illustration of the duplication issue, a survey carried out in 2005 revealed that out of roughly 5000 published mouse lines, about 40% were replicated 1–3 times [44].

Detailed protocols of most of the techniques described in this chapter can be found online at http://card.medic.kumamoto-u.ac.jp/card/multilingual/index.html (Reproductive engineering techniques of mice: Manual Information Center). The aim of this page, managed by Kumamoto University, is to advance biomedical science and promote the principles of 3Rs. Furthermore, the manual is available for download in various languages, such as English, Spanish, French, Chinese, and Japanese.

In conclusion of this chapter, the cryopreservation of mouse and rat germplasm stands as a fundamental chapter in the ongoing effort of advancing biomedical research. With its potential to reduce the use of live animals, promote resource sharing, and refine experimental methodologies, cryopreservation not only preserves genetic diversity but also aligns seamlessly with the principles of 3Rs. As technology continues to evolve, this invaluable tool opens new avenues for collaboration and knowledge dissemination. Given the rise in both the quantity and quality of GA rodent models worldwide, there is an imperative to streamline space and resources. Simultaneously, preserving these valuable lines becomes crucial to prevent genetic drift and otential losses resulting from pathogen outbreaks. Progress in cryobiology, evolving over decades, now provides knowledge and tools that enable the application of this technology in standard animal facilities. This advancement aligns with the pressing goals of optimizing research spaces, conserving resources, and safeguarding the integrity of invaluable genetically unique animal models.

References

1. Abedpour N, Rajaei F. Vitrification by Cryotop and the maturation, fertilization, and developmental rates of mouse oocytes. Iran Red Crescent Med J. 2015;17:e18172.
2. Agca Y, Bauer BA, Johnson DK, Critser JK, Riley LK. Detection of mouse parvovirus in Mus musculus gametes, embryos, and ovarian tissues by polymerase chain reaction assay. Comp Med. 2007;57:51–6.
3. Amos-Landgraf J, Franklin C, Godfrey V, Grieder F, Grimsrud K, Korf I, Lutz C, Magnuson T, Mirochnitchenko O, Patel S, Reinholdt L, Lloyd KCK. The mutant mouse resource and research center (MMRRC): the NIH-supported National Public Repository and distribution archive of mutant mouse models in the USA. Mamm Genome. 2022;33:203–12.
4. Arav A. Cryopreservation of oocytes and embryos. Theriogenology. 2014;81:96–102.
5. Argmann CA, Dierich A, Auwerx J. Uses of forward and reverse genetics in mice to study gene function. Curr Protoc Mol Biol. 2006;Chapter 29:Unit 29A 1.
6. Azkona G. Implementing the 3Rs in laboratory animal research-from theory to practice. Animals (Basel). 2023;13:3063.
7. Bath ML. Inhibition of in vitro fertilizing capacity of cryopreserved mouse sperm by factors released by damaged sperm, and stimulation by glutathione. PLoS One. 2010;5:e9387.
8. Berendzen KM, Sharma R, Mandujano MA, Wei Y, Rogers FD, Simmons TC, Seelke AMH, Bond JM, Larios R, Goodwin NL, Sherman M, Parthasarthy S, Espineda I, Knoedler JR, Beery A, Bales KL, Shah NM, Manoli DS. Oxytocin receptor is not required for social attachment in prairie voles. Neuron. 2023;111:787–96.e4.
9. Boersma A, Olszanska O, Walter I, Rülicke T. Microsurgical and percutaneous Epididymal sperm

aspiration for sperm collection from live mice. J Am Assoc Lab Anim Sci. 2015;54:471–7.
10. Bojic S, Murray A, Bentley BL, Spindler R, Pawlik P, Cordeiro JL, Bauer R, De Magalhães JP. Winter is coming: the future of cryopreservation. BMC Biol. 2021;19:56.
11. Brinster RL, Chen HY, Trumbauer M, Senear AW, Warren R, Palmiter RD. Somatic expression of herpes thymidine kinase in mice following injection of a fusion gene into eggs. Cell. 1981;27:223–31.
12. Candy CJ, Wood MJ, Whittingham DG. Restoration of a normal reproductive lifespan after grafting of cryopreserved mouse ovaries. Hum Reprod. 2000;15:1300–4.
13. Cantatore C, George JS, Depalo R, D'Amato G, Moravek M, Smith GD. Mouse oocyte vitrification with and without dimethyl sulfoxide: influence on cryo-survival, development, and maternal imprinted gene expression. J Assist Reprod Genet. 2021;38:2129–38.
14. Castello D, Cobo A, Mestres E, Garcia M, Vanrell I, Alejandro Remohi J, Calderon G, Costa-Borges N. Pre-clinical validation of a closed surface system (Cryotop SC) for the vitrification of oocytes and embryos in the mouse model. Cryobiology. 2018;81:107–16.
15. Choi YH, Toyoda Y. Cyclodextrin removes cholesterol from mouse sperm and induces capacitation in a protein-free medium. Biol Reprod. 1998;59:1328–33.
16. Cobo A, Garcia-Velasco JA, Remohi J, Pellicer A. Oocyte vitrification for fertility preservation for both medical and nonmedical reasons. Fertil Steril. 2021;115:1091–101.
17. Costantini F, Lacy E. Introduction of a rabbit beta-globin gene into the mouse germ line. Nature. 1981;294:92–4.
18. Critser JK, Arneson BW, Aaker DV, Huse-Benda AR, Ball GD. Factors affecting the cryosurvival of mouse two-cell embryos. J Reprod Fertil. 1988;82:27–33.
19. Del Val GM, Robledano PM. In vivo serial sampling of epididymal sperm in mice. Lab Anim. 2013;47:168–74.
20. Donahue LR, Hrabe De Angelis M, Hagn M, Franklin C, Lloyd KC, Magnuson T, Mckerlie C, Nakagata N, Obata Y, Read S, Wurst W, Hörlein A, Davisson MT. Centralized mouse repositories. Mamm Genome. 2012;23:559–71.
21. Dos Santos Neto PC, Vilarino M, Barrera N, Cuadro F, Crispo M, Menchaca A. Cryotolerance of day 2 or day 6 in vitro produced ovine embryos after vitrification by Cryotop or Spatula methods. Cryobiology. 2015;70:17–22.
22. Eroglu A, Bailey SE, Toner M, Toth TL. Successful cryopreservation of mouse oocytes by using low concentrations of trehalose and dimethylsulfoxide. Biol Reprod. 2009;80:70–8.
23. Fernandez A, Josa S, Montoliu L. A history of genome editing in mammals. Mamm Genome. 2017;28:237–46.
24. Foxworth WB, Carpenter E, Kraemer DC, Kier AB. Nonsurgical and nonlethal retrieval of mouse spermatozoa. Lab Anim Sci. 1996;46:352–4.
25. Fray MD. Biological methods for archiving and maintaining mutant laboratory mice. Part I: conserving mutant strains. Methods Mol Biol. 2009;561:301–19.
26. Fujiwara K, Kamoshita M, Kato T, Ito J, Kashiwazaki N. Generation of rats from vitrified oocytes with surrounding cumulus cells via in vitro fertilization with cryopreserved sperm. Anim Sci J. 2017;88:180–4.
27. Glenister PH, Whittingham DG, Lyon MF. Further-studies on the effect of radiation during the storage of frozen 8-cell mouse embryos at −196-degrees-C. J Reprod Fertil. 1984;70:229–34.
28. Gordon JW, Ruddle FH. Integration and stable germ line transmission of genes injected into mouse pronuclei. Science. 1981;214:1244–6.
29. Guan M, Bogani D, Marschall S, Raspa M, Takeo T, Nakagata N, Fray M. Conservation of mouse models through embryo freezing. Curr Protoc Mouse Biol. 2014;4:205–27.
30. Hall B, Limaye A, Kulkarni AB. Overview: generation of gene knockout mice. Curr Protoc Cell Biol. 2009;Chapter 19:Unit 19 12 19 12 1–17.
31. Hart-Johnson S, Mankelow K. Archiving genetically altered animals: a review of cryopreservation and recovery methods for genome edited animals. Lab Anim. 2022;56:26–34.
32. Hino C, Ueda J, Funakoshi H, Matsumoto S. Defined oocyte collection time is critical for reproducible fertilization in rats of different strains. Theriogenology. 2020;144:146–51.
33. Honda A, Tachibana R, Hamada K, Morita K, Mizuno N, Morita K, Asano M. Efficient derivation of knockout and knock-in rats using embryos obtained by in vitro fertilization. Sci Rep. 2019;9:11571.
34. Hsu PD, Lander ES, Zhang F. Development and applications of CRISPR-Cas9 for genome engineering. Cell. 2014;157:1262–78.
35. Iussig B, Maggiulli R, Fabozzi G, Bertelle S, Vaiarelli A, Cimadomo D, Ubaldi FM, Rienzi L. A brief history of oocyte cryopreservation: arguments and facts. Acta Obstet Gynecol Scand. 2019;98:550–8.
36. Jadoon S, Adeel M. Cryopreservation of oocytes. J Ayub Med Coll Abbottabad. 2015;27:22–8.
37. Jang TH, Park SC, Yang JH, Kim JY, Seok JH, Park US, Choi CW, Lee SR, Han J. Cryopreservation and its clinical applications. Integr Med Res. 2017;6:12–8.
38. Kawamata M, Ochiya T. Generation of genetically modified rats from embryonic stem cells. Proc Natl Acad Sci USA. 2010;107:14223–8.
39. Kim SS, Olsen R, Kim DD, Albertini DF. The impact of vitrification on immature oocyte cell cycle and cytoskeletal integrity in a rat model. J Assist Reprod Genet. 2014;31:739–47.
40. Kolbe T, Walter I, Rulicke T. Influence of graft size, histocompatibility, and cryopreservation on reproductive outcome following ovary transplantation in mice. J Assist Reprod Genet. 2019;36:2583–91.

41. Kuwayama M. Highly efficient vitrification for cryopreservation of human oocytes and embryos: the Cryotop method. Theriogenology. 2007;67:73–80.
42. Lee W, Zamudio-Ochoa A, Buchel G, Podlesniy P, Marti Gutierrez N, Puigros M, Calderon A, Tang HY, Li L, Mikhalchenko A, Koski A, Trullas R, Mitalipov S, Temiakov D. Molecular basis for maternal inheritance of human mitochondrial DNA. Nat Genet. 2023;55:1632–9.
43. Leibo SP, Sztein JM. Cryopreservation of mammalian embryos: derivation of a method. Cryobiology. 2019;86:1–9.
44. Lloyd K, Franklin C, Lutz C, Magnuson T. Reproducibility: use mouse biobanks or lose them. Nature. 2015;522:151–3.
45. Masrat-Un-Nisa EY, Malik AA, Sofi KA, Khatun A, Yousuf N. Recent advancements in vitrification cryodevices for gamete and gonadal tissue. Cryo Letters. 2022;43:129–39.
46. Mcgill TE, Coughlin RC. Ejaculatory reflex and luteal activity induction in Mus musculus. J Reprod Fertil. 1970;21:215–20.
47. Meikle MN, Schlapp G, Menchaca A, Crispo M. Minimum volume Spatula MVD vitrification method improves embryo survival compared to traditional slow freezing, both for in vivo and in vitro produced mice embryos. Cryobiology. 2018;84:77–81.
48. Menchaca A, Schlapp G, Noel Meikle M, Crispo M. Transgenesis and gene edition in mammals. In: Reference module in life sciences. Oxford: Elsevier; 2017.
49. Momozawa K, Matsuzawa A, Tokunaga Y, Abe S, Koyanagi Y, Kurita M, Nakano M, Miyake T. Efficient vitrification of mouse embryos using the Kitasato Vitrification system as a novel vitrification device. Reprod Biol Endocrinol. 2017;15:29.
50. Moreno-Del Val G, Munoz-Robledano P. Revitalizing genetically-modified mouse strains using frozen-thawed sperm after up to 192 h of refrigerated epididymis transportation. Lab Anim. 2017;51:526–9.
51. Moreno-Del Val G, Munoz-Robledano P, Caler AJ, Morante J. A method for multiple sampling mouse sperm in vivo. Biol Reprod. 2023;108:197–203.
52. Mukunoki A, Takeo T, Nakagata N. N-acetyl cysteine restores the fertility of vitrified-warmed mouse oocytes derived through ultrasuperovulation. PLoS One. 2019;14:e0224087.
53. Mullins JJ, Peters J, Ganten D. Fulminant hypertension in transgenic rats harbouring the mouse Ren-2 gene. Nature. 1990;344:541–4.
54. Nakagata N. High survival rate of unfertilized mouse oocytes after vitrification. J Reprod Fertil. 1989;87:479–83.
55. Nakagata N, Mikoda N, Nakao S, Nakatsukasa E, Takeo T. Establishment of sperm cryopreservation and fertilisation protocols for rats. Sci Rep. 2020;10:93.
56. Nakagata N, Nakao S, Mikoda N, Yamaga K, Takeo T. Observation of the in vitro fertilization process in living oocytes using frozen-thawed sperm in rats. Theriogenology. 2023;199:69–76.
57. Nakagata N, Takeshima T. Cryopreservation of mouse spermatozoa from inbred and F1-hybrid strains. Exp Anim. 1993;42:317–20.
58. Nakao S, Ito K, Sakoh K, Takemoto K, Watanabe H, Kondoh G, Irie T, Nakagata N, Takeo T. Dimethyl-alpha-cyclodextrin induces capacitation by removing phospholipids from the plasma membrane of mouse spermdagger. Biol Reprod. 2023;108:671–81.
59. Ostermeier GC, Wiles MV, Farley JS, Taft RA. Conserving, distributing and managing genetically modified mouse lines by sperm cryopreservation. PLoS One. 2008;3:e2792.
60. Polge C, Smith AU, Parkes AS. Revival of spermatozoa after vitrification and dehydration at low temperatures. Nature. 1949;164:666.
61. Pomeroy KO, Comizzoli P, Rushing JS, Lersten IL, Nel-Themaat L. The ART of cryopreservation and its changing landscape. Fertil Steril. 2022;117:469–76.
62. Rall WF, Fahy GM. Ice-free cryopreservation of mouse embryos at −196 degrees C by vitrification. Nature. 1985;313:573–5.
63. Rall WF, Schmidt PM, Lin X, Brown SS, Ward AC, Hansen CT. Factors affecting the efficiency of embryo cryopreservation and rederivation of rat and mouse models. ILAR J. 2000;41:221–7.
64. Remy S, Chenouard V, Tesson L, Usal C, Menoret S, Brusselle L, Heslan JM, Nguyen TH, Bellien J, Merot J, De Cian A, Giovannangeli C, Concordet JP, Anegon I. Generation of gene-edited rats by delivery of CRISPR/Cas9 protein and donor DNA into intact zygotes using electroporation. Sci Rep. 2017;7:16554.
65. Rulicke T, Autenried P. Potential of two-cell mouse embryos to develop to term despite partial damage after cryopreservation. Lab Anim. 1995;29:320–6.
66. Scavizzi F, Raspa M. Tissue distribution and duration of mouse hepatitis virus in naturally infected immunocompetent ICR (CD-1) and immunodeficient athymic nude-nu mouse strains used for ovarian transplantation and in vitro fertilization. Lab Anim. 2004;38:189–99.
67. Shao Y, Guan Y, Wang L, Qiu Z, Liu M, Chen Y, Wu L, Li Y, Ma X, Liu M, Li D. CRISPR/Cas-mediated genome editing in the rat via direct injection of one-cell embryos. Nat Protoc. 2014;9:2493–512.
68. Shaw JM, Jones GM. Terminology associated with vitrification and other cryopreservation procedures for oocytes and embryos. Hum Reprod Update. 2003;9:583–605.
69. Songsasen N, Leibo SP. Live mice from cryopreserved embryos derived in vitro with cryopreserved ejaculated spermatozoa. Lab Anim Sci. 1998;48:275–81.
70. Suzuki H, Yorozu K, Watanabe T, Nakura M, Adachi J. Rederivation of mice by means of in vitro fertilization and embryo transfer. Exp Anim. 1996;45:33–8.
71. Sztein J, Sweet H, Farley J, Mobraaten L. Cryopreservation and orthotopic transplantation of mouse ovaries: new approach in gamete banking. Biol Reprod. 1998;58:1071–4.
72. Sztein JM, Farley JS, Mobraaten LE. In vitro fertilization with cryopreserved inbred mouse sperm. Biol Reprod. 2000a;63:1774–80.

73. Sztein JM, Farley JS, Mobraaten LE. In vitro fertilization with cryopreserved inbred mouse sperm. Biol Reprod. 2000b;63:1774–80.
74. Sztein JM, Takeo T, Nakagata N. History of cryobiology, with special emphasis in evolution of mouse sperm cryopreservation. Cryobiology. 2018;82:57–63.
75. Taft R. In vitro fertilization in mice. Cold Spring Harb Protoc. 2017a;2017:pdb.prot094508.
76. Taft R. Mouse sperm cryopreservation using cryoprotectant containing monothioglycerol (MTG). Cold Spring Harb Protoc. 2017b;2017:pdb.prot094490.
77. Takeo T, Hoshii T, Kondo Y, Toyodome H, Arima H, Yamamura K, Irie T, Nakagata N. Methyl-beta-cyclodextrin improves fertilizing ability of C57BL/6 mouse sperm after freezing and thawing by facilitating cholesterol efflux from the cells. Biol Reprod. 2008;78:546–51.
78. Takeo T, Nakagata N. Combination medium of cryoprotective agents containing L-glutamine and methyl-β-cyclodextrin in a preincubation medium yields a high fertilization rate for cryopreserved C57BL/6J mouse sperm. Lab Anim. 2010;44:132–7.
79. Takeo T, Nakagata N. Reduced glutathione enhances fertility of frozen/thawed C57BL/6 mouse sperm after exposure to methyl-beta-cyclodextrin. Biol Reprod. 2011;85:1066–72.
80. Takeo T, Nakagata N. Superovulation using the combined administration of inhibin antiserum and equine chorionic gonadotropin increases the number of ovulated oocytes in C57BL/6 female mice. PLoS One. 2015;10:e0128330.
81. Takeo T, Nakao S, Mikoda N, Yamaga K, Maeda R, Tsuchiyama S, Nakatsukasa E, Nakagata N. Optimized protocols for sperm cryopreservation and in vitro fertilization in the rat. Lab Anim (NY). 2022;51:256–74.
82. Takeo T, Nakao S, Nakagawa Y, Sztein JM, Nakagata N. Cryopreservation of mouse resources. Lab Anim Res. 2020;36:33.
83. Takeo T, Sztein J, Nakagata N. The CARD method for mouse sperm cryopreservation and in vitro fertilization using frozen-thawed sperm. Methods Mol Biol. 2019a;1874:243–56.
84. Takeo T, Sztein J, Nakagata N. The CARD method for mouse sperm cryopreservation and in vitro fertilization using frozen-thawed sperm. Microinjection. 2019b;1874:243–56.
85. Takeshima T, Nakagata N, Ogawa S. Cryopreservation of mouse spermatozoa. Jikken Dobutsu. 1991;40:493–7.
86. Takizawa A, Eto T. Protocols for cryopreservation and rederivation of rat gametes. Methods Mol Biol. 2019;2018:131–49.
87. Taylor JH, Walton JC, Mccann KE, Norvelle A, Liu Q, Vander Velden JW, Borland JM, Hart M, Jin C, Huhman KL, Cox DN, Albers HE. CRISPR-Cas9 editing of the arginine-vasopressin V1a receptor produces paradoxical changes in social behavior in Syrian hamsters. Proc Natl Acad Sci USA. 2022;119:e2121037119.
88. Terren C, Fransolet M, Ancion M, Nisolle M, Munaut C. Slow freezing versus vitrification of mouse ovaries: from ex vivo analyses to successful pregnancies after auto-transplantation. Sci Rep. 2019;9:19668.
89. Tsang WH, Chow KL. Mouse embryo cryopreservation utilizing a novel high-capacity vitrification spatula. BioTechniques. 2009;46:550–2.
90. Ushigome N, Wakayama S, Yamaji K, Ito D, Ooga M, Wakayama T. Production of offspring from vacuum-dried mouse spermatozoa and assessing the effect of drying conditions on sperm DNA and embryo development. J Reprod Dev. 2022;68:262–70.
91. Wakayama T, Yanagimachi R. Development of normal mice from oocytes injected with freeze-dried spermatozoa. Nat Biotechnol. 1998;16:639–41.
92. Whittingham DG. Fertilization of mouse eggs in vitro. Nature. 1968;220:592–3.
93. Whittingham DG. Some factors affecting embryo storage in laboratory animals. Ciba Found Symp. 1977;52:97–127.
94. Whittingham DG, Leibo SP, Mazur P. Survival of mouse embryos frozen to −196 degrees and −269 degrees C. Science. 1972;178:411–4.
95. Windhofer L, Boersma A, Dahlhoff M, Rülicke T, Auer KE. The impact of percutaneous epididymal sperm aspiration on sperm quality in mice. Reprod Fertil. 2023;4(2):e230017.
96. Yoshimoto H, Takeo T, Nakagata N. Dimethyl sulfoxide and quercetin prolong the survival, motility, and fertility of cold-stored mouse sperm for 10 days. Biol Reprod. 2017;97:883–91.

Health Monitoring Programmes and International Standards

7

Werner Nicklas, Stephanie Buchheister, and André Bleich

Abstract

Health monitoring aims at investigating the microbiological status of laboratory animals and is therefore an important prerequisite to get reliable and reproducible results from animal experimentation. A proper health monitoring program must be established in order to evaluate the efficiency of hygienic measures. To obtain optimal information from testing, all the critical variables must be taken into account, and all the procedures need to be adapted to local and scientific needs. Variables include the various housing conditions (e.g., open cages and individually ventilated cages) and different types of samples (e.g., colony animals, sentinels, and environmental samples). Diverse agents such as bacteria, fungi, viruses, or parasites require suitable sampling strategies and specific test methods. The prevalence of agents and their importance must be considered. This chapter also provides some advice for the shipping of samples. The importance of comprehensive health reporting including test methods applied and the handling of unexpected or positive test results are stressed, and some suggestions are made on what to do in the case of an infectious outbreak. Finally, the authors give an outlook on future developments of health monitoring and the definition of the microbiological status of laboratory rodents.

Keywords

Rodents · Quality assurance · Host microbial interactions · Hygiene · Specific pathogen-free organisms

7.1 Introduction

Infections of laboratory animals used for research may not only compromise the animal's health but also have an impact on physiological parameters, thus increasing inter-individual variation and also variation between experiments. Prevention (and timely identification) of infections in laboratory animals by microbiological quality control is therefore an essential prerequisite not only to avoid animal losses but also to get reliable and reproducible results from animal experimentation.

To assess the microbiological quality of animals used for research, a proper health monitor-

W. Nicklas (✉)
Microbiological Diagnostics, German Cancer Research Centre, Heidelberg, Germany

S. Buchheister · A. Bleich
Institute for Laboratory Animal Science and Central Animal Facility, Hannover Medical School, Hannover, Germany
e-mail: Buchheister.Stephanie@mh-hannover.de; bleich.andre@mh-hannover.de

F. Benavides, A. K. Hansen (eds.), *Rodent Quality Control: Genes and Bugs*, Laboratory Animal Science and Medicine 2, https://doi.org/10.1007/978-3-031-59103-7_7

ing programme should be implemented in each animal facility aiming at the following goals: (1) definition of the hygienic status of individual animals and the population as a whole, (2) the detection of an infection as early as possible, (3) the evaluation of the efficiency of hygienic measures for the prevention of agent introduction. Systematic and scheduled laboratory testing is the most effective way to determine the colony status and to prevent or reveal possible influences on experiments. While the health monitoring procedure by itself does not affect the hygienic quality of animals, it enables to assess the adequacy of colony management practices.

Health monitoring can only provide optimal information when all critical variables such as the animal's species and strains, immune status, housing conditions, sampling strategies for testing, and test methodology, etc. are taken into account and when all procedures are adapted to local and scientific needs.

This chapter aims at providing basic information that should be considered when a health monitoring programme is established. Furthermore, we wish to give some guidance for the interpretation of health information from testing laboratories and for the understanding and evaluation of health reports accompanying animals from commercial breeders and from research institutions.

7.2 Microbiological Quality: The Hygienic Status of Laboratory Animals

Already in the 1950s, different terms were established to define different categories, which basically aim to describe the microbiological quality of the animals. The term *conventional* has been used for animals that are not housed behind strict physical barriers or which are not subject to comprehensive health monitoring, so that reliable information on the health status is not available. Such rodents are nowadays rarely used for scientific experiments. Specified pathogen-free (*"SPF"*) animals have been regularly tested for a certain panel of agents which are either capable to cause clinical disease in susceptible animals or are known confounders of study results. However, the rest of their microbiota is not defined and it is only known which agents were detected or not by the methods used. There is no uniform definition which agents should be included to use this term nor does a predetermined methodology exist how testing should be performed. Therefore, this term does not explicitly describe the microbiological status of an animal or a population or provide reliable information on the microbiological quality of animals; it rather stands for the concept how the quality control is performed. In contrast to that, the term *gnotobiotic* refers to all life forms in a host being fully defined and includes both germ-free animals and such colonized with defined microbial communities. *Germ-free* or axenic animals are theoretically free of all parasites, bacteria, fungi, and viruses. This status is only hypothetical because endogenous or previously uncharacterized viruses might be unknowingly present or integrated into the host genome. Defined colonized animals are intentionally associated with one or more known life forms (*"associated animals"*) [105]. Gnotobiotic animal models provide the strictest form of a standardized hygienic status, as their microbiological quality is fully defined. To maintain this status, animals have to be housed within completely isolated bio-compartments (typically in isolator systems with positive air pressure) and come in contact only with sterilized materials. Moreover, monitoring the gnotobiotic status is an enormously complex procedure, aiming at not only proving the absence of all living microorganisms but also the presence of the defined microbiota.

Besides those classical categories, many additional terms are used such as "barrier-reared", "virus-antibody-free" (VAF), "clean conventional", "pathogen-free", "murine pathogen-free" (MPF), "specific opportunistic pathogen-free" (SOPF), "optimal hygienic conditions" (OHC), or "health-monitored". They all represent some sort of variation of the SPF concept and should therefore never be used without providing a definition of the *individually listed* microorganisms which have been screened for by regular monitoring programmes using accepted and suitable

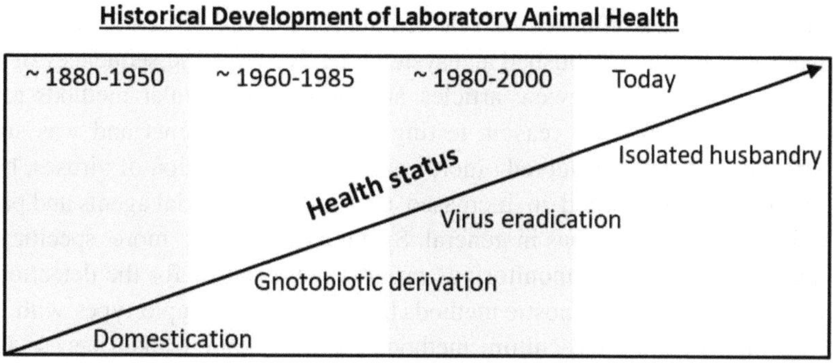

Fig. 7.1 Schematic timeline of the historical development of laboratory animal health

methods. In all these cases it is important to verify whether animals are indeed monitored for all relevant agents, especially for the ones that will not be tolerated. In this context it might happen, that certain agents (e.g., *Pneumocystis* sp.), for which the colony is positive, may be not included in the health reports (Fig. 7.1).

7.3 History of Hygienic Quality Control in Rodents

The detection of various infectious agents compromising the health of laboratory rodents already started in the early twentieth century. Disease resembling murine respiratory mycoplasmosis has been described already in 1915–1916 [51, 124] and Tyzzer's disease in 1917 [131, 134]. Ectromelia virus (ECTV) causing mousepox was the first virus discovered in rodents in 1930 [24, 88], followed by the lymphocytic choriomeningitis virus (LCMV) in 1934 [6, 10]. While some agents disappeared spontaneously from the colonies due to hygienically improved housing conditions and nutrition, active attempts to eradicate agents were not initiated before the 1950s. The primary motivation for conducting hygienic rederivation procedures was a generally higher demand for research animals after World War II. However, numerous research studies were negatively affected by infections. As a consequence, laboratory animal medicine evolved, aiming at improving the health situation, followed by foundation of the first laboratory animal science associations, such as American Association for Laboratory Animal Science (AALAS) 1950, the British Laboratory Animal Science Association (LASA) 1963, and Gesellschaft für Versuchstierkunde-Society of Laboratory Animal Science (GV-SOLAS) 1964. While complete eradication of infectious agents could not be achieved by use of traditional veterinary approaches such as vaccination or use of antibiotics, gnotobiotic rederivation techniques were established, which lead to an interruption of agent transmission from the mother to the offspring. Here, especially the implementation of hysterectomy/caesarean derivation was a big step ahead and allowed for the eradication of hitherto ineradicable agents like *Mycoplasma* (*M.*) *pulmonis* and other pathogens, especially viruses. Together with further improvement of the general hygiene concepts, it became possible to breed rodents for research purposes with improved health status that are and remain free of certain infectious agents including parasites, bacteria, and viruses. Despite those measures, some infections remained endemic in the colonies, as colonization with many agents was tolerated due to their lacking clinical manifestation.

With the implementation of systematic hygienic quality control of laboratory rodents by structured health monitoring programmes in the 1970s and increasing knowledge about molecular mechanisms during immune responses, it became more and more obvious that infections with certain agents may interfere with the host's physiological functions and thus severely influence the results of animal experiments. A symposium dealing with impacts of infectious agents on the validity of results from animal experiments was already held in 1971

[107] and known influences of selected microorganisms were later published and systematically summarized within review articles and books [9, 18, 102]). For this reason, testing for viral infections was considered more and more important and resulted in increased efforts to eradicate viral infections in general. Systematic and routine health monitoring programmes based on classical diagnostic methods like serology, microscopic, and culture methods for the detection of a broad panel of parasitic, bacterial, and viral agents were established.

Various agents such as *Clostridium* (*Cl.*) *piliforme* and *Pneumocystis* sp. were not or not easily detectable by these traditional methods. For their detection or exclusion, animals were immunosuppressed, e.g., by the application of cortisone. If silently infected, carriers developed disease or died, and agents became detectable in these animals by other methods, for example, by histopathology. This problem was solved by the introduction of molecular methods which enabled the detection of all agents by amplification of specific sequences of their nucleic acids. Use of molecular methods revolutionized the detection of agents and was soon used not only for the detection of viruses, but also for the detection of bacterial agents and parasites. These methods are faster, more specific, and more sensitive and allow for the detection of bacterial species even in sample types with complex microbial colonization like faeces, where numerous different agents occur. For example, the detection of *Helicobacter* sp. is possible by culture techniques but is much easier and faster by polymerase chain reaction (PCR) so that very soon PCR replaced culture methods for *Helicobacter* sp. and additional selected agents. Today, use of immunosuppressed or immunodeficient rodents for the detection of agents that usually lead to clinically silent infections becomes more and more obsolete, as the presence of these and all other agents, with the exception of prions, can be demonstrated by molecular methods (Fig. 7.2).

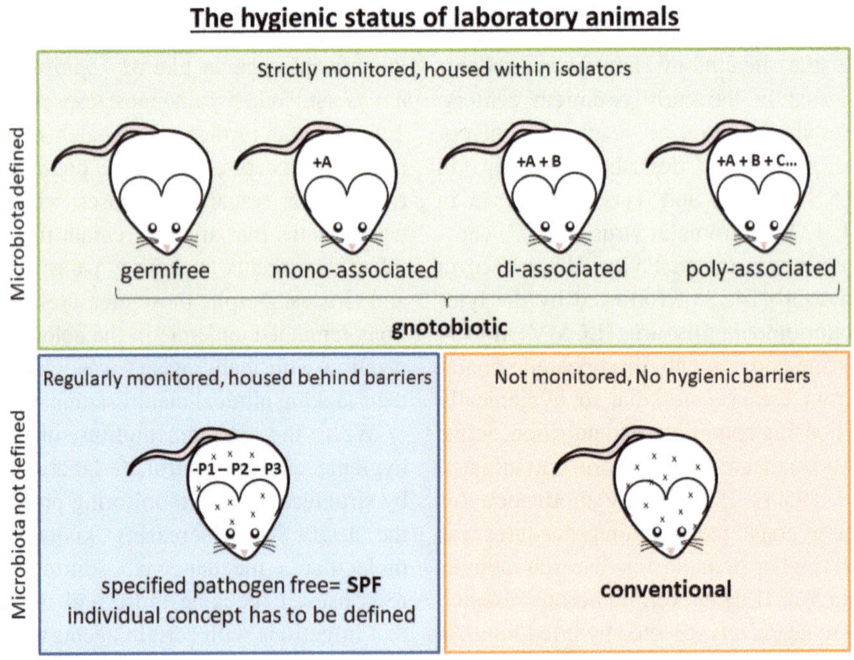

Fig. 7.2 Schematic overview about the hygienic status of laboratory animals

7.4 International Standards

Starting in the 1980s, many small breeding companies disappeared or were acquired by few large commercial vendors. These worldwide active companies were rather successful in eradicating viruses and other microorganisms and created uniform quality standards for their subsidiaries, setting the first international standards. Similarly, many multinational pharmaceutical companies later created their own worldwide standards, including for health monitoring [64, 99]. Therefore, purchasing animals from commercial vendors became more and more common due to their overall improved health status, which decreased the general breeding activity at research institutions and universities and reduced the need for health monitoring on-site. However, after the introduction of transgenic techniques, various institutions started producing and using genetically modified mice, reinstating local breeding colonies in research facilities. This growing availability of transgenic animals from external sources led to an increased animal exchange between research institutions. The worldwide transfer of genetically modified animals between research institutions resulted in re-emergence of many infectious agents, including various ecto- and endoparasites and other supposedly eradicated agents. It became very soon obvious that there is an increased need to define and standardize the microbial status of these animals through suitable health monitoring procedures. International efforts were initiated in the 1980s between the USA and Japan to create guidelines for microbiological monitoring of laboratory rodents [3]. In 1994 the World Health Organization (WHO) together with the International Council for Laboratory Animal Science (ICLAS) published guidelines for the care of laboratory animals and included suggestions for the definition of the health status [30, 40]. Participants at the 1999 national AALAS meeting discussed international harmonization [127]. A working group was formed with members from the USA, Europe, and Asia to consider the harmonization of health monitoring. The group concluded that such harmonization would be particularly valuable for the international transportation of laboratory rodents. Requirements for health status documentation were also discussed but a final statement was not published. Various authors stressed the importance of global harmonization which would enable the use of animals of the same microbiological status anywhere in the world [59]. suggested that standards should not only focus on animals but also on biological materials which derived from animals (e.g., transplantable tumours, sera) and claim reporting standards for a common language and format. In Europe, the GV-SOLAS Working Committee for Hygiene published information on the detection and identification of microorganisms, the selection of agents and the implementation of monitoring programmes including statistical consideration for sampling [69, 70] and in Great Britain very general rules were published by the Laboratory Animal Breeders Association [71]. Further details on international harmonization of health monitoring have been reviewed by [99].

In Europe, the Federation of European Laboratory Animal Science Associations (FELASA) established the first working group for the health monitoring of laboratory animals in 1989. It became obvious at that time that the exchange of animals within Europe would increase for political reasons (the Schengen agreement was already in preparation), and due to the diversity of cultures and traditions in Europe harmonization of health standards was considered essential. The first FELASA health monitoring recommendation for health monitoring of breeding colonies of rodents and rabbits was completed in 1992 and published in 1994 [68], but commercial breeders accepted most aspects of this recommendation only after initial hesitations. As a part of this recommendation, a format for a health certificate with suggestions for a layout and information to be included was also completed. The second working group [116] not only recommended health monitoring for rodent and rabbit experimental units but also considered it an update to the recommendation for breeding colonies as new agents [e.g., *Helicobacter* sp., "orphan" parvoviruses, *Corynebacterium* (*C.*) *bovis*] had been described since the completion of the first recommendation.

FELASA also prepared a guidance paper on quality assurance procedures for diagnostic laboratories [54]. In addition, a working group completed guidelines for the accreditation of health monitoring programmes and of laboratories that conduct health monitoring [104]. The aim of the FELASA accreditation scheme is to ensure that monitoring programmes and testing are conducted in accordance with FELASA recommendations. This programme was discontinued after a few years, as there was only little interest in the accreditation of health monitoring programmes. It was the aim of the FELASA Recommendation for the Health Monitoring of Rodent and Rabbit Colonies in Breeding and Experimental Units to combine the existing recommendations for rodents in a single document and to revise them concurrently [103]. A main objective was to stress the necessity of sufficient flexibility to ensure the adaptability of health monitoring programmes to specific needs. Furthermore, various explanations and details were included, for example, comments on agents and points to consider when monitoring animals from experimental units or various housing systems. This recommendation was widely accepted also outside Europe. It was revised in 2014 [83]. In the same year, a Working Group on Health Monitoring of Rodents for Animal Transfer which consisted of representatives from FELASA and AALAS published a report [113]. Both recommendations are similar and differ only in a few aspects. The FELASA recommendation provides more detailed information on the implementation of the testing programme, whereas the FELASA-AALAS recommendation puts forward the creation and formatting of a health monitoring report. An easy to recognize layout as well as a precise list of microorganisms to be specified in a health monitoring report are compiled [100].

7.5 Practical Aspects of Hygienic Quality Control: The Health Monitoring Programme

Laboratory animal colonies should undergo systematic health monitoring procedures to define the hygienic status of the animals and ensure that the animals have the anticipated microbiological quality. Therefore, the term "health monitoring" is misleading as not only those agents need to be detected or excluded that may have impact on the health of the animals. Most infections do not lead to overt disease but can have a considerable impact on the outcome of animal experiments. The underlying/causative agents must also be monitored systematically.

Each facility has to develop a routine health monitoring programme, adapted to the individual requirements and possibilities. In this context, suggestions were made by FELASA [83, 103] which factors should be considered. It is stated in these recommendations that "health monitoring is a complex issue". For this reason, "a person with sufficient understanding of the principles of health monitoring should be identified as the individual responsible for devising and maintaining a health monitoring policy". As every animal facility serves different needs and has its own structure (scientific aims, animal species, types of experiments, etc.) it is important that the monitoring programme is "tailored to individual and local needs". This also means that different programmes may be necessary for different units in the same facility.

Comprehensive health monitoring consists of several approaches. First of all, the responsible person will establish a scheduled programme which defines agents, frequency of testing, sampling and testing methods, etc. for every hygienic unit, depending on the structure of a facility (housing conditions, animal species, strains, types of research, immune status, breeding/research unit, procurement of animals from commercial breeders and/or from research institutions, etc.). It is the aim of this *active monitoring* to gain data and information specifically for a defined rodent population by systematic collection of data with predetermined methods in order to achieve the aim of the monitoring programme. However, this active monitoring approach should be supplemented by *passive monitoring*, i.e., by using data and information that were not specifically obtained for the purpose of defining the status of a population. This could be information that is acquired accidentally, for example, by clearing up an infectious aetiology of animals showing specific clinical signs or sudden death, or unexpected research results. Naturally, the sample collection and analysis of such data is per se an active process, but the initial data were collected for another purpose or were obtained randomly. The combination of both approaches may yield important information which can extend the information gained from scheduled and standardized active monitoring.

7.6 Considerations for Sampling: Housing Conditions

As already stated in the FELASA recommendations [83, 103], an important step in the establishment of a health monitoring programme is the definition of a microbiological or hygienic unit which is defined as "a self-contained microbiological entity, with separate space and traffic for animals, personnel and materials". Depending on various aspects, the size of such a hygienic unit can vary immensely. The health monitoring programme should be adapted to specificities of each unit within the facility. Naturally, the housing conditions will strongly influence the procedures for health monitoring.

7.6.1 Open Cage Systems

Traditionally, rodents have been housed in *open cages* within ventilated and air-conditioned rooms, and usually several rooms are combined to a hygienic unit. Animal caretakers and research personnel have access to such a hygienic unit, usually through sluices, to avoid introduction or spreading of agents. Hundreds or thousands of animals may be housed in such a hygienic unit. In contrast to rodent housing in individually ventilated cages, gloves are not generally changed or disinfected after the handling of every cage. Within such closed entities, all animals will eventually have gained the same microbiological status as agent exchange is, at least theoretically, possible by air, by aerosols, dust or fomites, and by handling.

In case of such open cage housing conditions, statistic approaches can be used to calculate the number of animals that should be tested to detect an infection with a defined probability and confidence level [25, 48, 58].

It is a common approach to select randomly *colony animals* for testing. This is an easy way to directly determine the microbiological status of a population when a sufficient number of animals is tested. However, testing of colony animals is not always feasible, as valuable individuals (e.g., certain genotypes within a complex breeding system of a genetically modified strain or animals intended for the use within experiments) are not available for health monitoring, especially when animals have to be sacrificed for the testing procedures in case of whole animal testing concepts.

7.6.2 Individually Ventilated Cage (IVC) Systems

During the last 20 years, more and more institutions house rodents in *individually ventilated cages* (IVC), and it likely represents the most frequently housing system being used today. Each individual cage receives filtrated air, which prevents that agents get access to a cage via the ventilation system or spread between cages. Therefore, the likelihood of spreading of accidentally introduced agents between cages (e.g., through infected animals) is substantially diminishced. This major advantage in terms of hygienic management unfortunately complicates the health monitoring procedures. In general, the stricter the handling procedures are the more unlikely it is that agents spread from one cage to another. Thus–theoretically—each IVC cage represents a separate hygienic unit, and the prevalence of an infection as well as the amount of cages with infected animals, respectively, may be very low. Therefore, reliable information on the microbiological status of the population might be easily obtained on sentinel than on colony mice. For many reasons molecular tests on environmental samples have become more and more common for testing the microbiological quality of animals housed in IVCs.

7.6.3 Isolator Systems

An isolator housing system normally consists of a completely sealed biocompartment, maintained under positive or negative air pressure. Inside this compartment, the animals are housed within open cages. Each *isolator* represents a separate hygienic unit and it is likely that all animals in an isolator have the same microbiological status as agent transmission within the isolator is similarly effective as in rooms with open cage housing systems. Only small animal numbers can be housed within common isolators due to space limitations. High numbers of animals are necessary to be tested, Naturally, molecular testing of environmental sample material is also possible. Compared to barrier units, the risk of introducing agents into an isolator is low when properly handled, so that less frequent testing, for example, once or twice a year, may be sufficient.

7.6.4 Filter Top Cages

Health surveillance in *filter top cages* is even more difficult. Depending on maintenance procedures, every cage may represent a separate hygienic unit. This means that samples should be tested from all or from the majority of cages to get reliable information on the microbiological status of a population. As filter top cages are only exceptionally used for long-term housing, systematic and repeated testing is frequently not performed. If necessary, dirty bedding sentinels or molecular tests (cage swabs, bedding) can be used.

7.7 Consideration for Sampling: Sources of Test Material

Traditionally, health monitoring programmes are based upon the testing of whole animals or samples derived from animals (e.g., blood/serum/plasma, oral swabs, organs, faecal samples). With development of highly specific and sensitive molecular test methods applicable in high throughput approaches, there is a shift towards sampling material of the animal's environment.

7.7.1 Animals

To define the microbiological status, health monitoring has relied primarily on testing animals and animal derived samples for decades. Here, either randomly selected colony animals or sentinels can be used for the testing of a closed colony.

Most animal facilities purchase animals from outside sources. These can be commercial breeders or research institutions and universities. To reduce the risk of agent transmission to existing colonies, critical inspection of the health reports (which should be requested prior to—and also accompany every animal shipment) should be ideally performed prior to the authorization to introduce animals from outside into the own facility. Moreover, it is generally recommended to test imported shipments thoroughly and house the animals under quarantine conditions before permission is given to introduce them into existing colonies. Animals from commercial breeders usually have a high microbiological quality. In contrast, animals derived from research institutions are more frequently colonized by "unwanted" agents. Independent from their origin, animals from outside should be thoroughly monitored during the quarantine period. In case of animals from commercial breeders, occasional testing may be sufficient, but regular and comprehensive testing is necessary especially for animals with unknown health status or which are derived from untrusted sources. In some cases, it is reasonable to fully quarantine and rederive animals upon arrival, e.g., by embryo transfer, as this is the best means to avoid introduction of agents.

When *different animal species* are housed within a hygienic unit it is important that every species is tested. Furthermore, animals belonging to *different strains* must be tested as susceptibility for certain agents may differ. As an example, C57BL/6 mice require a higher viral dose for seroconversion, whereas C3H/HeN, BALB/c or outbred mice more reliably produce antibodies against mouse parvovirus (MPV) and are therefore more suitable to detect a MPV infection in a population [15]. In contrast, the opposite is true for minute virus of mice (MVM) [62]. Examples of strain dependent susceptibility exist also for bacterial infections such as *Streptobacillus* (*Sb.*) *moniliformis* [42]. Generally, immunodeficient animals are more susceptible to bacterial and viral infections, *Pneumocystis* spp. and also certain parasites (e.g., *Spironucleus muris, Demodex musculi*) and may develop lesions and clinical disease in immunocompromised hosts. It is therefore important that such animals are submitted to testing for specific agents (see also "passive health monitoring"). It has to be pointed out that many animals with inherited or acquired defects in the immune system may not produce sufficient amounts of antibodies so that serological testing may result in false negative results. In this case, molecular methods should be used instead as many viral agents may persist in immunodeficient animals resulting in prolonged and increased shedding rates.

To get optimal diagnostic results, the *age* of animals must be taken into account. Since serology testing requires that individuals actively produce sufficient amounts of antibodies, animals should be at least 6 weeks old to serologically screen for viral infections which are endemic in a population. Moreover, seroconversion to bacterial agents such as *Pasteurellaceae*, *M. pulmonis*, *Streptobacillus* spp., and others is frequently detectable much later, usually 10–12 or even more weeks post-infection. The same minimum age of 6 weeks is advisable for traditional parasitological test methods since microscopic detection relies on the detection of parasite eggs. Their detectability depends on the prepatent period of nematodes, which is particularly important in case of *Aspiculuris tetraptera*. When sentinels are submitted to testing, the exposition period is more important than their age. When sick animals are tested, agents are usually directly detectable in tissues with lesions, independent from the animal's age.

7.7.2 Colony Animals

It is a common and quite simple approach to use randomly selected colony animals for the health monitoring, especially in rooms with open cage systems. A prerequisite for this concept is that infectious agents can easily spread between the cages within the hygienic unit. In general, the number of samples to be tested depends on the expected ratio of positive animals in a given population. Guiding principles for the calculation have been published by [25, 58]. However, this mathematical formula presumes that a positive sample will be detected, not taking the diagnostic sensitivity into account, which very much depends on the methodology and quality of sampling. More detailed information will be provided in Chap. 8.

In fact, methodology protocols vary between laboratories, which might influence the test accuracy. This is especially true for bacteriology testing, as the use of specific culture media as well as the number and sites of sampling may impact the sensitivity of bacterial culture.

7.7.3 Sentinel Animals

The concept of testing colony animals becomes problematic whenever small numbers of animals are housed in *isolators,* in *filter top* or in *individually ventilated cages (IVC).* Under these conditions, health monitoring "has become a challenging task" and requires more complex testing and sampling strategies [83]. To accomplish this, it became common to use *"sentinels"* for the health monitoring of colonies kept under various housing conditions.

Sentinels are animals with known hygienic status and are usually introduced into an existing unit and tested after a sufficiently long exposition period representatively for a colony. As a prerequisite, sentinel animals must be free from all agents for which the population is to be tested. Moreover, they should also be negative for all agents with any pathogenic potential [for example, *Staphylococcus (Staph.) aureus*, *Pneumocystis* spp.] as unintentionally introduced infections with those agents might compromise the health of immunodeficient animals. To be able to detect infectious agents via sentinels, the sentinel animals have to be either directly or indirectly exposed to colony animals. Direct exposure (= contact sentinels) will certainly allow transmission of agents very efficiently, however, this concept requires the use of relatively large number of sentinels. Most commonly indirect exposure concepts are used, such as exposing the sentinels to the exhaust air (= exhaust air sentinels) or to house the sentinels in the soiled or "dirty" bedding of the colony.

Exhaust air sentinels are not commonly used any more. When dirty bedding sentinels are used it is common to transfer small volumes of used bedding from several cages to the sentinel cage so that there is a risk of diluting microorganisms. The number of active or infectious agents transmitted may fall below the infectious dose so that sentinel animals may not get infected and do not seroconvert. It has been shown for various agents that agent transmission to sentinels by dirty bedding is efficient. Agents like murine hepatitis virus (MHV) [33, 129], *Cl. piliforme* [93, 135], parvoviruses [128], murine norovirus (MNV) and Helicobacter [86, 137] are usually detectable with an appropriate sentinel programme.

However, many agents do not survive long under the environmental conditions in bedding (e.g., low humidity) so that not all agents may be infectious for a sufficiently long time and may therefore not be detected by use of sentinels [82]. Particularly several respiratory agents like, e.g., Murine respirovirus (Sendai virus) [7, 37], *Filobacterium rodentium* (formerly *Cilia-associated-respiratory Bacillus)* [35], *Pasteurellaceae* [122] or mycoplasms and even agents with zoonotic potential like LCMV [57] may not be detected by the use of dirty bedding sentinel systems. Mites are frequently also not transmitted by dirty bedding and may therefore remain undetected [75, 130]. This has also been published for intestinal flagellates such as *Spironucleus muris* [109]. It has been shown for other agents like MPV [15] and mouse rotavirus (EDIM) [117] that older animals may develop a higher resistance to certain agents compared to younger animals.

Furthermore, the susceptibility for a certain agent may be strain-specific. It has been shown that BALB/c mice are resistant to an experimental infection by *Sb. moniliformis*, whereas C57BL/6 develop severe clinical signs followed by death within few days [42, 139]. As mentioned earlier, C57BL/6 mice are less susceptible to an infection with MPV so that a higher viral dose is necessary for an infection and for seroconversion compared to other mouse strains [15]. Vice versa, C57BL/6 raise antibody responses after experimental infection with MVM more reliably compared to outbred mice [62]. These examples clearly show that the selection of a mouse strain for use as sentinels can be crucial for the detectability of infections with certain agents. Moreover, some opportunistic agents (for example, *Pneumocystis* spp. or *C. bovis*) may be detectable only by use of immunodeficient sentinels. However, they may develop clinical signs which must be generally avoided. Finally, seroconversion requires a certain time period, depending on the agents. While antibodies to viruses are detectable usually within 2 weeks after infection, antibodies to agents living on mucous membranes (e.g., *Pasteurellaceae, Sb. moniliformis*, mycoplasms) are usually detectable not until 10–12 weeks after infection or even later, so that such infections may remain undetected by serological methods even after exposition periods of 12 weeks or more. For the above-mentioned reasons it is important that sentinels-based health reports are read and interpreted very critically and that supplementary measures (e.g., testing colony animals, testing sick animals, testing environmental samples) may be implemented to increase the sensitivity for the detection of many agents including parasites [63].

To determine the status of very small populations, e.g., breeding pairs of genetically modified animals from other research institutes during quarantine after importation, use of *contact sentinels* is possible [76, 94]. Female sentinels (to avoid fighting) can therefore be placed in the same cage together with the animals in question for 2–3 weeks and are tested after a sufficiently long exposition period of at least 8–10 weeks (usually in a separate cage after the contact period, additional use of dirty bedding). Due to the close contact, it is to be expected that all agents are reliably transmitted to susceptible sentinels. Alternatively, agents can also be detected via non-invasive sampling methods (e.g., faeces, fur samples, oral swabs) during quarantine directly by molecular methods [53].

7.7.4 Environmental Samples

With the increased availability of highly sensitive and specific molecular methods such as probe-based PCR [53], testing environmental samples especially in the widely used cage level biocontainment systems became very popular (dust from exhaust air, exhaust air filters or capture media from rack air ventilation systems, swabs from cages or cage surfaces). With increasing numbers of publications showing that environmental testing strategies might be superior compared to sentinels in terms of diagnostic sensitivity, it became obvious that sentinels are not ideal for all situations. Moreover, the number of animals sacrificed for health monitoring can massively be reduced by replacing sentinel animals by testing environmental samples, which is particularly in line with the 3Rs principle [84, 91, 110]. Shipping of live animals for diagnostic purposes is also not necessary, further improving animal welfare aspects. During the last decade, vendors of cage racks even developed special capture media to standardize the sampling procedure and make sampling and shipping easier for the users [80, 106]. Compared to traditional diagnostics based on sentinels, molecular methodology has markedly increased diagnostic sensitivity, as it is possible to detect very low concentrations of nucleic acids present in a broad majority of different sample matrices [32]. For soiled bedding sentinels an exposure time of 10–12 weeks is usually recommended, whereas it has been shown that testing environmental samples can detect a broad panel of different agents after a much shorter exposure period of 1–4 weeks [80, 84, 91, 106]. Moreover, the diagnostic specificity of PCR based methodology is also exceptionally high. While principally being beneficial, this primary advantage is automatically a major disadvantage of PCR based technology, as testing using specific primers can only detect a defined panel of agents for which the tests are performed. As another drawback, molecular assays simply detect the presence of nucleic acid of an agent in the sample material independent of its infectivity. As nucleic acid can be very stable in the environment, cleaning and decontamination of racks prior to testing environmental samples is crucial to avoid false positive results by the detection of residual DNA. Positive findings obtained by molecular methodology should therefore be verified by an independent diagnostic methodology such as serology, to proof infectivity of agents detected.

7.7.5 Biological Materials

When biological materials (tumour cell lines, tissues, serum, agents, parasites, etc.) are serially passaged in animals or applied to other animals during scientific studies it is important that these materials are tested for contamination as they may be contaminated by various viruses and other agents [17, 31, 36, 46, 77, 87, 101]). It has been shown that also materials from other donor species than rodents can be contaminated by agents that can be significant for rodents [34]. It is important that especially biological materials that have been frozen for years or decades are tested as storage at temperatures below −70 °C does not affect their viability, and also contaminants do not lose their infectivity.

7.8 Test Methods

In contrast to other fields (testing of e.g., pharmaceuticals), compulsory standards of test methodology do not exist for the health monitoring of laboratory animals. This means that all laboratories are free to decide which detection methods are used. Methods and procedures to determine the microbiological status of laboratory animals are therefore not uniform which can lead to confusing and contradictory results, especially regarding bacterial infections. Furthermore, direct and indirect detection methods are used. Direct methods detect the agent (or parts of it), whereas indirect methods detect substances, which are produced as a consequence of an infection. The most frequently used indirect approach is serology which detects agent-specific antibodies following adaptive immune responses. However, also enzymatic targets may be used, e.g., the increased activity of lactate dehydrogenase resulting from an infection with the lactate dehydrogenase elevating virus Gammaarterivirus lacdeh (LDV). Direct methods indicate the actual presence of an agent, whereas indirect methods may generate positive results even after clearance of the agent. In cases of unclear or contradictory results, the diagnostic laboratories should be contacted to find potential explanations before action is taken (e.g., euthanasia of animals) as a consequence of false positive (or false-interpreted) test results.

During the last years, traditional methods such as microscopy, microbiological culture, serology or histopathology have more and more been supplemented or even replaced by molecular methods which are highly sensitive. However, it must be considered that only agents can be detected for which specific tests are performed. Furthermore, the method of nucleic acid extraction is crucial, since especially environmental and faecal samples usually contain inhibitors of the DNA polymerase which may strongly affect the sensitivity of the test. In contrast, bacterial culture, microscopy, and histopathology may also detect other agents and yield additional information. Molecular methods can be used as standalone tests or together with traditional methods for the clarification of unclear or contradictory results.

It is generally important that diagnostic laboratories have broad spectrum methodology available for the detection of agents or infections so that unexpected positive reactions can be confirmed (or rejected) by an independent and alternative test method. This helps to reduce the risk of false positive reactions and is particularly valuable for serology and for molecular testing. In this context, results from direct detection methods can be confirmed by indirect detection methods and vice versa—whenever reasonable and possible.

7.8.1 Bacterial and Fungal Infections

Traditionally, culture methods have been used to test for bacterial and fungal agents. This approach is time consuming and expensive and requires highly skilled and experienced personnel of the testing laboratory.

Most commonly, agents are cultured by the use of liquid and solid culture media, and pure cultures are finally identified by appropriate methods. Testing for anaerobic bacteria is usually not performed routinely. The panel of bacterial species that are detectable during testing very much depends on the media or combination of media used. A good combination of selective and universal culture media is ideal but it may also slightly increase the expenses. Use of either selective or non-selective culture media or inappropriate combinations may severely affect the detectability of certain agents. As culture conditions [incubation atmosphere, incubation time, timepoint(s) of visual colony identification, enrichment procedures, etc.] might differ significantly between laboratories, multiple factors can lead to inconsistent test results. The major advantage of bacterial culture is that a large number of bacteria, including environmental bacteria, can be detected in one readout.

Direct detection of bacterial agents and their identification is also possible by molecular tests. In case of *Helicobacter* detection, it is a common approach to first use a PCR screening assay detecting all *Helicobacter* species and (in case of a positive result) subsequently identify relevant species by species specific PCR tests. Furthermore, it is possible to sequence the PCR products to confirm the specificity of the test, or for further specification.

Besides direct detection methods, (indirect) serological tests can also be used to detect bacterial infections in rodents. Serology has commonly been performed for the detection of agents that are not cultivable or difficult to detect by other available methods, like e.g., *M. pulmonis* or *Cl. piliforme.*

Historically, susceptible animals were used in diagnostic animal experiments for the detection of agents that were difficult to culture or cannot be cultured at all. To increase the susceptibility to infection, animals were immunosuppressed, for example, by cortisone, or animals with inherited immunodeficiency were used. Materials from sick or dead animals were injected into these animals which developed clinical disease if pathogenic agents were present, and the agents were identified after being cultured or by histopathology. In general, diagnostic animal experiments are not necessary any more for routine health monitoring of rodents as all agents can be detected by molecular methods much faster and cheaper and with a higher sensitivity. They should also be disclaimed for animal welfare reasons as these animals may suffer and even die during the course of the experiment. However, diagnostic experiments may be still valuable in specific situations, e.g., when infections with hitherto uncharacterized agents occur or in cases where modern test systems provide ambiguous results.

More detailed information will be provided in Chap. 10.

7.8.2 Viral Infections

The most common approach for the detection of viral infections is the detection of specific antibodies via serology. Traditional serological methods like, e.g., HI tests and complement fixation tests (CF) are nowadays only exceptionally performed and have been largely replaced by more sensitive and modern methods such as IFA and ELISA or MFIA.

As an alternative to serology, direct virus detection via molecular methods is applicable.

The time point of sampling and expected phase of infection substantially determines the suitability of the applied test method. PCR for the detection of viral nucleic acid will probably be successful in the first days after infection whereas antibodies might be detectable even weeks after clearance of the virus from the host. It is optimal to combine direct and indirect methods, as the phase of infection usually remains unknown.

More detailed information will be provided in Chap. 9 (Fig. 7.3).

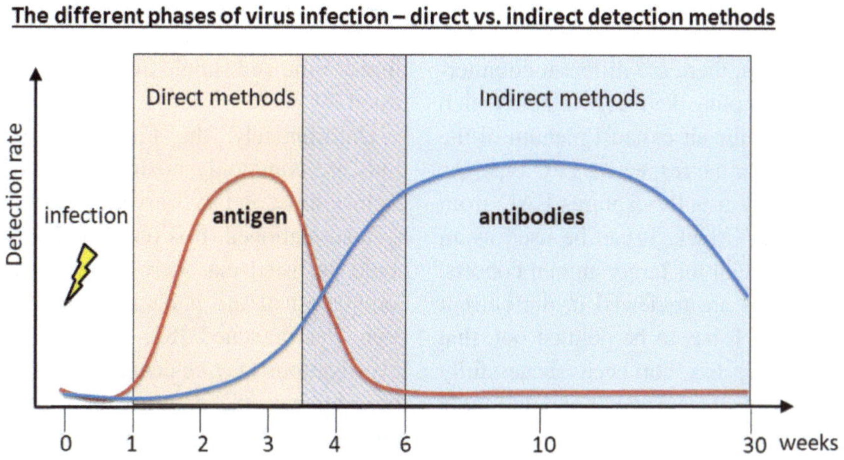

Fig. 7.3 Rough timeline summarizing the different phases of virus infection and implications for the use of viral detection methods. Antigen is usually detectable during the second half of the first week after infection (p. i.) and usually cleared during week 3 p.i. Antibodies are usually detectable during the beginning of week 2 p.i. for several months or lifelong. The time points chosen in this figure are only examples and significantly differ between viruses

7.8.3 Parasitic Infections

Most ecto- and also endoparasites are relatively large in size and can be directly visualized and identified by light microscopy (protozoan parasites, ectoparasites) or indirectly identified by detecting their respective eggs (metazoan endoparasites).

Testing for ecto- and endoparasites is nowadays also frequently performed by molecular tests. However, PCR tests are usually not available and therefore not applied for all known protozoans so that microscopy may yield positive results despite negative PCR results.

More detailed information will be provided in Chap. 11.

7.8.4 Environmental Sampling Strategy

Due to its high sensitivity molecular testing enables using environmental material, thus reducing the number of sacrificed animals necessary for routine health monitoring. Here, different kinds of environmental sample types can be used for the testing procedures, including bedding/nesting material, swabs taken from the animal's home-cages or even exhaust air dust (EAD) material. The latter can be sampled on different levels of the exhaust air filtration procedure, e.g., by swabbing the plenum or the pre-filter of IVC systems. Nowadays, there are different commercially available sampling devices available which can be placed into the air exhaust plenum of the IVCs, depending on the respective IVC type. As this testing strategy usually captures EAD from all cages of the IVC rack, it can be used as an optimal screening tool for larger animal cohorts. Validated protocols are reviewed in the current literature [23, 32]. It has to be pointed out, that EAD based testing has not been successfully implemented for all relevant agents and that some agents, e.g., murine parvoviruses are very hard to detect in EAD material due to very low antigen concentrations. Therefore, PCR based techniques should always be combined with traditional methodology.

7.8.5 Screening of Biological Materials

Biological materials, such as sera, tumour cells or stem cells might be contaminated with certain pathogens, especially after serial passages in animals, and should be screened for contaminants to prevent an infectious disease outbreak [101]. Historically, the sample to be tested was initially transferred to animals which were finally tested for specific antibodies (Mouse or rat antibody production test/MAP or RAP test) [73, 119]. Nowadays, traditional MAP/RAP tests have been replaced by molecular methods, as they are much faster, cheaper, and more sensitive [12, 72, 79].

7.9 Agents and Their Importance

7.9.1 Choice of Agents

Ideally, each facility should provide information on the agents that are included in the HM programme and are therefore regularly tested. Suggestions about the panel of relevant agents have been made by FELASA [83] as well as FELASA-AALAS [113]; however, it is necessary to adapt such a list to the local situation. As a consequence, the panel of agents for which animals are monitored might differ among animal research facilities or commercial breeders. Furthermore, novel agents are described from time to time and should also be considered to be tested for.

Unfortunately, the FELASA recommendations are commonly misinterpreted as a simple exclusion list and as "carved in stone" regarding agents mentioned. It is obvious that only agents could be listed that were known at the time of completion of the publication. It has therefore been clearly stated [83, 103] that "additional investigations may be deemed necessary. Should these indicate the presence of an agent which, although not listed in these recommendations, is suspected of being important, this agent should be mentioned in successive reports and treated as are listed agents". Furthermore, "monitoring for additional agents and their declaration in a health report is advised under specific circumstances, e.g., when associated with lesions; when associated with clinical signs of disease; when there is evidence of perturbation of physiological parameters or breeding performance; when using immunodeficient animals".

It is the decision of the "person with sufficient understanding" to decide if consequences are taken in case one or several agents were detected. It is already stated that "it is not a requirement of these recommendations that animals tested are free from all of the microorganisms listed" [103].

Testing is expensive and it may therefore not be possible and also not necessary to test for all known agents on a regular basis. In many cases, testing will focus on agents that may have impact on specific research disciplines and which are still prevalent whereas testing for agents that have not been found in laboratory rodents for several decades may be omitted or testing is performed only occasionally.

The type of a laboratory animal facility determines the agents for which testing is reasonable. Commonly, animals in breeding units, especially from commercial breeders, should be free from agents considered relevant. The testing regime may therefore be more complex compared to research units or mixed facilities with breeding and research in the same unit.

7.9.2 Prevalence of Agents

Decades ago, severe animal losses were caused by various bacterial pathogens such as *C. kutscheri* or *M. pulmonis*. Consequently, many agents, especially those with pathogenic potential, were eradicated from many populations when efficient rederivation methods became available. However, novel agents were detected as test methodology evolved, mainly viruses such as mouse norovirus (MNV), but also bacterial agents such as various *Helicobacter* species. Nevertheless, the microbiological quality was constantly improved over the last 6 decades, and many microorganisms are nowadays not detected any more.

Up to the 1980s, a broad variety of infectious agents were common in colonies of laboratory rodents, even in breeding populations from commercial breeders [26, 61, 67]. Lower prevalence of agents was later reported by [78, 125] who found a significant decrease in prevalence compared to the early 1990ies. In their studies, some infections such as murine pneumotropic virus (MPtV, formerly known as newborn mouse pneumonitis virus or K virus) and murine polyomavirus (MPyV) had been eliminated from laboratory animal colonies, whereas other infections [e.g., Mouse hepatitis virus (MHV) and parvoviruses] remained at a high rate [81]. Also reported low prevalence for most viruses but they frequently found antibodies to MNV in mice and parvoviruses, primarily rat minute virus (RMV) in rats which had both been newly detected less than a decade ago. Similar results were reported by [112].

In the USA, several surveys were performed in which animal facility directors were queried regarding the presence of infections in their experimental rodent colonies. In the first survey, the presence of many different viruses, bacterial agents, and parasites was reported in the majority of institutions, even in "SPF" colonies [60]. In a similar survey, MPV was reported to be endemic in more than 90% of mouse colonies [27]. The most frequent bacteria reported were *Helicobacter* sp. and *Pasteurellaceae* in almost 80% of mouse colonies. According to this survey there is evidence of an increase of parasitic infections compared to the 1998 survey, since pinworms were reported in almost 75% of institutions and mites in 40%. The results of the third survey demonstrated that the rate of viral outbreaks was further decreasing, whereas the prevalence of parasites has remained constant [89].

The prevalence of infections may differ between different parts of the world. For example, [74] reported positive findings for pneumonia virus of mice (PVM), Theiler's mouse encephalomyelitis virus (TMEV), LCMV, and ECTV in mouse colonies in Taiwan [138]. Reported findings of Hantaviruses in laboratory rodents in Korea, and [95] found, also in Korea,

MCMV-1, MPyV, and MPtV in commercial breeding colonies. All these viruses are extremely uncommon in European and North American colonies [27, 81, 89]. Higher prevalence rates for several viruses were reported in India compared to the situation in Europe and the USA [85]. In contrast, prevalence rates for various viruses, bacterial agents, and parasites were reported in Australia that are comparable to European and American laboratory rodent colonies [90].

Compared to commercial breeders, infections are generally more prevalent in animals from experimental colonies. In particular, genetically modified animals are frequently contaminated, as they are usually bred and housed on-site in research institutes. Due to the frequent worldwide exchange of these animals between research institutions, agents can easily be transmitted to rodents in other parts of the world [57, 140].

7.9.3 Zoonotic Agents

Several agents that are found in rodents can be zoonotic [96]. Among viral agents, LCMV plays the most significant role as it may cause neurological disease in humans. LCMV has been detected only rarely in laboratory mice [29, 57]. Hantaviruses can also be transmitted to humans by various rodent species but infections arising from laboratory rodents are very rare in Europe [21]. Bacterial zoonoses obtained from laboratory rodents became nowadays also relatively rare. Among zoonotic bacteria is the causative agent of the so-called rat bite fever [*Sb. moniliformis* and other species belonging to this genus] which are only occasionally found in laboratory rats and mice; however, they are highly prevalent in pet and in wild rats. Other bacterial pathogens like *Salmonella* spp. or *Leptospira* spp. are not found any more in laboratory rodents. Opportunistic bacteria [e.g., *Staph. aureus*, *Streptococcus* (*S.*) *pneumoniae*, and *S. agalactiae*] are still prevalent. Parasitic zoonoses transmitted by laboratory rodents do not play a significant role in laboratory rodents any more.

7.9.4 Viruses

There is general agreement that known viruses should not commonly be tolerated in laboratory rodents. However, novel viruses have again and again been characterized in laboratory rodents during the last decades, and it is very likely that hitherto unknown viruses will be detected in the future.

Some viruses became more and more relevant with the increased use of immunocompromised animals, as these animals are prone to develop clinical disease or even death when infected by agents with low or even lacking pathogenicity. The use of genetically modified rodents also drove the discovery of new viruses. As an example, MNV was detected when genetically modified and severely immunocompromised mice sporadically succumbed to a systemic disease [65]. Novel molecular techniques like metagenomic analyses also allow for systematic screening for hitherto unknown viruses [38, 111, 121] and demonstrated that rodents are carrying various additional viruses [e.g. murine astrovirus (MuAstV) [111]] or mouse kidney parvovirus (MKPV) [118].

Some of the known viruses infecting rodents were not found in laboratory rodent populations for decades. Parvoviruses are certainly among the most relevant viral agents that should also be included in health monitoring programmes for laboratory mice and rats. In general, those that have been known for several decades [MVM, Rat virus (Kilham rat virus (RV), Toolan's H-1 virus (H-1) in rats] are found at a much lower prevalence rate compared to those that were described later [MPV, MKPV, RMV, rat parvovirus (RPV)]. Some of these parvoviruses also play a role as contaminants of transplantable tumours and other cell lines [101].

Various newly detected viruses are far more prevalent. As an example, MuAstV has been shown to be widespread in mice housed in research facilities worldwide [39, 97, 111, 123]. Similarly, a high prevalence in contemporary rat colonies has been shown for rat polyomavirus (RatPyV2) which has been described in 2017 [14].

7.9.5 Bacteria and Fungi

Numerous bacterial species are found as commensals in all complex organisms, and many of them have important physiological functions. However, some bacterial species may have some pathogenic potential. For this reason, it may be difficult to decide which bacterial agents should be tolerated or not in laboratory animals.

As already stated for viruses, most of the classical "pathogens" that lead to clinical signs of disease do not play a significant role anymore and have been eradicated from laboratory rodent populations. Various bacterial agents such as *Pseudomonas aeruginosa*, *C. bovis*, and *Staph. aureus* may be accepted in immunocompetent animals but can cause serious health problems in animals with immunodeficiency. *Staph. xylosus* is commonly isolated from animals in numerous institutions, nonetheless, colonization may lead to disease under certain conditions [1, 22, 47, 120]. Zoonotic bacterial species like *S. pneumoniae* [133] or *S. agalactiae* (serogroup B) [66, 126] can be detected occasionally, even in colonies from commercial breeders. [*Pasteurella*] *pneumotropica* which has historically been considered a species with some pathogenicity has recently been re-classified into the novel genus *Rodentibacter* (R.) [2, 13]. A large complex of various *Helicobacter* species has been studied in detail, showing that some species not only affect the health of animals but also research results [43]. The list of bacteria that can be pathogenic for rodents under certain circumstances is endless, and virtually all bacteria may cause lesions or disease in certain immunodeficient or genetically modified rodents or under certain disease conditions [28, 41]. This means that it is impossible to define a simple and definite exclusion list for bacterial agents which fit to all situations.

With the increasing use of molecular methodology, researchers became more and more aware of the striking influence of the whole microbiota of the animals and that common commensals may contribute to disease phenotypes, thus impacting the overall research results [e.g., segmented filamentous bacteria (SFB), *Bifidobacteria*, *Clostridia*, *Akkermansia*]. Comprehensive lists of these bacteria are reviewed elsewhere [50].

As repeated rederivation and barrier housing led to eradication of pathogens, it also massively reduced the overall microbial diversity in laboratory animals, especially mice, since it has been shown that key microorganisms that are crucial to the function of the animals have been lost. Thus, laboratory animals frequently have kind of an artificial microbiota so that they may not be suitable models for research when certain autochthonous agents are missing [16, 115].

7.9.6 Parasites

Testing for endo- and ectoparasites is an integral part of a health monitoring programme, as many protozoan and metazoan parasites, among them *Syphacia* sp. and *Aspiculuris tetraptera*, can still be found in laboratory rodents. As commonly used dirty bedding sentinel systems will probably fail to detect most ectoparasites, mites (*Myobia* sp., *Radfordia* sp., *Myocoptes* sp.) are probably underdiagnosed.

Among protozoans, the most prevalent species of flagellates (e.g., trichomonads or *Chilomastix* sp.,) and *Entamoeba* are nonpathogenic, however, also pathogenic species (*Spironucleus* sp., *Giardia* sp.) have occasionally been found in rodents.

7.10 Shipping of Samples

The vast majority of research institutions and animal facilities, including commercial breeders, do not have access to the diagnostic infrastructure to perform all necessary tests inhouse, which makes shipping of samples to diagnostic laboratories inevitable. Finally, also test laboratories need to ship samples to other laboratories, e.g., for confirmatory purposes or when test results are unclear. To avoid problems or even failures during shipments as well as unnecessary delays, it is generally helpful to discuss shipping procedures with the testing laboratories beforehand and follow their instructions.

Submitting living animals to the testing laboratories requires additional compliance with animal welfare aspects. Here it makes sense to take professional couriers into service, which is expensive though, but courier services have knowledge of all information on legal requirements at hand and are very helpful and care for all organizational steps. When submitting living animals, all testing procedures, including necropsy and pathology are possible. Whenever testing of living animals is not possible or necessary, shipping of samples is less complicated.

7.10.1 Serology (Serum + Dried Blood Spots)

Shipping *serum* samples for serology is very common as serology testing is a main pillar of health surveillance. Serum samples may be stored and shipped at ambient temperature; however, cooling (ice packs) is recommended when high environmental temperatures and/or prolonged shipping time are to be expected. Serum samples should not be pooled as there is a risk of diluting antibodies under a detectable level. It is important to inform the diagnostic laboratory if samples are already diluted or otherwise pretreated or not. As an alternative, the so-called *dried blood spot* (DBS) samples can be used. With this technique, whole blood samples are directly transferred to a piece of filter paper and subsequently dried at room temperature. In contrast to serum, the dried whole blood specimen are stable at relatively high temperatures and offer a good opportunity for long-term storage of blood specimen without the need of centrifugation and cooling.

7.10.1.1 Bacterial Culture (Swab Material, Faeces, Tissues)

Swabs for bacterial culture should ideally be shipped in specific transportation media to extend the survival time of microorganisms and thus enable their detection. All materials designated for cultural testing may be shipped at ambient conditions, as long as very high environmental temperatures and/or prolonged shipping time are not to be expected. Otherwise, cooling is here also recommended.

7.10.2 Molecular Pathogen Detection (Swab Material, Faeces, Tissues, Environmental Sample Materials)

Swabs for molecular detection methods like PCR assays should not contain any media, as presence of chemicals may interfere with the methodology. Commercial producers of racks for individually ventilated cage systems offer devices for sampling dust and microorganisms in exhaust air, and these samples can easily be shipped to diagnostic laboratories. As molecular tests are very sensitive and specific, pooling of a small number of samples is possible to reduce expenses for testing. However, caution should be paid to the fact that low antigen concentrations in single faecal pellets might be diluted which may lead to false negative results.

7.10.3 Parasitology (Faeces, Skin Scrapings)

Faecal samples for parasitology testing must be shipped in sealed tubes or vials to prevent desiccation. However, some protozoans may lose their viability and may therefore not be detectable easily by microscopy. In this case, molecular methods may be used supplementary. Skin scrapings and hair samples may be tested for ectoparasites, dermatophytes or bacterial pathogens by microscopy or PCR.

7.10.4 Histopathology (Fixated Organs and Tissues)

Organs for histopathology can easily be shipped in fixative solutions (e.g., 4% paraformaldehyde/PFA or buffered formaldehyde) or after fixation.

When shipped after complete fixation for a few days, the fixative can be removed to reduce the weight or to avoid regularities for the shipment of dangerous substances.

7.10.5 Biological Substances

It may occasionally be necessary to ship life microorganisms for identification by external laboratories. Bacteria growing aerobically usually survive at ambient temperatures for several days. It is best to ship overnight cultures of bacteria on agar plates at the beginning of the week (Monday or Tuesday) so that they arrive in the testing laboratory during the same week. Swabs of samples taken from pure cultures can also easily be shipped in specific transportation media. In this case, shipping regulations must be followed (Table 7.1)!

7.11 Health Reporting

Details on information that should be given in a health report were already published in the first FELASA health monitoring recommendation [68] to describe the health status of rodents for customers. As stated in another FELASA health monitoring recommendation [116], a main purpose of health monitoring and reporting is "to supply experimenters with data on variables that might influence the outcome of an experiment. These data are part of the experimental work and have to be considered during the interpretation of the experimental results by the experimenter and by the reader of a publication. Results of health monitoring should therefore be included in scientific publications" [116], which is also mandatory according to the current guidelines on Animal Research Reporting of In Vivo Experiments ("ARRIVE") [108].

Table 7.1 Overview about different sample types, test methods, and recommended shipping and storage conditions

Diagnostic field		Sample type	Test method	Storage/shipping conditions
Specific diagnostic fields	Bacterial and fungal infections—*microbiology*	Swabs	Culture	Ambient or cooled; use transportation media if possible
		Faeces, tissues		Ambient or cooled
		Swabs	PCR, qPCR	Frozen; do not use transportation media
		Faeces, tissues		
	Viral infections—*virology*	Serum	IFA MFIA ELISA	Ambient or cooled
		Dried blood spots		Ambient
		Faeces, tissues	PCR, RT-PCR qPCR, qRT-PCR	Frozen
	Parasitic infections—*parasitology*	Faeces	Microscopy	Ambient or cooled; prevent desiccation
		Fur samples, skin scrapings		Ambient
		Faeces	PCR, qPCR	Frozen
		Fur samples, skin scrapings		
Broad spectrum diagnostics	*Pathology*	Tissues	Histopathology; Immunohistology	Ambient; prevent desiccation, special requirements for shipment of dangerous substances
	Screening of *environmental samples*	Filter material, swabs, bedding material, etc.	PCR, RT-PCR qPCR, qRT-PCR	Ambient or cooled
	Screening of *biological substances*	Bacterial cultures, cell cultures, etc.	Diverse	Depends on the test method; special requirements for shipment of biological substances

As requested in the FELASA recommendations [83, 103], a competent person is necessary in the consigner's institution to prepare a health report which enables the reader to assess the health status or the microbiological quality of animals. A competent person is also necessary to appraise the health report in the institution receiving the animals. The recipient of a health report must read the information critically. Is the number of animals tested sufficient to get reliable information? Are the test methods used appropriate for the detection or exclusion of a defined agent? Are all agents listed that are not accepted in the receiving facility? To have easy access to additional information when required, contact data of a competent person in the shipping institution must be provided.

It is meanwhile common that health information is provided prior to animal transfer from commercial breeders as well as from research institutions, however, there is no uniform or standardized form used worldwide. As health monitoring programmes and reporting often vary between institutions, critical reading and interpretation of information is often challenging and time consuming. As mentioned above, FELASA recommendations give some assistance regarding number of animals to be tested and agents, frequency of testing and reporting, but similar recommendations do not exist in most other countries outside Europe.

Health reporting has to inform about the agents that were included in the testing procedures and the respective test results. However, the diagnostic sensitivity depends on the applied serological methods and the preparation of the antigens used (e.g., purified viral particles, recombinant antigens, type of viral protein). Serology detects antibodies and thus past infections, whereas molecular tests demonstrate the presence of viral nucleic acid and thus actual infections. Viruses may persist for different periods of time in different organs so that it is helpful to know which organs were analysed by molecular tests and at which location a virus was found or not. As an example, MPV is detectable by PCR in faeces for several days while demonstration of MPV DNA is possible for weeks in mesenteric lymph nodes [11, 15]. For the evaluation of test results in parasitology it is crucial to know exactly how sampling was performed, including sampling location(s). Traditionally, bacteriology is performed by microbiological culture, and this is also today still the method of choice for the detection of most bacterial agents. Bacterial culture is a very critical part of testing as the detectability of agents strongly depends on the methodology. Since most testing laboratories usually report only those bacterial species that are somehow considered relevant, other laboratories include all detected agents in their report, which may be very confusing for the recipients of test reports. Disagreeing results can be obtained also when testing for bacterial agents is performed by PCR, as different primer sets can be selected that detect different target sequences.

Basically, there are two categories of health information. Frequently, simple "*test reports*" are provided when animals are shipped from research institutions. They were prepared by the diagnostic laboratory and include only the test results for the animals that were submitted for testing, without containing detailed information about the animals (strain, age, immune status, research use, exact location in the originating animal facility, etc.) In contrast, a "*health report*" summarizes all relevant health information. This includes the results from regular and repeated testing of animals housed in a defined hygienic unit (active health monitoring) as well as results obtained from passive health monitoring (e.g., diagnostic follow-up from sick or deceased animals). The health report should be also supplemented with additional basic information like clinical observations, maintenance practices, and a description of the monitoring programme. Occasionally, testing laboratories offer to prepare a more comprehensive report by summarizing all test results from an institution over a certain time period. However, this is usually not reasonable as the testing laboratory will not have all relevant information about the animals at hand necessary to provide robust health data. As mentioned earlier, a health report must be prepared by a competent person comprehending all details about the animals (e.g., their origin and use, and history of infections).

7.11.1 Basic Information/Animals

First of all, the health report should name the exact location or designation of the colony (hygienic unit), for example, the room number. Furthermore, information about the housing conditions should be provided, for example, experimental barrier unit with an open cage system or strict breeding barrier with an IVC system. Moreover, the testing laboratories involved in monitoring should be declared, supplemented with the date of restocking or rederivation of the colony and also the date of the last monitoring. A health report should include the number of animals monitored since the date of restocking or during a certain period (FELASA recommends a period of 18 months).

It is important to know details on husbandry practices. Only few persons have usually access to breeding colonies whereas more personnel must have access to experimental units leading to a higher risk of agent introduction. Also, information on the type of research may be important when this may lead to an increased risk for the animals, for example, the use of biological materials or special equipment which may result in cross contamination between animals from different experiments or colonies (e.g., imaging units). It is naturally important to provide the reader of a health report with details on the monitoring programme. Were resident animals tested or were sentinels used? How is the sentinel programme organized? Were any clinical observations made? Were any infections detected in the unit in the past?

7.11.2 Test Methods

The methodology of testing should be included, for example, morphological examination/clinical signs, microscopy, serology, microbiological culture, histopathology, PCR, and so on. To be able to understand and correctly interpret the information provided, details on the testing procedures (e.g., which culture media are used in bacteriology testing, test methods for parasites) would be helpful, but they are usually not provided.

7.11.3 Tested Agents and Test Results

Naturally, the complete panel of agents tested should be listed in the health report, including all results (positive AND negative). Besides the results of the most recent testing, also historic test results should be summarized, including all measures conducted as a consequence of infection outbreaks (e.g., drug treatments, rederivation strategies). Occasionally, animals also receive drugs (e.g., anthelmintics, antibiotics) prior to shipment. Information about any treatment or vaccinations is in these cases also necessary. Additional information which is useful for health testing or for the understanding and interpretation of a health report and also specific experimental procedures may have impact on tests and should be included [44].

Information to be generally given in a health report

- Exact location (designation) of the colony (e.g., room no.) or the hygienic unit
- Housing conditions (e.g., barrier, isolator, IVC)
- Name(s) of laboratory/ies involved in monitoring
- Date of restocking/rederivation of the colony
- Date of last monitoring
- No. of animals monitored since date of restocking or during a certain period (e.g., 18 months)
- Methods (clinical signs, microscopy, serology, microbiological culture, histopathology, molecular)
- Name(s) of all tested agents
- Test results (positive and negative)
- Treatments, vaccinations, etc.
- Contact person

7.12 What to Do in Case of an Infectious Outbreak

7.12.1 Confirmation of Positive Results

Clinical signs of disease or indirect signs like reduced food or water consumption, poor breeding performance, or altered research results could be indicators of an infection, and it is important that investigations are initiated to find their cause. However, in most cases laboratory test results first indicate an infection. In this situation it is important to avoid hasty panic reactions, as it is very likely that a potential infection has already happened several weeks prior to its detection. Unexpected test results could also occur when samples are not clearly labelled, have been mixed up, or due to pipetting errors. Unexpected results from bacterial culture could be due to a sample contamination or misidentification. In serology, false positive reactions occur due to cross reactions with other agents or unspecific signals. False positive reactions may also happen in PCR tests, but also contamination of samples or reagents can be a reason. It is therefore important that samples with unexpected positive findings, especially from serology or molecular tests, are re-tested and confirmed (or disproved) by the same and optimally also by one or more alternative methods. Laboratories usually have backup methods available, so that confirmation is usually possible within 1 or 2 days. Simultaneously, additional samples from the same location or shipment should be tested and additional laboratories be involved for confirmation. If re-testing confirms the initially positive results there is strong evidence that they are true positives, however, false positive reactions or other artefacts are still possible [44] and must be excluded.

As mentioned before, direct detection methods demonstrate the actual presence of an agent, while indirect methods, such as serology, do not automatically imply that an agent is still present or shed. Whenever reasonable and possible, direct detection methods should be applied to confirm unexpected positive findings before action is taken to fight an infection.

7.12.2 Measures in Case of an Infection Outbreak

In case of true positives, various factors must be considered to contain and eliminate an infection. Concurrently, it is reasonable to analyse additional samples to find out when and how the infection was introduced and how far the agent has spread in the particulate unit or facility. It is particularly important to determine the source of an infection in order to avoid similar breaks in future. As some agents can infect several rodent species it needs to be determined which animal species are at risk.

Some agents are more infectious than others and spread fast (e.g., MHV) or only slowly (e.g., MPV) within a given population. The prevalence of an infection should be determined, as this parameter can be used to estimate the necessary measures. If there are only few positives in a distinct area, a more conservative or less rigorous regimen may be implemented. However, in case of a more widespread infection, prompt and rigorous measures involving the entire facility may be implemented. Housing conditions (open cages, IVCs, isolators) also have impact on quarantine measures. Furthermore, all factors, including the economic or scientific value of an infected colony (donor animals for embryo transfer versus transgenic breeding of numerous lines, short- versus long-term experiments) play a role in evaluating which regimen could be put in place for the eradication of an agent. In any case, policies to contain and eradicate an infection must be decided cooperatively between animal health professionals, facility managers, and researchers. It is crucial that each decision regarding approaches to eliminate an infection are adapted to the specific situation.

7.12.3 Rederivation of Infected Colonies/Eradication of Infections

Different decisions regarding the fate of an infected colony are possible. Only very specific and severe situations would require an immediate cull of an infected colony, as highly pathogenic or zoonotic agents are nowadays very rare. In most cases it is reasonable that no radical action is taken and that the infected population or parts of it will be retained in isolation until the end of the animal's intended purpose. In case of an infection in a breeding population, the number of infected animals should be reduced to a necessary minimum before rederivation is performed.

Agents that do usually neither cause clinical signs nor are considered to have obvious impact on experiments while having a high risk of reintroduction, may be tolerated (e.g., *Klebsiella oxytoca, Staph. aureus*, *Pseudomonas aeruginosa* or certain protozoans such as trichomonads). However, it has to be considered that these agents may cause serious health problems for susceptible genetically modified lines or immunodeficient animals.

Various viruses with very low pathogenicity, low infectivity, and spreading tendency like MPV and also certain bacterial agents like, e.g., *Rodentibacter* spp. that may have mild impact on certain experiments are frequently tolerated. This is the case also for other viruses such as MNV, MuAstV, and others although they are more infectious and have a higher tendency to spread. Isolation of infected animals in individually ventilated cages is sometimes possible and restocking or rederivation may be considered when infected animals have been used up.

A number of agents are sensitive to pharmacological substances, and their elimination might be considered by medication. Eradication is usually successful in case of pinworms when sufficiently high doses of fenbendazole (or similar compounds) are orally administered (e.g., via food) for a sufficiently long period [55]. Successful eradication of fur mites with ivermectin has also been reported [4, 5] and a combination of ivermectin therapy with cross fostering is also possible [56]. Therapeutic elimination of selected bacterial agents such as *Helicobacter* spp. or *Rodentibacter* spp. [45, 98] has been described in the literature, however, attempts are often not successful [43, 132]. Infections by *Pneumocystis* spp. can be treated orally (trimethoprime/sulfonamide) via drinking water [92], but complete elimination is usually not possible. Whenever drugs are used, it is important that side effects on animals and on experiments are evaluated and excluded as far as possible.

More drastic action will usually be taken when infections are detected that are known to spread rapidly or when moderate or significant clinical signs or effects on research results are to be expected, e.g., after a MHV outbreak or infections with pathogenic bacteria (e.g., *C. kutscheri*), or in cases of zoonotic infections (LCMV, *Sb. moniliformis*). Here, more strict isolation procedures are necessary, which usually means that animals with lower importance may be eliminated to reduce the number of infected animals to a necessary minimum. Breeding cessation together with other measures has been reported to be successful in eradicating a mouse rotavirus infection [52] and has repeatedly been reported also for other virus infections [e.g., MHV [136]]. However, attempts to eliminate infections by breeding cessation can be expected to be successful only after prolonged restrictions for several months. Most situations require more secure and also more rapid agent eradication, e.g., by embryo transfer, hysterectomy or cross fostering as described for specific agents (e.g., *Helicobacter* sp., MNV) [8].

In specific situations, a "test and cull" strategy can be conducted, meaning that all animals of the infected colony are frequently tested and either eliminated or maintained in the husbandry based on their test result. However, this approach is extremely time and cost consuming and requires strict separation of negative animals and very high test frequencies. For these reasons it is generally not recommended, since this approach is also relatively uncertain and often not completely successful (Fig. 7.4).

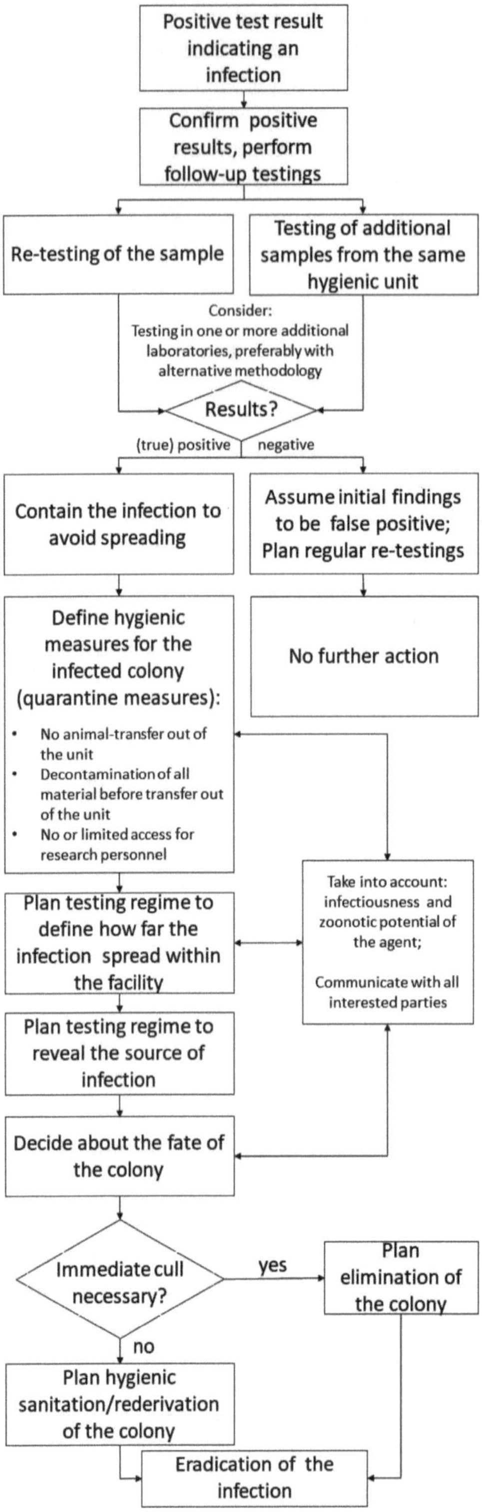

Fig. 7.4 Schematic flowchart depicting the course of action after positive test results indicating an infection is received

7.13 Outlook

Molecular methods are more and more used instead of traditional diagnostics. This development not only comes along with increased diagnostic sensitivity but also with a larger spectrum of detectable agents, especially since sequencing techniques became broadly available. In this context, besides classical pathogens, an increasing number of commensal bacteria moves into focus, which can be associated with certain (disease) phenotypes, depending on the animal model [50]. However, not only single agents might influence study results, since researchers become increasingly aware of the striking influence of the whole microbiome on the outcome of animal experiments [19, 20, 49]. The composition of these complex communities of more than 100 billion of different microorganisms (consisting of bacteria, protozoa, fungi, archaea and viruses) per gram of intestinal content massively depends on numerous host factors (e.g., strain, age, gender) but also on the animal's environment (housing type, food, bedding material, sterilization procedures) and therefore significantly differs between laboratory animal facilities. When comparing the hygienic status of two distinct animal colonies it is therefore of key importance that researchers also take the overall microbiome composition of the animals into account.

Since most pathogens were eradicated from laboratory rodent colonies due to repeated rederivation procedures and strict barrier housing, also the diversity of the microbiome was reduced, and most commercially produced mouse colonies nowadays have a limited microbiome composition compared to wild rodents or conventionally housed colonies [23]. This development towards "very clean" colonies may lead to an artificial composition of their microbiota and, as an example, to aberrant immune reactions [16, 115], thus compromising the translational value of the study outcome. Moreover, results obtained from distinct colonies housed in different animal facilities (or even different units within one facility) might not be reproducible due to major differences in the microbiome compositions [114]. To control this huge variable within in vivo experiments while achieving a maximum level of hygienic standardization, project planning has to involve a strategy on how to best respect this factor. For example, the use of appropriate control groups (e.g., cage-littermates or co-housed animals) help to minimize microbiome effects between animal groups, multicenter studies in different microbiome conditions give insights into the replicability of results [23]. Here it is of key importance that regular microbiome analyses should find their way into routine health monitoring programmes to fill this gap of knowledge and help researchers characterizing not only specific agents/known study confounders but also the animal's microbiome composition. Particularly, metagenome analyses (e.g., EAD material) could be a key methodology to combine targeted agent-detection with microbiome analyses. Though this approach would not compensate for the analyses of individual microbiomes, it could give valuable information about major shifts in microbiome compositions of closed colonies (e.g., indicating a hygienic break). Nevertheless, traditional diagnostic methods still serve as valuable instruments for the detection of infectious agents, indispensable to reveal the etiopathogenesis of disease development and confirmation of positive results. Optimally, basic veterinary skills (clinical examination, necropsy procedures) and classical microbiological, serological, and parasitological methods should build the solid foundation, supplemented with well-aimed molecular methodology for the highly sensitive screening of targeted agents or the systematic analyses of microbiome compositions.

References

1. Acuff NV, LaGana M, Nagy T, Watford WT. Severe dermatitis associated with spontaneous *Staphylococcus xylosus* infection in Rag(−/−) Tpl2(−/−) mice. Comp Med. 2017;67(4):344–9.
2. Adhikary S, Nicklas W, Bisgaard M, Boot R, Kuhnert P, Waberschek T, Aalbaek B, Korczak B, Christensen H. *Rodentibacter* gen. nov. including *Rodentibacter pneumotropicus* comb. nov., *Rodentibacter heylii* sp. nov., *Rodentibacter myodis* sp. nov., *Rodentibacter ratti* sp. nov., *Rodentibacter heidelbergensis* sp. nov., *Rodentibacter trehalosifermentans* sp. nov., *Rodentibacter rarus* sp. nov., *Rodentibacter mrazii* and two genomospecies. Int J Syst Evol Microbiol. 2017;67(6):1793–806.
3. Allen AM, Nomura T. Manual of microbiologic monitoring of laboratory animals. Bethesda: US Department of Health and Human Services, Public Health Service; 1986.
4. Arbona RJR, Lipman NS, Wolf FR. Treatment and eradication of murine fur mites: II. Diagnostic considerations. J Am Assoc Lab Anim Sci. 2010a;49(5):583–7.
5. Arbona RJR, Lipman NS, Wolf FR. Treatment and eradication of murine fur mites: III. Treatment of a large mouse colony with ivermectin-compounded feed. J Am Assoc Lab Anim Sci. 2010b;49(5):633–7.
6. Armstrong C, Lillie R. Experimental lymphocytic choriomeningitis of monkeys and mice produced by a virus encountered in studies of the 1933 St. Louis encephalitis epidemic. Publ Health Rep. 1934;49:1019–27.
7. Artwohl JE, Cera LM, Wright MF, Medina LV, Kim LJ. The efficacy of a dirty bedding sentinel system for detecting Sendai virus infection in mice: a comparison of clinical signs and seroconversion. Lab Anim Sci. 1994;44(1):73–5.
8. Artwohl JE, Purcell JE, Chrusciel K, Lang M, Fortman J. Assessment of cross-foster rederivation in the elimination of mouse norovirus and *Helicobacter*. J Am Assoc Lab Anim Sci. 2007;46(4):84.
9. Baker DG. Natural pathogens of laboratory mice, rats, and rabbits and their effects on research. Clin Microbiol Rev. 1998;11(2):231–66.
10. Barthold SW, Smith AJ. Lymphocytic choriomeningitis virus. In: Fox JG, Barthold SW, Davisson MT, et al., editors. The mouse in biomedical research. Vol. 2: diseases. Amsterdam: Elsevier Academic; 2007. p. 179–214.
11. Bauer BA, Riley LK. Antemortem detection of mouse parvovirus and mice minute virus by polymerase chain reaction (PCR) of faecal samples. Lab Anim. 2006;40(2):144–52.
12. Bauer BA, Besch-Williford CL, Riley LK. Comparison of the mouse antibody production (MAP) assay and polymerase chain reaction (PCR) assays for the detection of viral contaminants. Biologicals. 2004;32(4):177–82.
13. Benga L, Nicklas W, Lautwein T, Verbarg S, Gougoula C, Engelhardt E, Benten WPM, Kohrer K, Sager M, Christensen H. Rodentibacter haemolyticus sp. nov. isolated from laboratory rodents. Int J Syst Evol Microbiol. 2021;71(8):e004947.
14. Besch-Williford C, Pesavento P, Hamilton S, Bauer B, Kapusinszky B, Phan T, Delwart E, Livingston R, Cushing S, Watanabe R, Levin S, Berger D, Myles M. A naturally transmitted epitheliotropic polyomavirus pathogenic in immunodeficient rats: characterization, transmission, and preliminary epidemiologic studies. Toxicol Pathol. 2017;45(5):593–603.
15. Besselsen DG, Wagner AM, Loganbill JK. Effect of mouse strain and age on detection of mouse parvovi-

rus 1 by use of serologic testing and polymerase chain reaction analysis. Comp Med. 2000;50(5):498–502.
16. Beura LK, Hamilton SE, Bi K, Schenkel JM, Odumade OA, Casey KA, Thompson EA, Fraser KA, Rosato PC, Filali-Mouhim A, Sekaly RP, Jenkins MK, Vezys V, Haining WN, Jameson SC, Masopust D. Normalizing the environment recapitulates adult human immune traits in laboratory mice. Nature. 2016;532(7600):512–6.
17. Bhatt PN, Jacoby RO, Barthold SW. Contamination of transplantable murine tumors with lymphocytic choriomeningitis virus. Lab Anim Sci. 1986a;36(2):136–9.
18. Bhatt PN, Jacoby RO, Morse HC, New A. Viral and mycoplasm infections of laboratory rodents: effects on biomedical research. New York: Academic; 1986b.
19. Bleich A, Fox JG. The mammalian microbiome and its importance in laboratory animal research. ILAR J. 2015;56(2):153–8.
20. Bleich A, Hansen AK. Time to include the gut microbiota in the hygienic standardisation of laboratory rodents. Comp Immunol Microbiol Infect Dis. 2012;35(2):81–92.
21. Bleich A, Nicklas W. [Zoonoses transmitted by mouse and rat maintained as laboratory or pet animals]. Berl Munch Tierarztl Wochenschr. 2008;121(7–8):241–55.
22. Bradfield JF, Wagner JE, Boivin GP, Steffen EK, Russell RJ. Epizootic fatal dermatitis in athymic nude-mice due to *Staphylococcus-xylosus*. Lab Anim Sci. 1993;43(1):111–3.
23. Buchheister S, Bleich A. Health monitoring of laboratory rodent colonies-talking about (r)evolution. Animals. 2021;11(5):1410.
24. Buller RML, Fenner F. Mousepox. In: Fox JG, Barthold SW, Davisson MT, et al., editors. The mouse in biomedical research. Vol. 2: diseases. Amsterdam: Elsevier Academic; 2007. p. 67–92.
25. Cannon R, Roe R. Livestock disease surveys: a field disease manual for veterinarians. Canberra: Australian Government Publishing Service; 1986.
26. Carthew P, Verstraete A. A serological survey of accredited breeding colonies in the United Kingdom for common rodent viruses. Lab Anim. 1978;12(1):29–32.
27. Carty AJ. Opportunistic infections of mice and rats: Jacoby and Lindsey revisited. ILAR J. 2008;49(3):272–6.
28. Casey KM, Johnson AL, Hunrath MN, Fraser JK, McCowan NC, Wasson K, Doty RA, Griffey SM, Imai DM. Proliferative typhlocolitis with multinucleated giant cells: a nonspecific enteropathy in immunodeficient sentinel mice. Vet Pathol. 2019;56(1):157–68.
29. CDC. Notes from the field: lymphocytic choriomeningitis virus infections in employees of a rodent breeding facility – Indiana, May–June 2012. MMWR Morb Mortal Wkly Rep. 2012;61(32):622–3.
30. Coates M, Cooper J, Heine W, Kraft V, Hedrich H, Fujikura T, Govell G, Hänninen O, Pelkonen K. Germ-free, gnotobiotic and specified pathogen-free (SPF) animals. In: Guidelines for the breeding and care of laboratory animals. Rome: World Health Organization (WHO) and International Council for Laboratory Animal Science (ICLAS); 1994. p. 42–51, ISS/WHO/FAO-CC/IZSTe/94.23.
31. Collins MJ Jr, Parker JC. Murine virus contaminants of leukemia viruses and transplantable tumors. J Natl Cancer Inst. 1972;49(4):1139–43.
32. Compton SR. PCR and RT-PCR in the diagnosis of laboratory animal infections and in health monitoring. J Am Assoc Lab Anim Sci. 2020;59(5):458–68.
33. Compton SR, Ball-Goodrich LJ, Paturzo FX, Macy JD. Transmission of enterotropic mouse hepatitis virus from immunocompetent and immunodeficient mice. Comp Med. 2004;54(1):29–35.
34. Criley JM, Carty AJ, Besch-Williford CL, Franklin CL. Coxiella burnetii infection in C.B-17 scid-bg mice xenotransplanted with fetal bovine tissue. Comp Med. 2001;51(4):357–60.
35. Cundiff DD, Riley LK, Franklin CL, Hook RR Jr, Besch-Williford C. Failure of a soiled bedding sentinel system to detect *cilia-associated respiratory bacillus* infection in rats. Lab Anim Sci. 1995;45(2):219–21.
36. Dick EJ Jr, Kittell CL, Meyer H, Farrar PL, Ropp SL, Esposito JJ, Buller RM, Neubauer H, Kang YH, McKee AE. Mousepox outbreak in a laboratory mouse colony. Lab Anim Sci. 1996;46(6):602–11.
37. Dillehay DL, Lehner NDM, Huerkamp MJ. The effectiveness of a microisolator cage system and sentinel mice for controlling and detecting Mhv and Sendai virus-infections. Lab Anim Sci. 1990;40(4):367–70.
38. Ehlers B, Kuchler J, Yasmum N, Dural G, Voigt S, Schmidt-Chanasit J, Jakel T, Matuschka FR, Richter D, Essbauer S, Hughes DJ, Summers C, Bennett M, Stewart JP, Ulrich RG. Identification of novel rodent herpesviruses, including the first gammaherpesvirus of Mus musculus. J Virol. 2007;81(15):8091–100.
39. Farkas T, Fey B, Keller G, Martella V, Egyed L. Molecular detection of novel astroviruses in wild and laboratory mice. Virus Genes. 2012;45(3):518–25.
40. Festing M, Nomura T, Fujikura T, Govell G, Hänninen O, Pelkonen K. Animal quality control. In: Guidelines for the breeding and care of laboratory animals. Rome: World Health Organization (WHO) and International Council for Laboratory Animal Science (ICLAS); 1994. p. 68–72, ISS/WHO/FAO-CC/IZSTe/94.23.
41. Foreman O, Kavirayani AM, Griffey SM, Reader R, Shultz LD. Opportunistic bacterial infections in breeding colonies of the NSG mouse strain. Vet Pathol. 2011;48(2):495–9.
42. Fornefett J, Krause J, Klose K, Fingas F, Hassert R, Eisenberg T, Schrodl W, Grunwald T, Muller

U, Baums CG. Comparative analysis of clinics, pathologies and immune responses in BALB/c and C57BL/6 mice infected with Streptobacillus moniliformis. Microbes Infect. 2018;20(2):101–10.
43. Fox JG, Whary MT. *Helicobacter* infections in mice. In: Fox JG, Barthold SW, Davisson MT, et al., editors. The mouse in biomedical research. Vol. 2: diseases. Amsterdam: Elsevier Academic; 2007. p. 407–36.
44. Gaertner DJ, Batchelder M, Herbst LH, Kaufman HL. Administration of vaccinia virus to mice may cause contact or bedding sentinel mice to test positive for orthopoxvirus antibodies: case report and follow-up investigation. Comp Med. 2003;53(1):85–8.
45. Goelz MF, Thigpen JE, Mahler J, Rogers WP, Locklear J, Weigler BJ, Forsythe DB. Efficacy of various therapeutic regimens in eliminating *Pasteurella pneumotropica* from the mouse. Lab Anim Sci. 1996;46(3):280–5.
46. Goto K, Ishihara KI, Kuzuoka A, Ohnishi Y, Itoh T. Contamination of transplantable human tumor-bearing lines by *Helicobacter hepaticus* and its elimination. J Clin Microbiol. 2001;39(10):3703–4.
47. Gozalo AS, Hoffmann VJ, Brinster LR, Elkins WR, Ding L, Holland SM. Spontaneous *Staphylococcus xylosus* infection in mice deficient in NADPH oxidase and comparison with other laboratory mouse strains. J Am Assoc Lab Anim Sci. 2010;49(4):480–6.
48. Hansen AK. Statistical aspects of health monitoring of laboratory animal colonies. Scand J Lab Anim Sci. 1993;20:11–4.
49. Hansen AK, Franklin C. Microbiota, laboratory animals, and research. Lab Anim. 2019;53(3):229–31.
50. Hansen AK, Nielsen DS, Krych L, Hansen CHF. Bacterial species to be considered in quality assurance of mice and rats. Lab Anim. 2019;53(3):281–91.
51. Hektoen L. Observations on pulmonary infections in rats. Trans Chicago Pathol Soc. 1915–1916;10:105–9.
52. Held N, Hedrich HJ, Bleich A. Successful sanitation of an EDIM-infected mouse colony by breeding cessation. Lab Anim. 2011;45(4):276–9.
53. Henderson KS, Perkins CL, Havens RB, Kelly MJ, Francis BC, Dole VS, Shek WR. Efficacy of direct detection of pathogens in naturally infected mice by using a high-density PCR array. J Am Assoc Lab Anim Sci. 2013;52(6):763–72.
54. Homberger FR, Boot RF, Kornerup-Hansen A, van der Logt J. FELASA guidance paper for the accreditation of laboratory animal diagnostic laboratories – report of the Federation of European Laboratory Animal Science Associations (FELASA) Working Group on accreditation of diagnostic laboratories. Lab Anim. 1999;33:19–38.
55. Huerkamp MJ, Benjamin KA, Zitzow LA, Pullium JK, Lloyd JA, Thompson WD, Webb SK, Lehner NDM. Fenbendazole treatment without environmental decontamination eradicates *Syphacia muris* from all rats in a large, complex research institution. Contemp Top Lab Anim Sci. 2000;39(3):9–12.
56. Huerkamp MJ, Zitzow LA, Webb S, Pullium JK. Cross-fostering in combination with ivermectin therapy: a method to eradicate murine fur mites. Contemp Top Lab Anim Sci. 2005;44(4):12–6.
57. Ike F, Bourgade F, Ohsawa K, Sato H, Morikawa S, Saijo M, Kurane I, Takimoto K, Yamada YK, Jaubert J, Berard M, Nakata H, Hiraiwa N, Mekada K, Takakura A, Itoh T, Obata Y, Yoshiki A, Montagutelli X. Lymphocytic choriomeningitis infection undetected by dirty-bedding sentinel monitoring and revealed after embryo transfer of an inbred strain derived from wild mice. Comp Med. 2007;57(3):272–81.
58. ILAR. Committee on long-term holding of laboratoryrodents. Long-term holding of laboratory rodents. ILAR News. 1976;19:L1–L25.
59. Jacoby RO, Homberger FR. International standards for rodent quality. Lab Anim Sci. 1999;49(3):230.
60. Jacoby RO, Lindsey JR. Risks of infection among laboratory rats and mice at major biomedical research institutions. ILAR J. 1998;39(4):266–71.
61. Jakubik J, Skoda R. Studies on the occurrence of latent virus infections in various European breeds of conventional, SPF and germ-free mice. Z Versuchstierkunde. 1972;14(6):281–9.
62. Janus LM, Mahler M, Kohl W, Smoczek A, Hedrich HJ, Bleich A. Minute virus of mice: antibody response, viral shedding, and persistence of viral DNA in multiple strains of mice. Comp Med. 2008;58(4):360–8.
63. Jensen ES, Allen KP, Henderson KS, Szabo A, Thulin JD. PCR testing of a ventilated caging system to detect murine fur mites. J Am Assoc Lab Anim Sci. 2013;52(1):28–33.
64. Kagiyama N. Necessity of genetic and microbiologic quality network from the pharmaceutical industry's perspective. In: National Research Council. Microbial status and genetic evaluation of mice and rats: proceedings of the 1999 US/Japan conference. Washington, DC: National Academy Press; 2000. p. 77–84.
65. Karst SM, Wobus CE, Lay M, Davidson J, Virgin HW t. STAT1-dependent innate immunity to a Norwalk-like virus. Science. 2003;299(5612):1575–8.
66. Koga T, Aoki W, Fujii M, Satou K, Ikeda Y. Spontaneous infection caused by *Streptococcus agalactiae* in KK-A(y) mice. Comp Med. 2017;67(5):416–9.
67. Kraft V, Meyer B. Seromonitoring in small laboratory animal colonies. A five year survey: 1984–1988. Z Versuchstierkunde. 1990;33(1):29–35.
68. Kraft V, Deeny A, Blanchet H, Hansen A. FELASA recommendations for the health monitoring of mouse, rat, hamster, guineapig and rabbit breeding colonies. Lab Anim. 1994;28:1–12.
69. Kunstyr I. List of pathogens for specification in SPF laboratory animals. Publication no. 2. Biberach: Society for Laboratory Animal Science; 1988.
70. Kunstyr I. Diagnostic microbiology for laboratory animals: viruses, bacteria, chlamydia, fungi, and parasites. Stuttgart: G. Fischer; 1992.

71. LABA. (Laboratory Animal Breeders Association). Accredited health monitoring scheme manual (LABAHMS). 2000. http://www.laba-uk.com/docs/LABAHMS%20Manual.pdf
72. Labelle P, Hahn NE, Fraser JK, Kendall LV, Ziman M, James E, Shastri N, Griffey SM. Mousepox detected in a research facility: case report and failure of mouse antibody production testing to identify *Ectromelia* virus in contaminated mouse serum. Comp Med. 2009;59(2):180–6.
73. Lewis VJ, Clayton DM. An evaluation of the mouse antibody production test for detecting three murine viruses. Lab Anim Sci. 1971;21(2):203–5.
74. Liang CT, Shih A, Chang YH, Liu CW, Lee YT, Hsieh WC, Huang YL, Huang WT, Kuang CH, Lee KH, Zhuo YX, Ho SY, Liao SL, Chiu YY, Hsu CN, Liang SC, Yu CK. Microbial contaminations of laboratory mice and rats in Taiwan from 2004 to 2007. J Am Assoc Lab Anim Sci. 2009;48(4):381–6.
75. Lindstrom KE, Carbone LG, Kellar DE, Mayorga MS, Wilkerson JD. Soiled bedding sentinels for the detection of fur mites in mice. J Am Assoc Lab Anim Sci. 2011;50(1):54–60.
76. Lipman NS, Homberger FR. Rodent quality assurance testing: use of sentinel animal systems. Lab Anim. 2003;32(5):36–43.
77. Lipman NS, Perkins S, Nguyen H, Pfeffer M, Meyer H. Mousepox resulting from use of ectromelia virus-contaminated, imported mouse serum. Comp Med. 2000;50(4):426–35.
78. Livingston RS, Riley LK. Diagnostic testing of mouse and rat colonies for infectious agents. Lab Anim. 2003;32(5):44–51.
79. Mahabir E, Jacobsen K, Brielmeier M, Peters D, Needham J, Schmidt J. Mouse antibody production test: can we do without it? J Virol Methods. 2004;120(2):239–45.
80. Mahabir E, Durand S, Henderson KS, Hardy P. Comparison of two prevalent individually ventilated caging systems for detection of murine infectious agents via exhaust air particles. Lab Anim. 2019;53(1):84–8.
81. Mähler M, Köhl W. A serological survey to evaluate contemporary prevalence of viral agents and *Mycoplasma pulmonis* in laboratory mice and rats in western Europe. Lab Anim (NY). 2009;38(5):161–5.
82. Mähler M, Nicklas W. Health management and monitoring. In: The laboratory mouse. London: Elsevier Academic; 2012. p. 601–20.
83. Mähler M, Berard M, Feinstein R, Gallagher A, Illgen-Wilcke B, Pritchett-Corning K, Raspa M. FELASA recommendations for the health monitoring of mouse, rat, hamster, Guinea pig and rabbit colonies in breeding and experimental units. Lab Anim. 2014;48(3):178–92.
84. Mailhiot D, Ostdiek AM, Luchins KR, Bowers CJ, Theriault BR, Langan GP. Comparing mouse health monitoring between soiled-bedding sentinel and exhaust air dust surveillance programs. J Am Assoc Lab Anim Sci. 2020;59(1):58–66.
85. Manjunath S, Kulkarni PG, Nagavelu K, Samuel RJ, Srinivasan S, Ramasamy N, Hegde NR, Gudde RS. Sero-prevalence of rodent pathogens in India. PLoS One. 2015;10(7):e0131706.
86. Manuel CA, Hsu CC, Riley LK, Livingston RS. Soiled-bedding sentinel detection of murine norovirus 4. J Am Assoc Lab Anim Sci. 2008;47(3):31–6.
87. Manuel CA, Bagby SM, Reisinger JA, Pugazhenthi U, Pitts TM, Keysar SB, Arcaroli JJ, Leszczynski JK. Procedure for horizontal transfer of patient-derived xenograft tumors to eliminate *Corynebacterium bovis*. J Am Assoc Lab Anim Sci. 2017;56(2):166–72.
88. Marchal J. Infectious ectromelia, a hitherto undescribed virus disease of mice. J Pathol Bacteriol. 1930;33:713–8.
89. Marx J, Gaertner DJ, Smith AL. Results of survey regarding prevalence of adventitial infections in mice and rats at biomedical research facilities. J Am Assoc Lab Anim Sci. 2017;56(5):527–33.
90. McInnes EF, Rasmussen L, Fung P, Auld AM, Alvarez L, Lawrence DA, Quinn ME, del Fierro GM, Vassallo BA, Stevenson R. Prevalence of viral, bacterial and parasitological diseases in rats and mice used in research environments in Australasia over a 5-y period. Lab Anim (NY). 2011;40(11):341–50.
91. Miller M, Brielmeier M. Environmental samples make soiled bedding sentinels dispensable for hygienic monitoring of IVC-reared mouse colonies. Lab Anim. 2018;52(3):233–9.
92. Mossmann H, Nicklas W, Hedrich HJ. Management of immunocompromised and infected animals. In: Kaufmann SHE, Kabelitz D, editors. Immunology of infection, vol. 32. San Diego: Elsevier Academic; 2002. p. 183–231.
93. Motzel SL, Riley LK. Subclinical infection and transmission of Tyzzers disease in rats. Lab Anim Sci. 1992;42(5):439–43.
94. Myers DD, Smith E, Schweitzer I, Stockwell JD, Paigen BJ, Bates R, Palmer J, Smith AL. Assessing the risk of transmission of three infectious agents among mice housed in a negatively pressurized caging system. Contemp Top Lab Anim Sci. 2003;42(6):16–21.
95. Na YR, Seok SH, Lee HY, Baek MW, Kim DJ, Park SH, Lee HK, Park JH. Microbiological quality assessment of laboratory mice in Korea and recommendations for quality improvement. Exp Anim. 2010;59(1):25–33.
96. Newcomer CE, Fox JG. Zoonoses and other human health hazards. In: Fox JG, Barthold SW, Davisson MT, et al., editors. The mouse in biomedical research. Vol. 2: diseases. Amsterdam: Elsevier Academic; 2007. p. 719–46.
97. Ng TFF, Kondov NO, Hayashimoto N, Uchida R, Cha Y, Beyer AI, Wong W, Pesavento PA, Suemizu H, Muench MO, Delwart E. Identification of an astrovirus commonly infecting laboratory mice in the US and Japan. PLoS One. 2013;8(6):e66937.

98. Nicklas W. Pasteurellaceae. In: Fox JG, Barthold SW, Davisson MT, et al., editors. The mouse in biomedical research. Vol. 2: diseases. Amsterdam: Elsevier Academic; 2007. p. 469–506.
99. Nicklas W. International harmonization of health monitoring. ILAR J. 2008;49(3):338–46.
100. Nicklas W, Seidel K. Expert information from the working group on hygiene: harmonisation of health monitoring reports. Lab Anim. 2019;53(2):208–9.
101. Nicklas W, Kraft V, Meyer B. Contamination of transplantable tumors, cell lines, and monoclonal antibodies with rodent viruses. Lab Anim Sci. 1993;43(4):296–300.
102. Nicklas W, Homberger FR, Illgen-Wilcke B, Jacobi K, Kraft V, Kunstyr I, Mahler M, Meyer H, Pohlmeyer-Esch G. Implications of infectious agents on results of animal experiments. Lab Anim. 1999;33:39–87.
103. Nicklas W, Baneux P, Boot R, Decelle T, Deeny AA, Fumanelli M, Illgen-Wilcke B. FELASA recommendations for the health monitoring of rodent and rabbit colonies in breeding and experimental units. Lab Anim. 2002;36(1):20–42.
104. Nicklas W, Deeny A, Diercks P, Gobbi A, Illgen-Wilcke B, Seidelin M. FELASA guidelines for the accreditation of health monitoring programmes and for testing laboratories involved in health monitoring. Lab Anim (NY). 2010;39(2):43–8.
105. Nicklas W, Keubler L, Bleich A. Maintaining and monitoring the defined microbiota status of gnotobiotic rodents. ILAR J. 2015;56(2):241–9.
106. Niimi K, Maruyama S, Sako N, Miyata K, Yoshimoto T, Bileckei B, Henderson KS, Takahashi E. The Sentinel (TM) EAD(R) program can detect more microorganisms than bedding sentinel animals. Jpn J Vet Res. 2018;66(2):125–9.
107. Pakes SP, Benirschke K. Diseases of laboratory animals complicating biomedical research – introduction. Am J Pathol. 1971;64(3):624.
108. Percie du Sert N, Hurst V, Ahluwalia A, Alam S, Avey MT, Baker M, Browne WJ, Clark A, Cuthill IC, Dirnagl U, Emerson M, Garner P, Holgate ST, Howells DW, Karp NA, Lazic SE, Lidster K, MacCallum CJ, Macleod M, Pearl EJ, Petersen OH, Rawle F, Reynolds P, Rooney K, Sena ES, Silberberg SD, Steckler T, Wurbel H. The ARRIVE guidelines 2.0: updated guidelines for reporting animal research. PLoS Biol. 2020;18(7):e3000410.
109. Perdue KA, Copeland MK, Karjala Z, Cheng LI, Ward JM, Elkins WR. Suboptimal ability of dirty-bedding sentinels to detect *Spironucleus muris* in a colony of mice with genetic manipulations of the adaptive immune system. J Am Assoc Lab Anim Sci. 2008;47(5):10–7.
110. Pettan-Brewer C, Trost RJ, Maggio-Price L, Seamons A, Dowling SC. Adoption of exhaust air dust testing in SPF rodent facilities. J Am Assoc Lab Anim Sci. 2020;59(2):156–62.
111. Phan TG, Kapusinszky B, Wang C, Rose RK, Lipton HL, Delwart EL. The fecal viral flora of wild rodents. PLoS Pathog. 2011;7(9):e1002218.
112. Pritchett-Corning KR, Cosentino J, Clifford CB. Contemporary prevalence of infectious agents in laboratory mice and rats. Lab Anim. 2009;43(2):165–73.
113. Pritchett-Corning KR, Prins JB, Feinstein R, Goodwin J, Nicklas W, Riley L. AALAS/FELASA working group on health monitoring of rodents for animal transfer. J Am Assoc Lab Anim Sci. 2014;53(6):633–40.
114. Rausch P, Basic M, Batra A, Bischoff SC, Blaut M, Clavel T, Glasner J, Gopalakrishnan S, Grassl GA, Gunther C, Haller D, Hirose M, Ibrahim S, Loh G, Mattner J, Nagel S, Pabst O, Schmidt F, Siegmund B, Strowig T, Volynets V, Wirtz S, Zeissig S, Zeissig Y, Bleich A, Baines JF. Analysis of factors contributing to variation in the C57BL/6J fecal microbiota across German animal facilities. Int J Med Microbiol. 2016;306(5):343–55.
115. Reese TA, Bi K, Kambal A, Filali-Mouhim A, Beura LK, Bürger MC, Pulendran B, Sekaly R-P, Jameson SC, Masopust D. Sequential infection with common pathogens promotes human-like immune gene expression and altered vaccine response. Cell Host Microbe. 2016;19(5):713–9.
116. Rehbinder C, Baneux P, Forbes D, VanHerck H, Nicklas W, Rugaya Z, Winkler G. FELASA recommendations for the health monitoring of mouse, rat, hamster, gerbil, guineapig and rabbit experimental units. Lab Anim. 1996;30(3):193–208.
117. Riepenhoff-Talty M, Offor E, Klossner K, Kowalski E, Carmody P, Ogra P. Effect of age and malnutrition on rotavirus infection in mice. Pediatr Res. 1985;19(12):1250–3.
118. Roediger B, Lee Q, Tikoo S, Cobbin JCA, Henderson JM, Jormakka M, O'Rourke MB, Padula MP, Pinello N, Henry M, Wynne M, Santagostino SF, Brayton CF, Rasmussen L, Lisowski L, Tay SS, Harris DC, Bertram JF, Dowling JP, Bertolino P, Lai JH, Wu W, Bachovchin WW, Wong JJ, Gorrell MD, Shaban B, Holmes EC, Jolly CJ, Monette S, Weninger W. An atypical parvovirus drives chronic tubulointerstitial nephropathy and kidney fibrosis. Cell. 2018;175(2):530–543.e24.
119. Rowe WP, Hartley JW, Estes JD, Huebner RJ. Studies of mouse polyoma virus infection. 1. Procedures for quantitation and detection of virus. J Exp Med. 1959;109:379–91.
120. Russo M, Invernizzi A, Gobbi A, Radaelli E. Diffuse scaling dermatitis in an athymic nude mouse. Vet Pathol. 2013;50(4):722–6.
121. Sachsenröder J, Braun A, Machnowska P, Ng TFF, Deng XT, Guenther S, Bernstein S, Ulrich RG, Delwart E, Johne R. Metagenomic identification of novel enteric viruses in urban wild rats and genome characterization of a group a rotavirus. J Gen Virol. 2014;95:2734–47.
122. Scharmann W, Heller A. Survival and transmissibility of *Pasteurella pneumotropica*. Lab Anim. 2001;35(2):163–6.
123. Schmidt K, Butt J, Mauter P, Vogel K, Erles-Kemna A, Pawlita M, Nicklas W. Development of a mul-

tiplex serological assay reveals a worldwide distribution of murine astrovirus infections in laboratory mice. PLoS One. 2017;12(10):e0187174.
124. Schoeb TR. *Mycoplasma pulmonis*, other murine mycoplasmas, and cilia-associated respiratory bacillus. In: Fox JG, Barthold SW, Davisson MT, et al., editors. The mouse in biomedical research. Vol. 2: diseases. Amsterdam: Elsevier Academic; 2007. p. 437–68.
125. Schoondermark-van de Ven EM, Philipse-Bergmann IM, van der Logt JT. Prevalence of naturally occurring viral infections, *Mycoplasma pulmonis* and *Clostridium piliforme* in laboratory rodents in Western Europe screened from 2000 to 2003. Lab Anim. 2006;40(2):137–43.
126. Shuster KA, Hish GA, Selles LA, Chowdhury MA, Wiggins RC, Dysko RC, Bergin IL. Naturally occurring disseminated group B *Streptococcus* infections in postnatal rats. Comp Med. 2013;63(1):55–61.
127. Smith AL. BULLETIN BOARD-experts address international standards for rodent quality. Comp Med. 2000;50(2):232.
128. Smith AL, Jacoby RO, Johnson EA, Paturzo F, Bhatt PN. Invivo studies with an orphan parvovirus of mice. Lab Anim Sci. 1993;43(2):175–82.
129. Smith PC, Nucifora M, Reuter JD, Compton SR. Reliability of soiled bedding transfer for detection of mouse parvovirus and mouse hepatitis virus. Comp Med. 2007;57(1):90–6.
130. Thigpen JE, Lebetkin EH, Dawes ML, Amyx HL, Caviness GF, Sawyer BA, Blackmore DE. The use of dirty bedding for detection of murine pathogens in sentinel mice. Lab Anim Sci. 1989;39(4):324–7.
131. Tyzzer EE. A fatal disease of the Japanese waltzing mouse caused by a spore-bearing bacillus (*Bacillus Piliformis*, N. SP.). J Med Res. 1917;37(2):307–338.5.
132. Ueno Y, Shimizu R, Nozu R, Takahashi S, Yamamoto M, Sugiyama F, Takakura A, Itoh T, Yagami K. Elimination of *Pasteurella pneumotropica* from a contaminated mouse colony by oral administration of enrofloxacin. Exp Anim. 2002;51(4):401–5.
133. van der Linden M, Al-Lahham A, Nicklas W, Reinert RR. Molecular characterization of pneumococcal isolates from pets and laboratory animals. PLoS One. 2009;4(12):e8286.
134. Waggie KS. Clostridial species. In: Fox JG, Barthold SW, Davisson MT, et al., editors. The mouse in biomedical research. Vol. 2: diseases. Amsterdam: Elsevier Academic; 2007. p. 349–63.
135. Waggie KS, Hansen CT, Ganaway JR, Spencer TS. A study of mouse strains susceptibility to *Bacillus piliformis* (Tyzzer's disease): the association of B-cell function and resistance. Lab Anim Sci. 1981;31(2):139–42.
136. Weir EC, Bhatt PN, Barthold SW, Cameron GA, Simack PA. Elimination of mouse hepatitis-virus from a breeding colony by temporary cessation of breeding. Lab Anim Sci. 1987;37(4):455–8.
137. Whary MT, Cline JH, King AE, Hewes KM, Chojnacky D, Salvarrey A, Fox JG. Monitoring sentinel mice for *Helicobacter hepaticus*, *H rodentium*, and *H bilis* infection by use of polymerase chain reaction analysis and serologic testing. Comp Med. 2000;50(4):436–43.
138. Won YS, Jeong ES, Park HJ, Lee CH, Nam KH, Kim HC, Hyun BH, Lee SK, Choi YK. Microbiological contamination of laboratory mice and rats in Korea from 1999 to 2003. Exp Anim. 2006;55(1):11–6.
139. Wullenweber M, Kaspareit-Rittinghausen J, Farouq M. Streptobacillus moniliformis epizootic in barrier-maintained C57BL/6J mice and susceptibility to infection of different strains of mice. Lab Anim Sci. 1990;40(6):608–12.
140. Yamamoto H, Sato H, Yagami K, Arikawa J, Furuya M, Kurosawa T, Mannen K, Matsubayashi K, Nishimune Y, Shibahara T, Ueda T, Itoh T. Microbiological contamination in genetically modified animals and proposals for a microbiological test standard for national universities in Japan. Exp Anim. 2001;50(5):397–407.

Statistical Aspects of Health Monitoring in Rodents

8

Axel Kornerup Hansen

Abstract

To make a valid diagnosis for specific infections in animals, the sample size has to be calculated based upon the expected prevalence, i.e., the fraction of animals that it is reasonable to assume, that this specific agent will infect in the colony, and the sensitivity of the assay, i.e., how the fraction of the infected animals are likely to become positive in the assay. In smaller colonies, the size of the colony also has an impact on the sample size

Keywords

Rodents · Statistics · Quality assurance

8.1 Prevalence and Incidence and How It Leads to the Sample Size

If an infection enters a colony of animals, it may of course infect all the animals of that colony, but in most cases only some of the animals will be infected. The fraction or percentage of animals infected is known as *the prevalence (p)* (Fig. 8.1), while the fraction or percentage of new animals catching the infection within a certain period, is called *the incidence* (*i*). So, the prevalence is a measure of how widespread the infection is, while the incidence is a measure of, how rapidly the infection is spreading.

The prevalence is influenced by environmenal, host and agent factors. If the animals are maintained highly stocked and with current exchange of animals between the cages, the prevalence will be higher than if the animals are maintained with certain degrees of isolation from one another, e.g., in individually ventilated cages (IVC) changed under a strict laminar air flow regime. Some animals may have a genetic constitution, which makes them more or less sensitive to infection. For example, SJL mice seem to be resistant to infection with mouse hepatitis virus, which seems to be related to genetic differences in a receptor [4]. Most important is, however, how infective the agent is. Mouse hepatitis virus is easily transmitted through the faeco-oral route and contaminated equipment is highly contagious, and therefore mouse hepatitis virus will balance at high prevalences [5], while lymphocytic choriomeningitis primarily spreads vertically from mother to off-spring in a low rate, and, therefore, it will balance at much lower prevalences [7, 8].

The sampler, who samples animals for testing from a colony, therefore needs to ask the basic question: Is this specific infection present in animal of the colony, i.e., can I find one animal about

A. K. Hansen (✉)
Department of Veterinary and Animal Sciences, University of Copenhagen, Frederiksberg, Denmark
e-mail: akh@sund.ku.dk

F. Benavides, A. K. Hansen (eds.), *Rodent Quality Control: Genes and Bugs*, Laboratory Animal Science and Medicine 2, https://doi.org/10.1007/978-3-031-59103-7_8

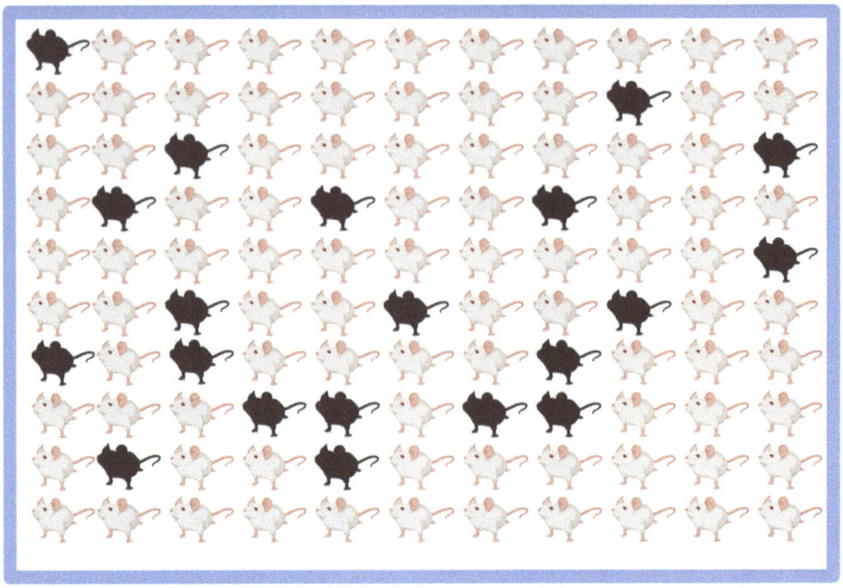

Fig. 8.1 The prevalence. An infection has entered a mouse colony with 110 mice. All the dark mice, i.e., 20 mice, have caught the infection, while the white mice are still uninfected. The prevalence p is, therefore, 20/110 = 0.18 or 18%. If one mouse and one mouse only is sampled the chance of sampling a non-infected mouse, i.e., a mouse which erroneously when examined will indicate that the colony is not infected, is $1 - p = 82\%$. (Created with Biorender.com)

which, I can say that this animal for sure is infected? To answer this question, the sampler needs to estimate, which p is expected for the specific infection to be tested for, in case it have infected animals of the colony. At every sampling of a mouse for testing from an infected colony the chance that the particular mouse is infected will be p, and the risk of sampling a non-infected mouse will be $1 - p$. If a sample of a certain size, S, is sampled then the confidence limit c, i.e., the risk that the sampler will not find the infection, is

$$c = (1 - p)^S \tag{8.1}$$

The sampler needs to define, which c is acceptable, e.g., if it is acceptable to run a risk of not finding the infection of 5% (0.05). Eq. 8.1 can be logarithmized

$$\log(c) = S \times \log(1 - p) \tag{8.2}$$

Then comes that S, i.e., the sample size needed to have a risk lower than c, has to be calculated as

$$S \geq \frac{\log(c)}{\log(1 - p)} \tag{8.3}$$

Using ≥ instead of = simply comes from that the number of animals to be sampled must be an integer. This equation has generally been known as the ILAR equation [6].

8.2 Specificity and Sensitivity and How It Affects the Sample Size

Very few assays would find all infected animals without also claiming some non-infected animals to be infected. Therefore more or less all assays will result in not only true positives and negatives, i.e., animals which in accordance with reality are diagnosed as infected, but also a fraction of false positives, i.e., animals being diagnosed as infected in spite of this not being the case, and false negatives, i.e., infected animals for which the assay fails in diagnosing the infection (Table 8.1). This can be used to define the *sensitivity* (SENS) and the *specificity* (SPEC) (Table 8.1).

If the sensitivity of an assay is

- 100%, this means that the assay applied will make all infected animals tested come out as positive in the assay
- 50%, this means that 50% of the infected animals will come out as positive in the assay
- 0%, this means that the assay will never be able to make an infected animal come out as positive in the assay.

If the specificity of an assay is

- 100%, this means that all positive animals in the assay are really infected with the agent searched for
- 50%, this means that 50% of the positive animals in the assay are really infected with the agent searched for, while the remaining 50% are positive for reasons not related to the specific agent
- 0%, it means than none of the positive results in the assay are caused by the specific agent.

The sensitivity and specificity depend on several factors. One of them is the *cut-off value* defined for the assay (Fig. 8.2). The cut-off value is the value of a sample tested in the assay above which, the sample is considered positive (Fig. 8.2). The sensitivity and the specificity relate to one another, as striving for a high sensitivity may lower the specificity (Fig. 8.2). Also, sensitivity and specificity relate to the biological phenomenon actually looked for in the assay. In a rat colony infected with *Clostridium piliforme*, after an incubation period more than 50% of the rats will elicit an antibody response [3], while only a very low number, i.e., a few out of thousand animals, will have pathological changes in the liver, and, therefore, as a screening method for this agent histopathology of the liver has a very low and serology has a very high sensitivity [2]. Some epitopes in viruses are induced by genes, which are conserved within a family of viruses, such as parvoviridae and coronaviridae, while other epitopes are products of genes very specific to well-defined species of viruses. As the consequence of revealing an infection in a colony of laboratory mice are most frequently the same, no matter the virus, assays for laboratory animal health monitoring are often set up to verify the presence of all sorts of, e.g., parvoviruses. However, this does not imply that the assay is considered unspecific, as far as the test question raised is: "Is the colony infected with parvoviruses." In this case, the specificity relates to being able to differ infection with parvoviridae from other families of viruses. In disease diagnostics,

Table 8.1 Definition of assay sensitivity and specificity

An assay attempts to describe reality. In reality, some animals are infected with an agent, while other animals are not. Of those animals, which are in reality infected, only some are also positive in the assay, and they are known as true positives. The remaining infected animals, which are negative in the assay, are false negatives. The *sensitivity* is defined as

$$Sensitivity\left(\text{SENS}\right) = \frac{\text{True positives}}{\text{Infected}}$$

Out of those animals, which come out as positive in the assay, only some are infected. The remaining animals, which come out as positives without being infected, are known as false positives. The specificity is defined as

$$Specificity\left(\text{SPEC}\right) = \frac{\text{True positives}}{\text{Positive}}$$

		Reality	
		Infected	Not infected
Assay	Positive	**True positives**	False positives
	Negative	False negatives	**True negatives**

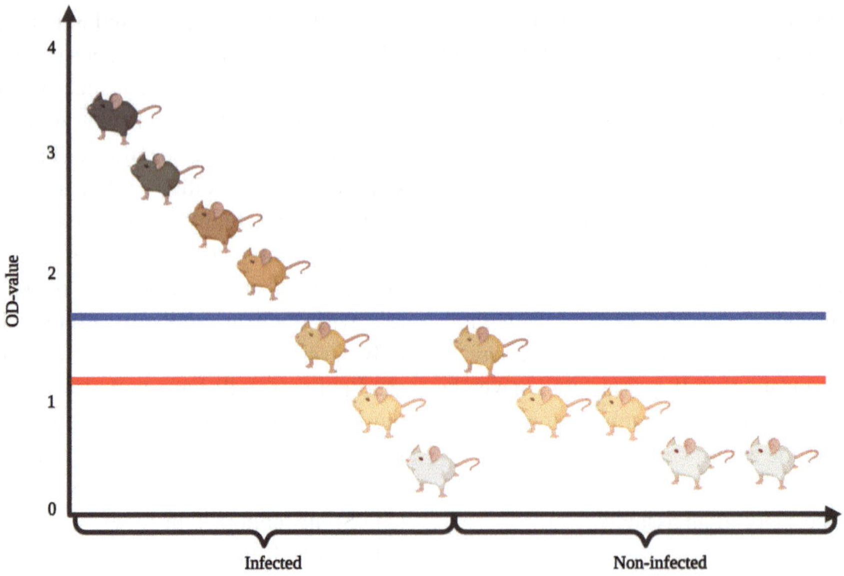

Fig. 8.2 Sensitivity and specificity. In ELISA for antibodies against an infectious agent the optical density (OD) is monitored, i.e., how colored does the microwell become on the ELISA plate. The scale is normally from 0 to 4, for which 0 is no color at all and 4 is color that high that all light is blocked on its way through the microwell. In this case serum from infected mice results in ODs from in entire range, while non-infected animals result in colors from just above 0 to just below 1. The laboratory must decide upon a *cut-off value*, i.e., for which they will consider the sample positive. If they set the cut-off value just above 1 (red line), the sensitivity of the assay will be 5/7 = 0.71 = 71%, because 5 out of 7 animals are defined as positive; probably because the two false negatives have not been infected long enough to produce antibodies. However, out of the six positive results, one is a false positive, i.e., the specificity is only 5 out of 6, i.e., 83.3%. If they instead set the cut-off just below 2 (blue line), the sensitivity will only be 4/7 = 0.57 = 57%, because one of the five infected mice is not defined as positive. If the prevalence is, e.g., 0.5 = 50% according to Eq. 8.3, a sample size of 7 for the red cut-off and a sample size of 9 for the blue cut-off will both result in a confidence limit of 5%, i.e., the risk of not finding the infection among the sampled animals is less than 5%. (Created with Biorender.com)

it can, in contrast, be important to be able to differ very precisely between species within a virus family. For instance, a COVID-19 antibody assay used in human medicine must be able to differ SARS-CoV2 antibodies from antibodies to common cold coronaviruses, while in rodent diagnostics it may be of less importance to differ the different strains of mouse hepatitis virus, if any of them are unacceptable.

It is difficult in health monitoring to work with assays, which are not very specific to the test question asked. At least, it must be expected that false positives will frequently occur and retest with other and more specific assays must be done. It is, however, not a statistical problem to work with assays with low sensitivity, as far as the task is to find at least one for sure infected animal. However, it needs to be included in the sample size calculation, which is quite simple. In Fig. 8.1 20 out of 110 mice are infected, i.e., the prevalence is 20/110 = 18%. However, if we use an assay with a sensitivity of 80%, the prevalence that the assay can detect is only 0.80 × 0.18 = 0.145. Therefore, Eq. 8.3 has to be corrected to

$$S \geq \frac{\log(c)}{\log(1-(\text{SENS} \times p))} \tag{8.4}$$

If we assume that the expected prevalence would be 18%, using this equation we will find that sampling 15, 20, or 82 mice will all give a risk of not finding the infection of less than 5% (c) for assays with sensitivities of 100%, 80%, or 20%, respectively. Therefore, the problem of low sensitivity is mostly a resource and ethics issue, i.e., it is not smart using 82 mice if one can do with 15 mice.

8.3 How the Colony Size Affects the Sample Size

The prevalence shown in Fig. 8.1 is 0.18. Let us assume that we use an assay with a sensitivity of 100% (which is not very likely). The calculated sample size will be 14.9, i.e., we need to sample 15 mice. However, if the first mouse we sample is negative, the prevalence changes into being 20 out of 109 animals, and with this prevalence the calculated sample size will be 14.8, i.e., our sampling has influenced the prevalence. After seven samples, the prevalence has become that low, that we only need to sample 14 mice. If instead the number of infected mice had been 200 out of 1100, then we could still calculate the sample size as 14.9, i.e., 15 mice, and although the prevalence drops to 200 out of 1099 this is still 0.18, and the sample size is not influenced. If we want to incorporate the total number of animals to be sampled from (*T*), we use this equation.

$$S \geq \left(1 - C^{1/(D\times\mathrm{SENS})}\right) \times \left(T - \left(\left(\left(\mathrm{SENS}\times D\right) - 1\right)/2\right)\right) \quad (8.5)$$

in which *D* is the absolute number of animals expected to be infected [1].

As can be seen from Fig. 8.3, Eq. 8.4 should be used for $T \geq 1000$ animals, while Eq. 8.5 should be used for $T < 1000$ animals.

8.4 How the Cut-off Value Is Set

In any laboratory assay there will always be some sort of low reaction from samples from uninfected animals. This is often referred to as background noise. In a serological assay, it will take a while before the infected animals have produced a sufficient amount of antibodies to respond in the assay. Therefore, the sensitivity will seldom be 100%. In an ELISA the optical density (OD), which is an expression of how stained the microwell is, must be above a certain value, i.e., the cut-off value, before we count the sample as positive. In a quantitative PCR the number of cycles needed to get a fluorescence signal at a certain levels must be below a certain number known as the cycle threshold (*Ct*) to count the sample as positive. The cut-off value or the cycle threshold must be set so that some sam-

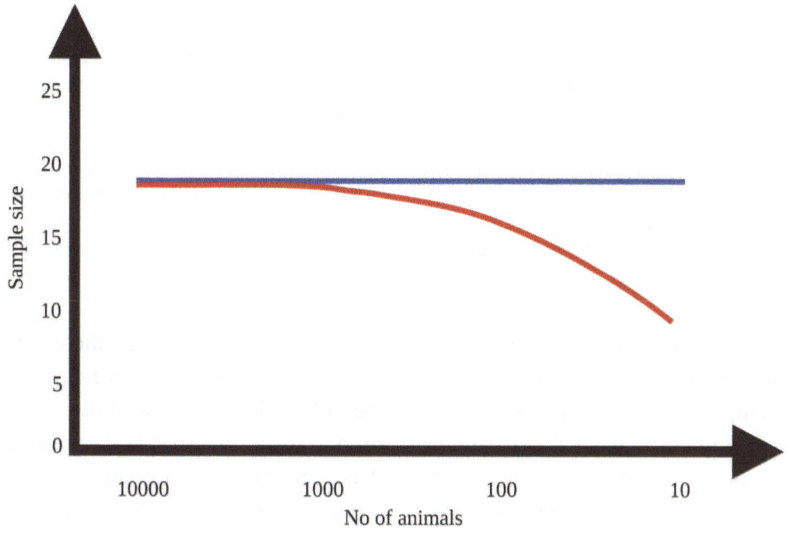

Fig. 8.3 Sample size calculation. The sample size can be calculated without (blue line) or with (red line) incorporation of the total number of animals to be sampled from. On the graph, the prevalence (*p*) is 0.18 and the sensitivity (SENS) is 0.80. If the total number of animals exceeds 1000 animals, the total number of animals does not influence the sample size (red line follows blue line). However, if the total number of animals is only 100, the sample size can be reduced to 19 animals (red line) instead of 20 animals (blue line)

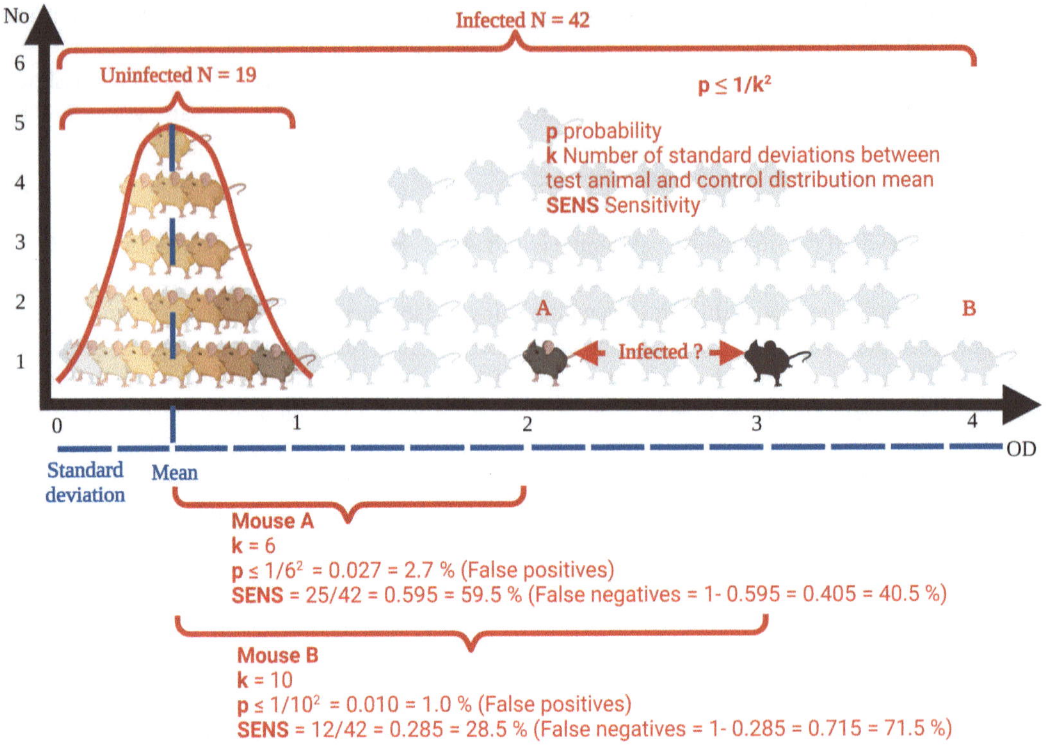

Fig. 8.4 The cut-off value. The outcome of an ELISA is the optical density (OD), which is an expression on how colored the microwell is, which corresponds to the amount of bound antibody. To set the cut-off value for an ELISA the laboratory will test a number (in this case 19) of mice, which are truly not infected. The uninfected animals will not all have an OD-value (see Fig. 8.2 legend) of 0, but will give a low response, which is likely to be normally distributed, because they have no antibodies, and all responses are due to unspecific binding in the wells. The mean and the standard deviation are calculated. The infected animals (gray mice) to be found with the assay give responses within the entire range, because they are in different stages of antibody production. Some (in this case 7 mice) respond within the range of the uninfected animals, and, therefore, the sensitivity cannot be 100%. According to Chebushev's theorem, the probability (*p*) that a response belongs to a described distribution, in this case the responses of test mice A and B compared to the distribution of uninfected mice, is $p \leq 1/k^2$, in which k is the number of standard deviations, which can be placed between the value of the test mouse and the mean of the distribution. A cut-off value just below test mouse A, will result in 2.7% false positives and a sensitivity of 59.5%, while a cut-off value just below test mouse B, will result in 1.0% false positives and a sensitivity of 28.5%. According to Eq. 8.3, 9 animals and 20 animals, respectively, must be sampled to obtain a confidence limit of 5% if the prevalence is 50%

ples are clearly positive and some samples are clearly negative, but there will in most assays be intermediate samples, which are not clearly positive or negative (Figs. 8.2 and 8.4). As outlined above in laboratory animal health monitoring the question raised is: Is this specific infection present in animal of the colony, i.e., can I find one animal about which, I can say that this animal for sure is infected. Therefore, in, e.g., an ELISA the cut-off value should be set at a relatively high level to avoid false positives. The higher it is set, the more false negatives there will be, i.e., the lower the sensitivity. However, as this can easily be solved by sampling more animals, the most important is to avoid false positives, which can have enormous economic and ethical consequences, which could have been avoided. Therefore, it should be noted that an assay validated to reveal at least one positive sample from a stack of samples from an infected colony, i.e., make a diagnosis on the colony level, is not necessarily applicable for diagnosing individual animals. This is in contrast to, e.g., a human medical laboratory, which often will use an automatized assay with a high sensitivity known to reveal very few false negatives, but unfortunately a high fraction of false positives, which hereafter need to be confirmed by a more laborious assay.

Cut-off values can be defined from the distribution of the positive sample or the distribution of the negative samples. Defining it from the positive samples is normally not a wise strategy, as the infected animals will cover the entire range from no antibodies in serology or no DNA in a PCR in the early phases of infection to very high loads. Therefore, the most applicable is to define the cut-off value from the distribution of the negative samples. As most of the variation in the samples of the uninfected animals come from the assay, e.g., unspecific binding in ELISA, these values will often be normally distributed, and the probability that a test animal belongs to the distribution of the negative samples can be calculated by Chebyshev's theorem [9] (Fig. 8.4).

8.5 Concluding Remarks

To be able to make statistically valid health monitoring in laboratory animal colonies, it is important to estimate the sample size for each agent from considerations on the expected prevalence in the colony, the sensitivity of the assay, and the colony size. In the dialogue with a testing laboratory, it is important to ask questions on how they validate their assays and calculate their cut-off values, and which specificity and sensitivity each of their assays have.

References

1. Cannon RM, Roe RT. Livestock disease surveys. A field manual for veterinarians. Canberra: Bureau of Resource Science, Department of Primary Industry. Australian Government Publishing Service; 1982.
2. Hansen A, Dagnaes-Hansen F, Møllegaard-Hansen K. Correlation between megaloileitis and antibodies to Bacillus piliformis in laboratory rat colonies. Lab Anim Sci. 1992;42(5):449–53.
3. Hansen AK, Andersen HV, Svendsen O. Studies on the diagnosis of Tyzzer's disease in laboratory rat colonies with antibodies against Bacillus piliformis (Clostridium piliforme). Lab Anim Sci. 1994;44(5):424–9.
4. Hirai A, Ohtsuka N, Ikeda T, Taniguchi R, Blau D, Nakagaki K, Miura HS, Ami Y, Yamada YK, Itohara S, Holmes KV, Taguchi F. Role of mouse hepatitis virus (MHV) receptor murine CEACAM1 in the resistance of mice to MHV infection: studies of mice with chimeric mCEACAM1a and mCEACAM1b. J Virol. 2010;84(13):6654–66.
5. Homberger FR. Enterotropic mouse hepatitis virus. Lab Anim. 1997;31(2):97–115.
6. ILAR. Committee on long-term holding of laboratoryrodents. Long-term holding of laboratory rodents. ILAR News. 1976;19:L1–L25.
7. Knust B, Stroher U, Edison L, Albarino CG, Lovejoy J, Armeanu E, House J, Cory D, Horton C, Fowler KL, Austin J, Poe J, Humbaugh KE, Guerrero L, Campbell S, Gibbons A, Reed Z, Cannon D, Manning C, Petersen B, Metcalf D, Marsh B, Nichol ST, Rollin PE. Lymphocytic Choriomeningitis virus in employees and mice at multipremises feeder-rodent operation, United States, 2012. Emerg Infect Dis. 2014;20(2):240–7.
8. Parker JC, Whiteman MD, Richter CB. Susceptibility of inbred and outbred mouse strains to Sendai virus and prevalence of infection in laboratory rodents. Infect Immun. 1978;19(1):123–30.
9. Tchebichef P. Des valeurs moyennes. Journal de Mathématiques Pures et Appliquées. 1867;2(12):177–84.

9 Viruses and Viral Diagnostics

Axel Kornerup Hansen,
Torben Sølbeck Rasmussen, and Lars Andresen

Abstract

Wild rodents carry viruses, but during the last 60 years the specific pathogen free technology has been applied to make laboratory rodents virus free, because they may interfere with research projects or cause zoonoses. Laboratory rodent colonies are, therefore, currently screened for virus infections, which can be done by serology or polymerase chain reaction (PCR). For more detailed characterization of the virome, i.e. all the viruses carried by an animal host, cultivation and sequencing can be applied, which also shows that commercial colonies in spite of the specific pathogen free technology, are not free of all viruses. Recently it has also been argued that the lack of viruses in laboratory rodents make these less translational to humans.

Keywords

Rodents · Viruses · Specific Pathogen-Free Organisms · Diagnosis · Host Microbial Interactions

9.1 The Virus

A virus is an infectious agent only to be seen on the electron microscopic level. It is not in itself a living organism in the traditional sense, as it can only replicate inside a living cell, which can have its origin in everything from microbes over plants to animals. Inside the infected host cell, the virus uses the cell organelles for a rapid production of a stack of copies, which burst out of the cell to infect new cells. Outside the cell the virus may be more or less viable and it cannot replicate. Some viruses, such as parvoviruses, are extremely stable in the external environment, while other viruses such as paramyxoviruses only survive in droplets. The virus consists of its genetic material, which may be a piece of either DNA or RNA, which may be either single-stranded or double-stranded. This is surrounded by a protein coat, the capsid, and for some viruses also an envelope of lipids. This makes up a structure, which can be simply helical or icosahedral forms, but which can also appear as rather complex. Systematically viruses are divided into realms based upon being DNA or RNA viruses and into

A. K. Hansen (✉) · L. Andresen
Department of Veterinary and Animal Sciences, University of Copenhagen, Frederiksberg, Denmark
e-mail: akh@sund.ku.dk; laran@sund.ku.dk

T. S. Rasmussen
Department of Food Sciences, University of Copenhagen, Frederiksberg, Denmark
e-mail: torben@food.ku.dk

F. Benavides, A. K. Hansen (eds.), *Rodent Quality Control: Genes and Bugs*, Laboratory Animal Science and Medicine 2, https://doi.org/10.1007/978-3-031-59103-7_9

kingdoms based upon, whether the nucleotide string is single-stranded or double-stranded and the nature of the capsid proteins (Figs. 9.1 and 9.2). Viruses are members of all ecosystems that can affect dynamics and function of microbial communities. Most of these viruses are represented by bacteriophages (phages) that infect bacteria in a host specific manner [9, 66, 109], while the remaining are eukaryotic and archaeal viruses [129]. There is emerging evidence that the viral community in the mammal environment is a key player in maintaining also the mammal bacterial community [118, 155, 157, 188].

9.2 The First Recognition of Rodent Viruses

Laboratory rodents back in the early twentieth century were wild caught, and some of them had zoonotic infections, which might infect and even kill research staff. Viruses exhibit a high degree of species specificity, and, therefore, such zoonotic infections were most frequently bacterial [152], but a few viruses do have a zoonotic potential. Although the common cold viruses of the type betacoronavirus 1 in humans probably have their origin in rodents, e.g., as the betacoronavi-

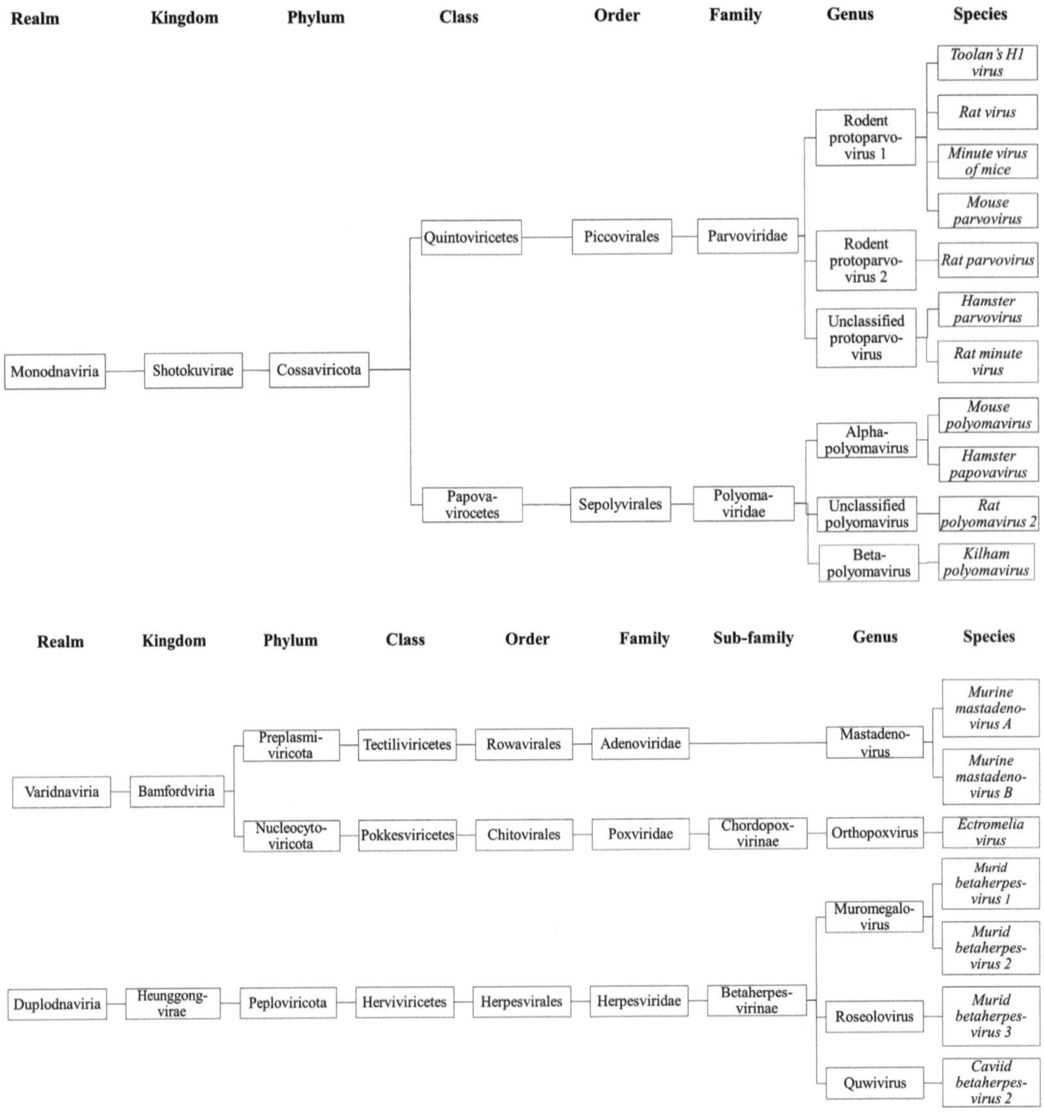

Fig. 9.1 Classification of the most common single-stranded (A. Monodnaviria) and double-stranded (B.Varidnaviria and Duplodnaviria) DNA viruses in colonies of laboratory rodents

rus HKU24, which comes from rats [87], these jumps occurred centuries ago, and today both virus and the human immune system have adapted to each other, and, therefore, rodent coronaviruses are not a very likely pandemic zoonosis. In the 1980s there were a few cases of disease in animal research staff due to human infection with Hantaan orthohantavirus from rats or rat cell cultures [44, 99]. Also lymphocytic choriomeningitis virus (LCMV) may transmit from mice and especially from hamsters [21, 42], and as late as in 2012 the US Center for Disease Control at commercial breeders isolated LCMV from eight mice, and found that 32% of the employees had antibodies to LCMV, and four employees had meningitis [81]. The first motivation to isolate infectious agents from laboratory rodents was mostly curiosity on their pathogenesis and pathology in order to elucidate their usefulness as models for humans. The first attention was given to bacteria [160], which were easier to see in the microscope and Robert Koch had already long before the millennium described the first culture media for bacterial isolation [82]. Even the Spanish flue in 1918, which was one of the major human pandemics and as all subsequent influenzas caused by influenza virus, was considered bacterial until influenza virus was described in 1933 [161]. Ectromelia virus, the first rodent virus to be described, had been described 3 years earlier [102]. In 1937 Max Theiler described the "mouse poliovirus," which became known as Theiler's murine encephalomyelitis virus (TMEV) [169], while working with mice as models for human polio. Also, mouse hepatitis virus (MHV), probably a distant cousin of some of the betacoronaviruses in humans, was first isolated in 1949 in connection with severe encephalomyelitis symptoms in mice [32].

9.3 The Specific Pathogen-Free Technology

The young veterinarian Henry Foster, who in 1947 named his breeding center in Boston (USA) after the Charles River, in the late 50 s reported the first use of caesarean section and barrier protection for large scale production of specific pathogen-free (SPF) rodents [54], which is an efficient way to eliminate virus infections. This created a new era, in which viruses as a rule of thumb were considered unwanted inhabitants of laboratory animals, and a number of commercial breeders popped up supplying animals free of many of those research disturbing infections, and it eradicated the small family breeders and their rodent colonies with all their virus infections (see Chap. 6). Currently, new viruses have been isolated, described, and eradicated (Tables 9.1, 9.2, and 9.3; Figs. 9.1 and 9.2).

Table 9.1 Some well-described DNA viruses in mice (M), rats (R), hamsters (H), and guinea pigs (GP)

Name		Species	Transmission	Disease, research impact, and pathology	References
Parvoviridae					
Rat virus (RV) (Kilham rat virus)		R	Fecal-oral contact vertical transmission for some types in female breeders without protective immunity	Normally none. It has been claimed that some RV-serotypes may cause jaundice and ataxia in rats less than 10 days of age, but this is not properly documented. Fetal mummifications at vertical transmissions	[39, 80, 7, 175, 16]
Toolan's H1 virus		R			
Rat parvovirus		R			
Rat minute virus		R			
Minute virus of mice		M			
Mouse parvovirus		M			
Hamster parvovirus		H			
Adenoviridae					
Murine mastadenovirus	A. Murine adenovirus type 1	M	Infects by oral or ocular transport, but close contact is required	In mice mostly clinically unnoticable, although myocarditis, nephritis, adenitis, encephalitis, and mortality have been observed after experimental inoculation of neonatal mice. In guinea pigs, necrotizing broncheoalveolitis is regularly observed	[63, 65, 85, 166]
	B. Murine adenovirus type 2	M			
Guinea pig adenovirus		GP			
Poxviridae					
Ectromelia virus		M	Ectromelia virus infects through skin lesions	Ectromelia or mouse pox; after app. 10 days edematous skin erosions and hyperplasias characterized by large eosinophilic cytoplasmatic inclusions in the epithelial cells; extensive necrosis in lymphatic organs and liver; most severe in DBA, C3H and BALB/c mice; C57BL/6 rare elatively resistant and may harbor latent infections; prevalence of overt disease may vary from few to all animals	[178, 19, 74, 45, 94]
Herpesviridae					
Murid betaherpesvirus 1	Mouse cytomegalovirus	M	Direct saliva or urine contact, transmission in utero cannot be excluded	Known as *salivary gland viruses*; normally asymptomatic; may induce megalic cells and nuclear inclusions in the glandular epithelium of the salivary glands	[31, 117, 24]
Murid betaherpesvirus 2	Rat cytomegalovirus	R			
Murid betaherpesvirus 3	Thymic virus	M		May be found in both the thymus and the salivary glands; asymptomatic	
Caviid betaherpesvirus 3	Guinea pig cytomegalovirus	GP		Abortions and variable mortality may be observed in breeding females	
Polyomaviridae					
Murine pneumotropic virus (K virus)		M	Urine-nasal route	Fatal interstitial pneumonia with congested and edematous lungs in mice <1 week	[53, 67, 57, 58]
Mouse/murine polyoma virus		M		Some strains are oncogenic in C3H mice, particularly in the parotid gland, but also in other tissues of newborn immune-deficient mice. Genetic susceptibility varies	[59, 135, 167]
Rat polyoma virus 2		R		Emaciation, labored breathing and mortality, particularly among breeding females and pups of immune-deficient strains	
Hamster papovavirus		H		Leukemia and lymphoma	[149]

Table 9.2 Some well-described RNA viruses in mice (M), rats (R), hamsters (H), and guinea pigs (GP)

Name		Species	Transmission	Disease, research impact and pathology	References
Arenaviridae Lymphocytic choriomeningitis virus		M GP H	Persistent tolerant infection in utero or within a few days after birth leads to life-long viremia and shedding of virus	Modest growth retardation; at 7–10 months of age immune complex glomerulonephritis; emaciation, ruffled fur, hunched posture, ascites and sometimes death; in guinea pigs, more severe symptoms may be observed with splenic necrosis and lymphocytic infiltrates in meninges, choroid and around vessels in the liver and kidney after infection with specific strains	[20, 21, 42, 123, 49, 112, 142, 11]
Arteriviridae Gammaarterivirus lacdeh (lactate dehydrogenase elevating virus)		M	Infection from transplanted cells; transplacental infection may occur	Elevates serum lactate dehydrogenase, as well as a range of other serum enzymes in mice; may cause poliomyelitis with clinical signs of paresis, especially due to the c strain, also called the Murphy strain, and especially in AKR and C58 mice	[40, 103, 116, 111, 150]
Bunyaviridae Hantaan orthohantavirus		R	Transmitted orally through contact with feces, urine, aerosols, and direct contact	No clinical signs	[89, 44, 88, 110]
Caliciviridae Murine norovirus		M	Fecal-oral route; persistent disease and shedding is possible in immunodeficient animals	Acute onsets with mild enteritis; rapid recovery in immunocompetent animals	[78, 69, 177, 126]
Coronaviridae	Mouse hepatitis virus	M	Easily spreads in rodent colonies; prevalences close to of 100% within 4 weeks; some strains infect through the gastrointestinal system, while others infect through the respiratory system. MHV-4 has a specific affinity for the nervous tissues; may be transported passively between facilities by staff or equipment	Tropism for all tissues; clinical symptoms are rare; some strains are more virulent; BALB/c and C57BL/6 mice seem to be more sensitive; immune-deficient mice has high mortality; most common symptom is diarrhea, eventually lethal, in suckling mice; the liver may be pale with multiple white, yellow, or hemorrhagic foci; the spleen may be enlarged, while the thymus may be reduced in size; widespread necrosis of lymphoid tissue	[153, 25, 165, 10, 125, 68]
Murine coronavirus	Rat coronavirus/ Sialodacryadenitis virus	R		Often asymptomatic; mild necrotizing rhinotracheitis; interstitial pneumonia with severe necrosis and swelling of salivary and lacrimal glands; red-brown porphyrin rings around their eyes; full recovery within 5 weeks	[75]

Table 9.3 Some well-described RNA viruses in mice (M), rats (R), hamsters (H), and guinea pigs (GP)

Name	Species	Transmission	Disease, research impact and pathology	References
Paramyxoviridae Murine respirovirus (Sendai virus)	M R	Unstable; non-persisting infection in immune-competent rodents; persisting in immune-deficient rodents; high prevalences; transmitted mainly by the respiratory route; does not spread efficiently to sentinels from dirty bedding; Guinea pig parainfluenza virus type 3 probably a lineage of human parainfluenza virus type 3 introduced into colonies via infected humans	Pathological signs are catarrhal bronchitis, eventually extending into the alveoli of the lung, but infections are mostly silent. High mortality may be observed in young mice before or around weaning, and DBA/2 and 129 mice may be more sensitive than other strains	[72, 124, 29, 30 6]
Murine pneumonia virus	M R H		Pathological symptoms absent in immune-competent mice; eventually minor pathological changes in rats and nude mice	[29]
Guinea pig parainfluenza type 3	GP		No clinical disease	[120, 159]
Picornaviridae Theiler's murine encephalomyelitis virus GDVII group (GDVII and FA)	M	Transmitted orally through contact with feces and urine, in infected colonies, prevalences are generally high, and the virus may be transmitted intrautero	In most cases asymptomatic; certain strains may invade the central nervous system and produce polio-like neurodegenerative disease; GDVII may cause acute encephalitis, and paralysis develops when the virus from the gray matter infects the white matter, thereby damaging the upper motor neuron system CD-1; DBA/2, SJL and SWR mice seem to be most susceptible than other strains;	[95, 96, 1, 183]
Cardiovirus B Theiler's murine encephalomyelitis virus Theiler's original (TO4, TO-Yale, BeAn, WW, and DA)	M		Less virulent; biphasic; early phase acute encephalomyelitis; long-term phase demyelinating disease	
Rat theilovirus	R		Probably asymptomatic; Sprague-Dawley may be more prone for infection than CD	[48, 50]
Guinea pig cardiovirus	GP		Paralytic and mortal disease (Guinea pig lameness) mostly observed in pets	[62, 61]
Reoviridae Reovirus type 3	M R H GP	Transmitted via fecal-oral and aerosol routes; highly contagious; high prevalences	No clinical disease; antibodies have been detected in mice, rats and Guinea pigs, but the virus has only been isolated from mice	
Rotavirus group A: Mouse rotavirus	M		Epizootic diarrhea of infant mice (EDIM)	[4]
Rotavirus group A: Rat rotavirus	R		Infectious diarrhea of infant rats (IDIR)	[176, 70]
Retroviridae Mouse mammary tumor virus	M	Both exogenous and endogenous viruses; mouse mammary tumor virus infects through mothers milk	Mammary tumors	[22]
Gammaretrovirus (leukemia viruses)	M R H GP		Leukemia	[134]

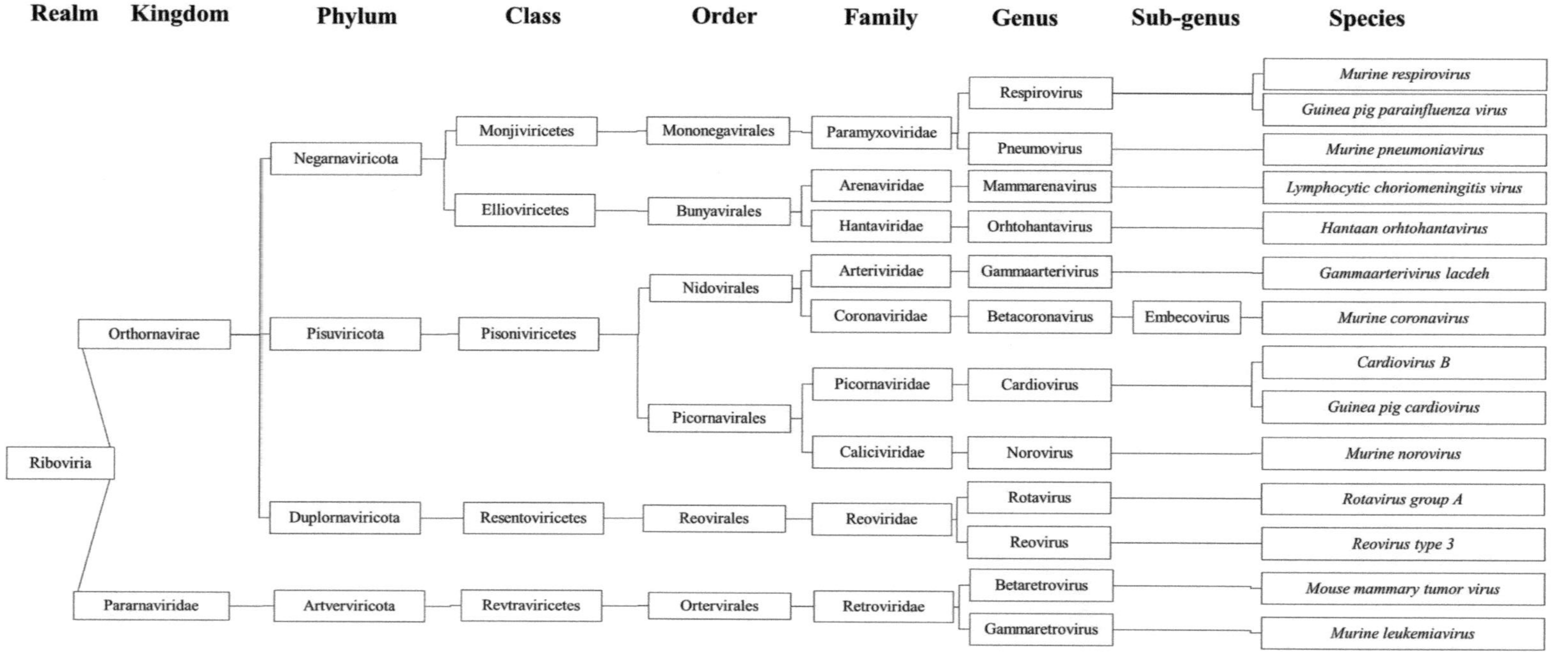

Fig. 9.2 Classification of the most common RNA viruses (Ribiviria) in colonies of laboratory rodents

9.4 Virus Screening

For many years the term "virus antibody-free animal" was used about animals documented to be free of antibodies to viral pathogens by a battery of serologic tests [37]. From the first [84] to the current [100] issue of the guidelines for rodent health monitoring from the Federation of European Laboratory Animal Associations (FELASA) there has been a list of viruses to be tested; some every 3 months and some annually. There is no doubt that the viruses recommended by FELASA have been currently diagnosed in rodent research facilities over the last 20 years [93, 101, 108, 128, 151] (Table 9.4), and most of the mouse viruses have also been found in pet shops [41, 51, 64, 137] and wild mice [13] (Table 9.4). In contrast, the few surveys available

Table 9.4 Viruses recommended by FELASA for routine screening in breeding colonies of rats and mice[100] and the prevalences with which they have been diagnosed globally in recent surveys from laboratory animal colonies and pet shops

Screening recommended by FELASA	Mice	Rats	Global observations in laboratory rats (R) and mice (M)[a]	Global observations in pet shop/wild rats (R) and mice (M)[b]	Zoonotic
Hantaviruses		Annually	R	R	+
Lymphocytic choriomeningitis virus	Annually		M	M	+
Adenoviruses general screening			MR	M	
Mouse adenovirus type 1	Annually			M	–
Mouse adenovirus type 2	Annually			M	–
Mouse hepatitis virus	Every 3 months		M	M	–
Mouse rotavirus	Every 3 months		M		–
Murine norovirus	Every 3 months		M	M	–
Murine pneumonia virus	Annually	Every 3 months	MR	M	–
Ectromelia virus	Annually		M	M	–
Parvoviruses general screening			MR	M	–
Minute virus of mice	Every 3 months		M	M	–
Mouse parvovirus	Every 3 months		M	M	–
Rat virus		Every 3 months	R		–
Rat minute virus		Every 3 months	R		–
Rat parvovirus		Every 3 months	R		–
Toolan's H-1 virus		Every 3 months	R		–
Rat coronavirus/ Sialodacryoadenitis virus (RCV/SDAV)		Every 3 months	R		–
Rat theilovirus		Every 3 months	R		–
Reovirus type 3	Annually	Annually	MR	M	–
Murine respirovirus (Sendai virus)	Annually	Annually	MR	M	–
Theiler's murine encephalomyelitis virus	Every 3 months		M	M	–

[a]Refs. [151, 93, 128, 108, 101]
[b]Refs. [41, 137, 64, 17, 51]

on wild rats seem to indicate that the FELASA recommended viruses apart from Hantaan orthohantavirus [36] are more common in laboratory than in wild rats [28, 141].

9.5 Virus Impact on Research

Although an issue for immune-deficient and transgenic animals [18] the argument against viruses is not only related to disease. Biological samples from the animals may if contaminated spread virus to other laboratories [116]. For example, major economic impact on two US laboratories was reported after import of samples infected with ectromelia virus [45, 94]. Viruses such as MHV exhibit a range of immune-modulating traits [83], and viruses may modulate the immune system as part of their strategy for invading the host [5]. Some physiological parameters may be changed in virus infected animals, e.g., when gammaarterivirus lacdeh, also known as lactate dehydrogenase (LDV) elevating virus, inhibits the clearance of LDV and a number of other enzymes in infected mice [119]. Some viruses may cross the placental barrier, although it is only common for lymphocytic choriomeningitis virus [123] and parvoviruses [80]. Certain parvoviruses may decrease the incidence of specific tumors in rodents [171, 172].

Most DNA viruses cause persistent infection. In general, they do not produce overt disease with the important exception of poxviruses, but they may influence research in other ways. RNA viruses are normally non-persisting in immune-competent animals, again with some important exceptions. Retroviruses are able to incorporate themselves into the genome of the host, and if this includes the germ cells, they will be endogenous, i.e., they are transmitted vertically with the genome. Lymphocytic choriomeningitis virus only generates insufficient immune responses, and it will, therefore, cause life-long infections. In immune-deficient animals most viruses should be considered as being able to cause persisting infections. Most viruses are rather species-specific, but infection in different species may have been caused by substrains, which as a historical event has jumped from one species to another, such as it is well known for coronaviruses, both in humans and rodents. Major species un-specific exceptions are the zoonotic viruses lymphocytic choriomeningitis virus and Hantaan orthohantavirus.

The number of viruses with a capability to infect rodents seems endless, and historically all sorts of viruses have been isolated from rodents. Due to the high mutation rates that some viruses exhibit mostly any new isolate is to be regarded as a new strain. However, within a certain species of virus there will be a high level of serological cross-reaction between these different strains; especially if conserved epitopes are used for diagnosis. So, therefore, the potentially pathogenic viruses likely to infect laboratory rodent colonies can be down-scaled to a limited number (Tables 9.1, 9.2, and 9.3).

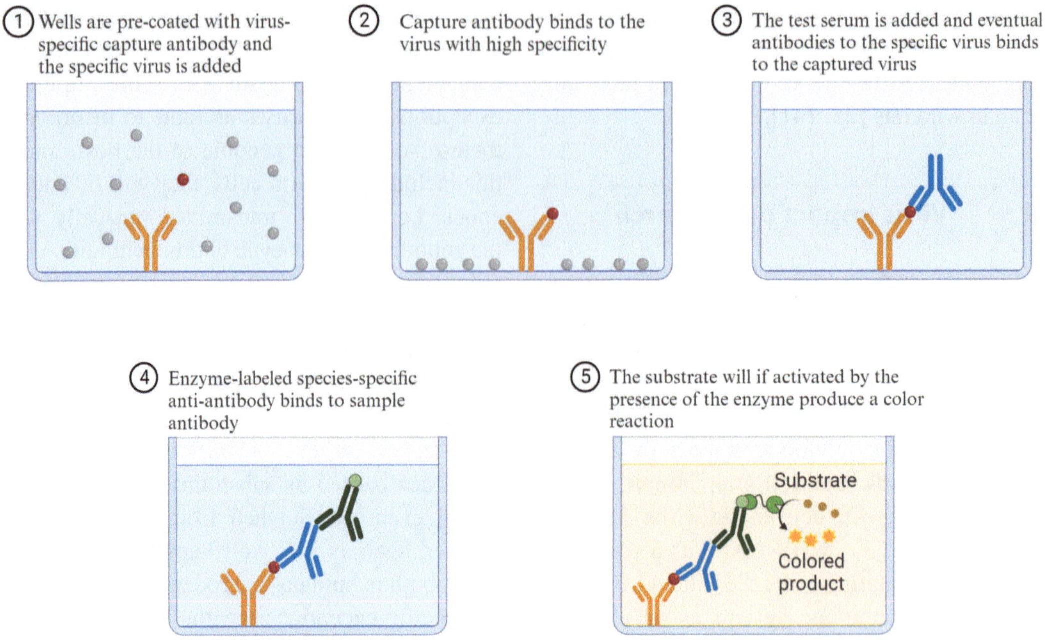

Fig. 9.3 The principles of sandwich ELISA used for verifying the presence of antibodies against a specific virus in rodent health monitoring. (Created with BioRender.com)

9.6 The Virus Infected Laboratory Rodent in Research

During the last decade the concept of virus-free laboratory animals has been challenged. A minor issue is that sequencing has shown that if properly examined viruses not listed by FELASA, such as herpesviruses, can be found in barrier protected commercial laboratory mice, although some colonies are closer to being virus free than others [131]. A major issue is the lack of immune stimulation in animals unexposed to common virus infections, which is a consequence of the barrier breeding and SPF housing technology [17, 168]. It has been argued that this may be a main reason why rodents studies fail to reproduce and translate into humans [104]. In humans, childhood disease viruses and common colds activate a lasting memory T cell response [179]. This is totally lacking in SPF rodents, which are more comparable to newborn babies in their abundancies of activated memory T cells [17]. Laboratory mice exposed to virus infections by co-housing with pet shop mice [17] or born after embryo transfer to wild mice [140] in contrast to SPF mice display vaccine-induced transcriptional signatures similar to humans, and the same declining antibody response [52]. They also survive far better after influenza infection, which is more comparable to humans [139]. Also, mice pre-immunized with inactivated murine viruses generates a full memory T cell response, and later exhibits a higher therapeutic window in a skin inflammation model [51]. It has, therefore, been proposed that, e.g., vaccines are tested both in SPF rodents as well as in rodents harboring a selection of their natural pathogens [104].

9.7 Diagnostic Tools

9.7.1 Serology

Shortly after an animal is infected with a virus or other pathogen, antiviral antibodies will start to be produced. First as IgM antibodies and later in the immune response as IgG antibodies. Measurement of these antiviral antibodies was one of the first diagnostic methods for specific diagnosis of a viral infection and remains to be an important tool in modern diagnostics.

9.7.1.1 Binding Assay

Assays that directly measure the binding of serum-derived antibodies to viral antigens are most widely used and appear in many different forms. The core principle is simple: a viral antigen is coated on the bottom of a plastic well, and serum from the test subject is applied to the surface. If present in the sample, antiviral antibodies will bind to the antigen. After a washing step to remove unbound antibodies, the well is incubated with a detection antibody specific for antibodies from the test subject the amount of which is determined after a second washing step. In 1960 the Radioimmunoassay (RIA) was the first binding assay to be published [180]. Here, the detection antibody (also referred to as the secondary antibody) is linked to a radioactive label. However, the many problems with the use of radioactive reagents 10 years later led to the development of an enzyme immunoassay (EIA)/enzyme linked immunosorbent assay (ELISA) in which the radioactive label is replaced by an enzyme [90]. The double antibody sandwich technique is a very highly used EIA/ELISA assay for the detection of viral antibodies (Fig. 9.3). Initially, the wells are precoated with antibodies against a viral antigen and then incubated with the inactivated virus, and then the test sample with the eventual virus-specific-antibodies. The well is now incubated with an anti-antibody specific for the species linked to an enzyme, which will make a substrate change color. Adding more layers in the "sandwich" provides more sensitivity to the assay, but at the cost of specificity.

9.7.1.2 Functional Assay

Another type of serum-based diagnostics are the functional assays, which are based on the detection of viral activity in the sample. One example is the neutralization assay in which serial dilutions of serum from the test subject is measured for ability to block viral infectivity in an in vitro cell line culture. If the test subject has developed antiviral antibodies, the number of infected cells will be decreased at a certain dilution titer compared to a control sample without serum. Another example is the hemagglutination inhibition assay that has been in use since the 1940s [162]. The assay is used to titrate the antibody response to a viral infection, such as rodent parvovirus infections. The virus can bind to the surface of red blood cells and cause the cells to agglutinate and precipitate. Antibodies against the virus will bind to the virus and therefore the virus cannot bind to the red blood cell and agglutination is blocked. This can easily be seen by the naked eye making this assay cheap and rapid.

9.7.2 Nucleic Acid Amplification Test (NAAT)

9.7.2.1 Conventional PCR

The Polymerase Chain Reaction (PCR) has since its development in the early 1980s [113, 114, 146] revolutionized almost every aspect of biological sciences and diagnostics. The reaction is easy to set up, cheap and is completed within a few hours (Fig. 9.4a). A specific stretch of DNA sequence can be amplified into millions of copies, and thereby easily analyzed, accounting for the high analytic sensitivity of this method. The key ingredients are a pair of single-stranded DNA oligoes (Primers), a purified DNA sample (Template), an enzyme blend containing a thermostable DNA polymerase [35, 77, 147], and a reaction buffer providing nucleotides, ions, and pH stability. The PCR reaction consists of repetitive cycles of denaturation of the template followed by annealing/extension of the primers. These steps take place at different temperatures controlled by a thermal cycler instrument. At high temperature, the double-stranded DNA tem-

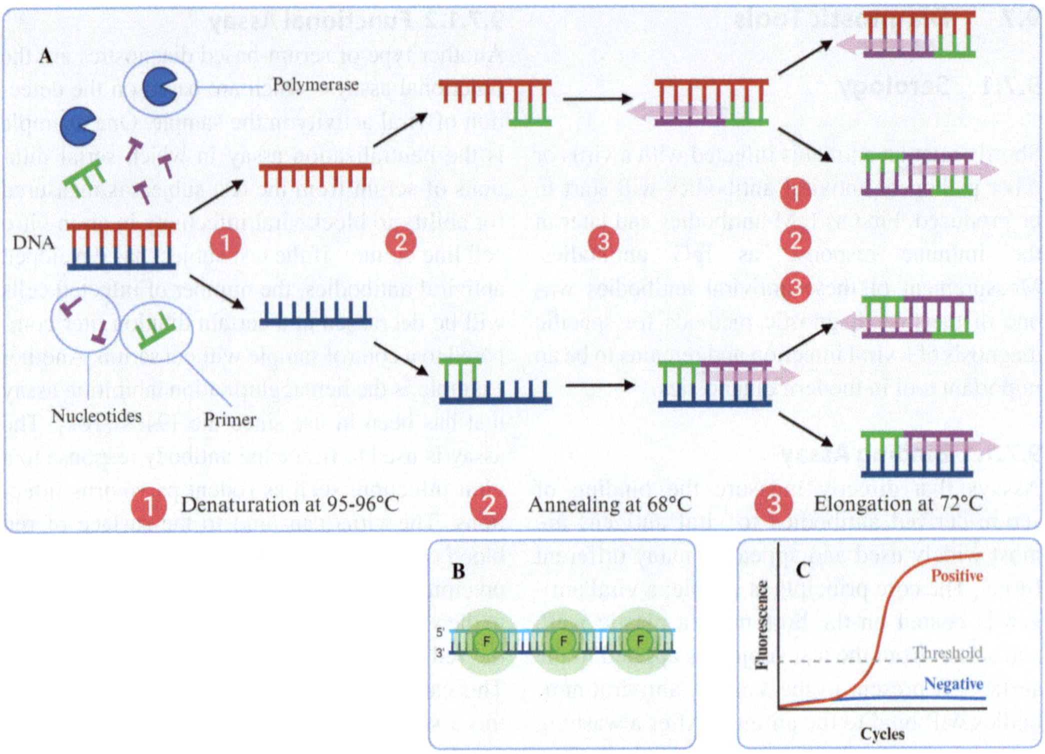

Fig. 9.4 Principles of polymerase chain reaction. (**a**) Thermocycling, (**b**) DNA staining with a fluorescent dye for quantitative PCR, (**c**) A cycle graph showing how the sample becomes positive, when the fluorescence becomes more of intensive than the cycle threshold. The number of cycles to obtain this is called the ct-value

plate will dissociate and become single stranded. Lowering the temperature will allow the two primers to anneal to their complementary sequence, which subsequently primes the DNA polymerase to extend the template. The resulting PCR product, which serves as template in the following cycles, will double for every repetition of the temperature cycle. After 35 temperature cycles, it is possible to detect as little as 1–10 copies of target, making PCR assays the most sensitive analytical method currently in use. After the completion of the PCR reaction, the product is visualized by gel electrophoresis. At some point during the progression of the temperature cycles, the production of PCR product is no longer exponential due to rate limiting factors such as availability of nucleotides, polymerase shortage and steric hindrance between enzyme and template. Therefore, conventional PCR cannot be used for quantitative measurements.

9.7.2.2 Real-Time PCR

Currently, most PCRs run in diagnostic laboratories are carried out using the real-time PCR technique also known as quantitative PCR (qPCR) (Fig. 9.4b, c). When performing qPCR, a special thermocycler is used that can monitor the production of the PCR product, while the reaction is ongoing. This has three major advantages (1): the need for doing gel electrophoresis is eliminated, and thereby (2): the final PCR product is not exposed to the environment, minimizing the risk of cross contamination between samples and accumulation of PCR products in the laboratory, and (3): it is possible to calculate the exact initial target copy number in the sample and thus an estimate of the viral load. To monitor the ongoing production of PCR products, two types of reporters can be used: DNA binding fluorescent dyes or ssDNA probes. In both cases the florescent entity is added to the enzyme blend. The most prevalent dye used, is SYBR Green, which emits florescence when it is bound to double-stranded DNA (Fig. 9.4b). As the PCR reaction progresses more and more PCR product is bound by the dye, making the intensity of emitted florescence proportional to the amount of generated PCR product. Since the thermocycler measure the amount of florescence after completion of each cycle, in real time, the data can be used to generate a plot of cycle number vs. florescence intensity (Fig. 9.4c). As the data are collected while the reaction is still in the exponential phase, it is possible to make an exact calculation of the initial relative amount of the DNA target. If standards with known copy numbers of template are included in the reaction setup, an estimate of the copy number in the samples (e.g., viral load) can be obtained. qPCR reactions can also be performed using a ssDNA probe as the reporter, which is also referred to as a Taqman assay. To work, the probe must be complementary to one of the strands in the amplicon and have the same annealing temperature as the primers. The probe is covalently linked to a florescent dye at the 5′-end and a matching fluorescent quencher at the 3′-end. As the dye and the quencher are physically close to each other, the overall fluorescence intensity in the reaction mixture is low. During the amplification cycles the probe will anneal to the template in the same way as the primers. Since the probe is annealed between the primer pair, the DNA polymerase will bump into the probe during the extension phase and cleave the probe into single nucleotides. This will subsequently release the dye from the quencher and increase the fluorescence intensity in the reaction mix proportionally to PCR product production. In the same way as with SYBR Green-based qPCRs, copy number/viral load in the sample can be calculated. An added benefit of Taqman probes is a much higher specificity of the reaction since both primers and the probe needs to be annealed to the DNA strand before fluorescence can be detected.

9.7.3 Sequencing

9.7.3.1 Sequencing of Viral Metagenomes and Its Challenges

The characterization of the viral metagenome is a key element for increasing the understanding of the complex interaction between bacteria, viruses, and the host. Rodents harbor far more viruses than those listed for health monitoring [131, 186]. The advancement in next-generation sequencing (NGS) technologies and computational methods has revolutionized research to allow researchers to study complex viral communities in a culture-independent and increasingly affordable way [60, 105]. However, our ability to produce sequence data, in this rapidly growing fields of genomics, has surpassed our ability to sufficiently extract meaningful information from the data [127]. The following five parameters are some of the main challenges associated to the characterization of viral metagenomes:

1. Complex biological samples with low input of viral genomes [127].
2. Biases associated with the methodologies used for preparing samples for viral metagenome sequencing [27].
3. Viruses do not hold a universal marker gene that can identify them as a viral genome in contrast to the 16S rRNA gene that is found in almost all bacterial genomes [14, 27, 127, 148].
4. It is estimated that a vast majority of viral taxa are yet to be described ("viral dark matter"). Hence, there is a limited availability of reference sequences in databases for most viral species [14, 92] [55, 181]. Recently, a Minimum Information about an Uncultivated Virus Genome (MIUViG) standards to overcome this issue has been suggested [138].
5. The high evolutionary rate (e.g., nucleotide mutations and methylations) of viruses leads to the current emergence of new virus lineages [143].

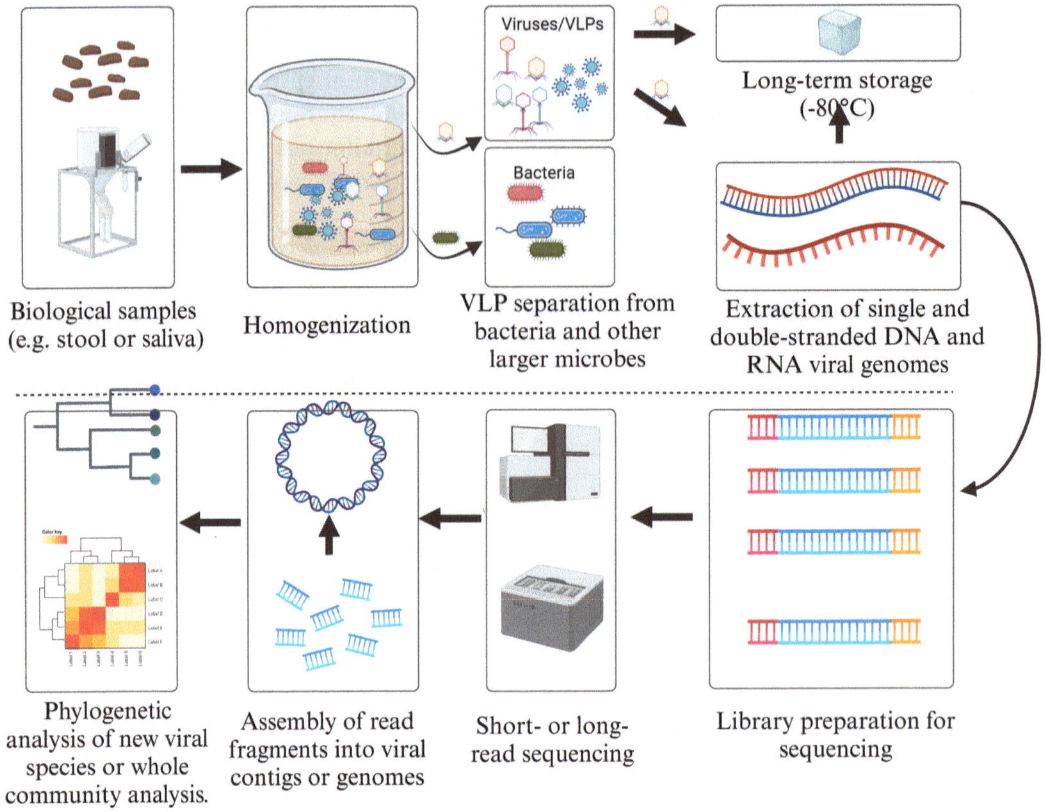

Fig. 9.5 Flow diagram illustrating the process from sample to the analysis of viral metagenomes

Tools and protocols have been developed to address these limitations [27, 105, 127, 143] (Fig. 9.5), which applies for both characterization of complex viral communities, and for screening samples for potential new pathogenic viruses.

9.7.3.2 Sample Preparation and Extraction of Viral Metagenomes

In most biological samples, the vast majority of genomic materials origins from bacteria, host cells, and other eukaryotic cells, while the viral genome biomass only constitute a minor fraction [79, 143]. For a high resolution and to avoid wasting sequencing capacity and money it is important to separate the viral-like particles (VLPs) [43, 156], which term is used, as it is uncertain whether all these particles are replication-competent viruses [92]. The separation process only isolates free viral particles regardless of their life cycle being lysogenic or lytic, thus inactive proviruses integrated in the bacterial (i.e., prophages) or mammal chromosome will not be isolated and sequenced in the downstream analysis [43, 97, 156]. The purification of VLPs is, therefore, one of the most critical steps in the quantitative and qualitative metagenomic analyses of the total viral population, which often include centrifugation and sterile filtration to remove large microbes, ultrafiltration to increase virus concentration and removal of metabolites, and nuclease treatment to remove free-floating non-viral DNA and RNA [43, 86, 156]. A semi-quantitative analysis of phages as well as eukaryotic viruses can be obtained if samples are spiked with known concentrations of exogenous viruses [156]. The preferred suspension buffer for homogenizing samples is SM-buffer, while PBS-buffer and other storage buffers should be avoided since it has been reported to inactivate some phages [3, 185]. Purified VLP samples should be stored at −80 °C for long-term storage, although the activity of phages appears relatively stable at 4 °C for shorter periods [156].

Next is the release of the genetic material encapsulated within the viral particles that either consists of single-stranded (ss)/double-stranded (ds)DNA or ss/dsRNA with either commercially available kits for viral DNA/RNA extraction or traditional methods using phenol/chloroform and formamide/cetyltrimethylammonium bromide (CTAB)-based protocols when high purity and yield of nucleic acids are important for downstream analysis [14, 27]. Reverse transcription (i.e., converting RNA to cDNA) is necessary for including the metagenome of RNA viruses, and during the RNA extraction it is crucial that the materials and reagents are RNase-free to avoid loss of viral RNA. Despite the effort of performing the reverse transcription, many studies find either no or few RNA viruses [27, 156], which is likely due to destabilization of the viral particle or as a result insufficient nucleic acid extraction [156]. The yield of viral nucleic acids from biological samples is often very low [148, 156, 158], hence multiple displacement amplification (MDA) is often used to ensure a sufficient cDNA/DNA amount required for library preparation kits. MDA is reported to preferentially amplify small and circular ssDNA [182], however, decreasing amplification time to 30 min. have shown to largely address the bias. Improved library preparation with commercial kits such as Illumina Nextera XT and Accel-NGS 1S have shown to be efficient for low-input samples of DNA/cDNA [27, 158, 184]. The short-read (250–300 bp) Illumina sequencing platform is the most widely used, but in spite of a lot (>1 Gbp per sample) of sequencing data, it remains a challenge to assemble the short-reads into full viral genomes [105, 127]. The rise of long-read (up to 200 kb) sequencing technologies, spearheaded by Oxford Nanopore and Pacific Biosciences, shows great potential to overcome this major issue associated with the short-read Illumina platform, but are currently limited by factors such as price and/or higher error rates [27, 184].

The available and verified protocols cited above [27, 43, 156, 158, 164] suggest different centrifugation speeds, pore sizes of sterile filters, amplification, reverse transcription, cleaning procedures, and library preparations, where the researchers have decided on a balance between contamination of non-viral content and the loss of viruses due to size or nucleotide structure. Thus, the choice of protocol should therefore be aligned with the properties and matrix associated to the investigated biological sample.

9.7.3.3 Analysis and Interpretation of Viral Metagenomes

The analysis of viral metagenomes involves step, such as quality control, assembly, protein annotation, classification, and estimation of relative abundance viral taxa in the samples [8, 27, 105, 127]:

- **Quality control:** First important step in viral metagenome analysis is the assessment of the raw sequencing data quality of the removal of any low-quality reads and adapter sequences to ensure that the subsequent analyses are based on high-quality data [79, 127, 181].
- **De novo assembly:** The sequencing read data can be assembled into contiguous genomic fragments or contigs based on database-independent de novo assembly [8, 130], which is, particularly useful for detecting previously unknown viruses that may be classified in downstream analyses [105]. Contigs that matches DNA with non-viral origin such as bacteria, the host, fungi, etc. are removed after assembly [130, 181]. The remaining contigs are expected to represent viruses and will therefore undergo protein annotation.
- **Protein annotation:** Once assembled and separated from non-viral sequences, the viral sequences can be annotated to identify coding sequences and functional elements such as integrases, lysins, tail-fibers or other virus-associated genes to provide insights into the potential functions of viral genes and the mechanisms for their host interaction [26, 79, 127].
- **Classification:** Several bioinformatics tools, such as Kraken2, MetaPhlan2, Virsorter2, and VirBot [34, 79, 122, 127] are available for classification of the reads based on their similarity to known viral genomes or reference databases.

- **Estimation of relative abundance**: The abundance of different viral species present in the sample can be estimated by comparing the number of reads assigned to each viral contig based on a contingency-table of reads per Kbp of contig sequence per million reads sample (RPKM) [181]. This information can provide insights into the viral diversity and dynamics of viral communities in different environments [122, 130, 131, 181]. If the samples were spiked with known exogenous viruses, a semi-quantitative analysis can be made [156].

Viral metagenome analysis can be useful to investigate the overall viral composition (β-diversity) and viral diversity (α-diversity) between different treatment groups [107, 132] or to show that eukaryotic viruses are either in high abundance or removed from samples [2]. With the increasing resolution of viral taxonomy [15, 154, 170], viral metagenome analysis can also highlight associations between certain viral taxa and different phenotypes [106], as shown for the bacteriome in numerous studies [46, 136, 173]. Also, hosts of the phages (not eukaryotic viruses) can be predicted based on among other CRISPR spacer databases, which increase the chances of isolating pure cultures of novel phages [46, 144, 187] that may have potential of valuable applications such as therapeutics. Although it should be noted that CRISPR-Cas activity in the gut might be limited [133] and the spacer-based predictions are not necessarily reflecting the most recent phage antagonists.

9.7.4 Cultivation

Cultivation methods are mostly used for bacterial diagnosis, although viral cultivation and isolation also have been used for viral testing and continues to be valuable as a research tool in characterization of newly discovered viruses. In viral diagnostics, however, these methods are increasingly replaced by modern molecular methods, i.e., qPCR and sequencing. Since viruses propagate in living systems a key element is some sort of living host. Most often a cultured cell line infectable by the virus is used but whole animals may be needed. After inoculation viral growth is monitored in several ways such as visually looking for appearance of cytopathic effects in a microscope or observation of interference, a behavior by which some viruses turn an infected cell resistant to infection by other viruses. Immunofluorescent staining of infected cells can also be used to confirm infection. A frequented method for viral culturing is the shell-vial culture assay first described in 1976 as a method to diagnose for cytomegalovirus infection [71], and soon to become the preferred diagnostic culture method in clinical virology laboratories [56]. The procedure is quite simple: after inoculation of the viral specimen in a tube containing a growing monolayer of cells, viral entry is enhanced by a centrifugation step. Next day, propagation is detected by immunofluorescence. The procedure was soon refined by growing different cell types in the same vial and combining several antibodies each labeled with unique florescent labels for staining. Since the ability for a virus to enter and propagate within a cell is highly cell type restricted, the approach of mixing several types of cells in the same vial enabled test for different viruses or subtypes in one assay [115]. Modern molecular techniques have enabled the engineering of culture cells to contain reporter systems easy to detect once activated. When a virus enters the cell the reporter gene activates the production of e.g. fluorescent proteins, enzymes such as beta-galactosidase or luciferase [47, 121, 163]. This eliminates the antibody staining step and makes the assay faster, cheaper, more convenient, and potentially gives a high-throughput. Today, several reporter cell lines for detection of specific viruses such as Influenza virus [91], Ebola virus [76], Rubella virus [174], or classical swine fever virus [33] to name a few, have been made. Until now, these novel cell lines are primarily used in research laboratories and not so much in diagnostics. One major limitation is the need for a specialized cell culturing facility at site, a requirement that many diagnostic laboratories cannot fulfill. The mouse antibody production (MAP) test [145] is an assay, which to a great extent has been used as a biosecurity test to screen for contamination of biologicals. The test involves homogenization of the testing material and inoculation into SPF mice either intraperitoneally and/or intranasally. Four to 6 weeks after inoculation the animals are tested for pathogen specific antibodies. This is resourceful and laborious and as PCR assays are cheaper, faster and more in accordance with the 3Rs, they have now replaced MAP testing [12, 23, 98].

9.8 Concluding Remarks

With the development of modern high-throughput sequencing techniques the amount of genomic information available for almost every circulating pathogen has increased enormously within recent years. Therefore, PCR assays available for detection of these pathogens have increased and replaced many of the traditional testing methods. Over 60 rodent infectious agents, including viruses, can now be monitored by PCR [38] that either can be used alone or in combination with serologic tests. Implementation of these new assays have many advantages such as speed, high sensitivity, and specificity, making it possible to discriminate between two closely related pathogenic and nonpathogenic virus strains. Therefore, PCR along with serology are today the go-to methods in laboratory animal diagnostics [38]. With the increasing cheaper costs for second and third generation whole genome sequencing methods as well as technological advances in bioinformatics, one might envision these methods as the next step forward in diagnostics [73]. Complete sequencing of a test sample could give information on all known pathogens present in the sample in one test, dramatically increasing diagnostic potential and throughput. Although diagnostic sequencing still needs some refinement to be established as reliable and proven method the perspective of this technology most likely will secure its place in future diagnostics. Even if certain viruses will be preferred in future laboratory rodents, this will not decrease the need for efficient viral diagnostics, as evidence for the presence of such viruses will be needed, as well as the animals still need to be free of zoonotic viruses.

References

1. Abzug MJ, Tyson RW. Picornavirus infection in early murine gestation: significance of maternal illness. Placenta. 2000;21(8):840–6.
2. Adamberg S, Rasmussen TS, Larsen SB, Mao X, Nielsen DS, Adamberg K. Reproducible chemostat cultures to eliminate eukaryotic viruses from fecal transplant material. bioRxiv. 2023; https://doi.org/10.1101/2023.03.15.529189.
3. Adams MH. The stability of bacterial viruses in solutions of salts. J Gen Physiol. 1949;32(5):579–94.
4. Adams WR, Kraft LM. Epizootic diarrhea of infant mice: identification of the etiologic agent. Science. 1963;141(3578):359–60.
5. Alcami A, Ghazal P, Yewdell JW. Viruses in control of the immune system. Workshop on molecular mechanisms of immune modulation: lessons from viruses. EMBO Rep. 2002;3(10):927–32.
6. Artwohl JE, Cera LM, Wright MF, Medina LV, Kim LJ. The efficacy of a dirty bedding sentinel system for detecting Sendai virus infection in mice: a comparison of clinical signs and seroconversion. Lab Anim Sci. 1994;44(1):73–5.
7. Ball-Goodrich LJ, Johnson E. Molecular characterization of a newly recognized mouse parvovirus. J Virol. 1994;68(10):6476–86.
8. Bankevich A, Nurk S, Antipov D, Gurevich AA, Dvorkin M, Kulikov AS, Lesin VM, Nikolenko SI, Pham S, Prjibelski AD, Pyshkin AV, Sirotkin AV, Vyahhi N, Tesler G, Alekseyev MA, Pevzner PA. SPAdes: a new genome assembly algorithm and its applications to single-cell sequencing. J Comput Biol. 2012;19(5):455–77.
9. Barrangou R, Yoon SS, Breidt F Jr, Fleming HP, Klaenhammer TR. Characterization of six Leuconostoc fallax bacteriophages isolated from an industrial sauerkraut fermentation. Appl Environ Microbiol. 2002;68(11):5452–8.
10. Barthold SW, Smith AL. Mouse hepatitis virus strain – related patterns of tissue tropism in suckling mice. ArchVirol. 1984;81(1–2):103–12.
11. Barton LL, Mets MB, Beauchamp CL. Lymphocytic choriomeningitis virus: emerging fetal teratogen. Am J Obstet Gynecol. 2002;187(6):1715–6.
12. Bauer BA, Besch-Williford CL, Riley LK. Comparison of the mouse antibody production (MAP) assay and polymerase chain reaction (PCR) assays for the detection of viral contaminants. Biologicals. 2004;32(4):177–82.
13. Becker SD, Bennett M, Stewart JP, Hurst JL. Serological survey of virus infection among wild house mice (Mus domesticus) in the UK. Lab Anim. 2007;41(2):229–38.
14. Beller L, Matthijnssens J. What is (not) known about the dynamics of the human gut virome in health and disease. Curr Opin Virol. 2019;37:52–7.
15. Benler S, Yutin N, Antipov D, Rayko M, Shmakov S, Gussow AB, Pevzner P, Koonin EV. Thousands

of previously unknown phages discovered in whole-community human gut metagenomes. Microbiome. 2021;9(1):78.
16. Besselsen DG, Myers EL, Franklin CL, Korte SW, Wagner AM, Henderson KS, Weigler BJ. Transmission probabilities of mouse parvovirus 1 to sentinel mice chronically exposed to serial dilutions of contaminated bedding. Comp Med. 2008;58(2):140–4.
17. Beura LK, Hamilton SE, Bi K, Schenkel JM, Odumade OA, Casey KA, Thompson EA, Fraser KA, Rosato PC, Filali-Mouhim A, Sekaly RP, Jenkins MK, Vezys V, Haining WN, Jameson SC, Masopust D. Normalizing the environment recapitulates adult human immune traits in laboratory mice. Nature. 2016;532(7600):512–6.
18. Bhan AK, Mizoguchi E, Smith RN, Mizoguchi A. Colitis in transgenic and knockout animals as models of human inflammatory bowel disease. Immunol Rev. 1999;169:195–207.
19. Bhatt PN, Jacoby RO. Mousepox in inbred mice innately resistant or susceptible to lethal infection with ectromelia virus. III. Experimental transmission of infection and derivation of virus-free progeny from previously infected dams. Lab Anim Sci. 1987;37(1):23–7.
20. Biggar RJ, Douglas RG, Hotchin J. Letter: lymphocytic choriomeningitis associated with hamsters. Lancet. 1975a;1(7911):856–7.
21. Biggar RJ, Woodall JP, Walter PD, Haughie GE. Lymphocytic choriomeningitis outbreak associated with pet hamsters. Fifty-seven cases from New York state. JAMA. 1975b;232(5):494–500.
22. Bittner JJ. Some possible effects of nursing on the mammary gland tumor incidence in mice. Science. 1936;84(2172):162.
23. Blank WA, Henderson KS, White LA. Virus PCR assay panels: an alternative to the mouse antibody production test. Lab Anim (NY). 2004;33(2):26–32.
24. Brabb T, Newsome D, Burich A, Hanes M. Chapter 23: Infectious diseases. In: Suckow MA, Stevens KA, Wilson RP, editors. The laboratory rabbit, guinea pig, hamster, and other rodents. Boston: Academic; 2012. p. 637–83.
25. Broderson JR, Murphy FA, Hierholzer JC. Lethal enteritis in infant mice caused by mouse hepatitis virus. Lab Anim Sci. 1976;26(5):824.
26. Callanan J, Stockdale SR, Shkoporov A, Draper LA, Ross RP, Hill C. Expansion of known ssRNA phage genomes: from tens to over a thousand. Sci Adv. 2020;6(6):eaay5981.
27. Callanan J, Stockdale SR, Shkoporov A, Draper LA, Ross RP, Hill C. Biases in viral metagenomics-based detection, cataloguing and quantification of bacteriophage genomes in human Faeces, a review. Microorganisms. 2021;9(3):524.
28. Camp JV, Desvars-Larrive A, Nowotny N, Walzer C. Monitoring urban zoonotic virus activity: are City rats a promising surveillance tool for emerging viruses? Viruses. 2022;14(7):1516.
29. Carthew P, Sparrow S. Persistence of pneumonia virus of mice and Sendai virus in germ-free (nu/nu) mice. Br J Exp Pathol. 1980a;61(2):172–5.
30. Carthew P, Sparrow S. Sendai virus in nude and germ-free rats. Res Vet Sci. 1980b;29(3):289–92.
31. Chatterjee A, Harrison CJ, Britt WJ, Bewtra C. Modification of maternal and congenital cytomegalovirus infection by anti-glycoprotein B antibody transfer in Guinea pigs. J Infect Dis. 2001;183(11):1547–53.
32. Cheever FS, Daniels JB, Pappenheimer AM, Bailey OT. A murine virus (JHM) causing disseminated encephalomyelitis with extensive destruction of myelin: I. Isolation and biological properties of the virus. J Exp Med. 1949;90(3):181–94.
33. Chen F, Yang X, Pang D, Peng Z, Ma T, Ouyang H, Ren L. A dark-to-bright reporter cell for classical swine fever virus infection. Antivir Res. 2015;117:44–51.
34. Chen G, Tang X, Shi M, Sun Y. VirBot: an RNA viral contig detector for metagenomic data. Bioinformatics. 2023;39(3):btad093.
35. Chien A, Edgar DB, Trela JM. Deoxyribonucleic acid polymerase from the extreme thermophile Thermus aquaticus. J Bacteriol. 1976;127(3):1550–7.
36. Clement J, LeDuc JW, Lloyd G, Reynes J-M, McElhinney L, Van Ranst M, Lee H-W. Wild rats, laboratory rats, pet rats: global Seoul hantavirus disease revisited. Viruses. 2019;11(7):652.
37. Committee_on_Infectious_Diseases_of_Mice_and_Rats. Infectious diseases of mice and rats. Institute of Laboratory Animal Resources, Commission on Life Sciences, National Research Council; 1991.
38. Compton SR. PCR and RT-PCR in the diagnosis of laboratory animal infections and in health monitoring. J Am Assoc Lab Anim Sci. 2020;59(5):458–68.
39. Crawford LV. A minute virus of mice. Virology. 1966;29(4):605–12.
40. Crispens CG Jr. Lactate dehydrogenase virus and mouse embryos. Nature. 1967;214(90):819.
41. Dammann P, Hilken G, Hueber B, Köhl W, Bappert MT, Mähler M. Infectious microorganisms in mice (Mus musculus) purchased from commercial pet shops in Germany. Lab Anim. 2011;45(4):271–5.
42. Deibel R, Woodall JP, Decher WJ, Schryver GD. Lymphocytic choriomeningitis virus in man. Serologic evidence of association with pet hamsters. JAMA. 1975;232(5):501–4.
43. Deng L, Silins R, Castro-Mejía JL, Kot W, Jessen L, Thorsen J, Shah S, Stokholm J, Bisgaard H, Moineau S, Nielsen DS. A protocol for extraction of infective Viromes suitable for metagenomics sequencing from low volume fecal samples. Viruses. 2019;11(7):667.
44. Desmyter J, LeDuc JW, Johnson KM, Brasseur F, Deckers C, C van Ypersele de Strihou. Laboratory rat associated outbreak of haemorrhagic fever with renal syndrome due to Hantaan-like virus in Belgium. Lancet. 1983;2(8365–66):1445–8.
45. Dick EJ, Kittell CL, Meyer H, Farrar PL, Ropp SL, Esposito JJ, Buller RML, Neubauer H, Kang YH,

McKee AE. Mousepox outbreak in a laboratory mouse colony. Lab Anim Sci. 1996;46(6):602–11.
46. Dion MB, Plante PL, Zufferey E, Shah SA, Corbeil J, Moineau S. Streamlining CRISPR spacer-based bacterial host predictions to decipher the viral dark matter. Nucleic Acids Res. 2021;49(6):3127–38.
47. Dolskiy AA, Grishchenko IV, Yudkin DV. Cell cultures for virology: usability, advantages, and prospects. Int J Mol Sci. 2020;21(21):7978.
48. Drake MT, Riley LK, Livingston RS. Differential susceptibility of SD and CD rats to a novel rat theilovirus. Comp Med. 2008;58(5):458–64.
49. Dutko FJ, Oldstone MB. Genomic and biological variation among commonly used lymphocytic choriomeningitis virus strains. J Gen Virol. 1983;64(Pt 8):1689–98.
50. Dyson MC. Management of an outbreak of rat theilovirus. Lab Anim. 2010;39(5):155–7.
51. Falkenberg C, Bartholdy C, Koch J, Toft MF, Skov S, Hansen CHF, Hansen AK. Induction of CD8+ immune memory and enhanced inflammation in a skin inflammation model through pre-immunization with inactivated pathogens. Clin Transl Sci. 2024;17(1):e13697.
52. Fiege JK, Block KE, Pierson MJ, Nanda H, Shepherd FK, Mickelson CK, Stolley JM, Matchett WE, Wijeyesinghe S, Meyerholz DK, Vezys V, Shen SS, Hamilton SE, Masopust D, Langlois RA. Mice with diverse microbial exposure histories as a model for preclinical vaccine testing. Cell Host Microbe. 2021;29(12):1815–1827.e1816.
53. Fisher ER, Kilham L. Pathology of a pneumotropic virus recovered from C3H mice carrying the Bittner milk agent. AMA Arch Pathol. 1953;55(1):14–9.
54. Foster HL. Large scale production of rats free of commonly occurring pathogens and parasites. Proc Anim Care Panel. 1958;8:92–100.
55. Geoghegan JL, Holmes EC. Predicting virus emergence amid evolutionary noise. Open Biol. 2017;7(10):170189.
56. Gleaves CA, Smith TF, Shuster EA, Pearson GR. Comparison of standard tube and shell vial cell culture techniques for the detection of cytomegalovirus in clinical specimens. J Clin Microbiol. 1985;21(2):217–21.
57. Greenlee JE. Pathogenesis of K virus infection in newborn mice. Infect Immun. 1979;26(2):705–13.
58. Greenlee JE. Effect of host age on experimental K virus infection in mice. Infect Immun. 1981;33(1):297–303.
59. Gross L. A filterable agent, recovered from Ak leukemic extracts, causing salivary gland carcinomas in C3H mice. Proc Soc Exp Biol Med. 1953;83(2):414–21.
60. Gwinn M, MacCannell DR, Khabbaz RF. Integrating advanced molecular technologies into public health. J Clin Microbiol. 2017;55(3):703–14.
61. Hager C, Glage S, Held N, Bleich EM, Burghard A, Mahler M, Bleich A. Detection of antibodies against Theiler's murine encephalomyelitis virus GDVII strain in experimental Guinea pigs. Lab Anim. 2016;50(5):400–3.
62. Hansen AK, Thomsen P, Jensen HJ. A serological indication of the existence of a guineapig poliovirus. Lab Anim. 1997;31(3):212–8.
63. Hansen AK, Velschow S, Clausen FB, Svendsen O, Amtoft-Neubauer H, Kristensen K, Jorgensen PH. New infections to be considered in health monitoring of laboratory rodents. Scand J Lab Anim Sci. 2000;27(2):65–83.
64. Hayashimoto N, Morita H, Ishida T, Uchida R, Tanaka M, Ozawa M, Yasuda M, Itoh T. Microbiological survey of mice (Mus musculus) purchased from commercial pet shops in Kanagawa and Tokyo, Japan. Exp Anim. 2015;64(2):155–60.
65. Heck FC Jr, Sheldon WG, Gleiser CA. Pathogenesis of experimentally produced mouse adenovirus infection in mice. Am J Vet Res. 1972;33(4):841–6.
66. Holmfeldt K, Middelboe M, Nybroe O, Riemann L. Large variabilities in host strain susceptibility and phage host range govern interactions between lytic marine phages and their Flavobacterium hosts. Appl Environ Microbiol. 2007;73(21):6730–9.
67. Holt D. Presence of K-virus in wild mice in Australia. Aust J Exp Biol Med Sci. 1959;37(2):183–92.
68. Homberger FR. Enterotropic mouse hepatitis virus. Lab Anim. 1997;31(2):97–115.
69. Hsu CC, Riley LK, Wills HM, Livingston RS. Persistent infection with and serologic cross-reactivity of three novel murine noroviruses. Comp Med. 2006;56(4):247–51.
70. Huber AC, Yolken RH, Mader LC, Strandberg JD, Vonderfecht SL. Pathology of infectious diarrhea of infant rats (IDIR) induced by an antigenically distinct rotavirus. Vet Pathol. 1989;26(5):376–85.
71. Hudson JB, Misra V, Mosmann TR. Cytomegalovirus infectivity: analysis of the phenomenon of centrifugal enhancement of infectivity. Virology. 1976;72(1):235–43.
72. Iida T. Experimental study on the transmission of Sendai virus in specific pathogen-free mice. J Gen Virol. 1972;14(1):69–75.
73. Irina Vlasova-St L, Andrew G, Mark P. Chapter 5: Diagnostic applications for RNA-Seq technology and transcriptome analyses in human diseases caused by RNA viruses. In: Vlasova-St LI, editor. Applications of RNA-Seq in biology and medicine. Rijeka: IntechOpen; 2021.
74. Jacoby RO, Bhatt PN. Mousepox in inbred mice innately resistant or susceptible to lethal infection with ectromelia virus. II. Pathogenesis. Lab Anim Sci. 1987;37(1):16–22.
75. Jacoby RO, Bhatt PN, Jonas AM. Pathogenesis of sialodacryoadenitis in gnotobiotic rats. Vet Pathol. 1975;12(3):196–209.
76. Kainulainen MH, Nichol ST, Albariño CG, Spiropoulou CF. Rapid determination of ebolavirus infectivity in clinical samples using a novel reporter cell line. J Infect Dis. 2017;216(11):1380–5.

77. Kaledin AS, Sliusarenko AG, Gorodetskiĭ SI. Isolation and properties of DNA polymerase from extreme thermophylic bacteria Thermus aquaticus YT-1. Biokhimiia. 1980;45(4):644–51.
78. Karst SM, Wobus CE, Lay M, Davidson J, Virgin HWT. STAT1-dependent innate immunity to a Norwalk-like virus. Science. 2003;299(5612):1575–8.
79. Kieft K, Adams A, Salamzade R, Kalan L, Anantharaman K. vRhyme enables binning of viral genomes from metagenomes. Nucleic Acids Res. 2022;50(14):e83.
80. Kilham L, Margolis G. Transplacental infection of rats and hamsters induced by oral and parenteral inoculations of H-l and rat viruses (RV). Teratology. 1969;2(2):111–23.
81. Knust B, Stroher U, Edison L, Albarino CG, Lovejoy J, Armeanu E, House J, Cory D, Horton C, Fowler KL, Austin J, Poe J, Humbaugh KE, Guerrero L, Campbell S, Gibbons A, Reed Z, Cannon D, Manning C, Petersen B, Metcalf D, Marsh B, Nichol ST, Rollin PE. Lymphocytic Choriomeningitis virus in employees and mice at multipremises feeder-rodent operation, United States, 2012. Emerg Infect Dis. 2014;20(2):240–7.
82. Koch R. Zur Untersuchung von pathogenen Organismen. Mitthdungen uus dem Kaiserlichen Gesundbeitsamte. 1881;1:1–48.
83. Körner RW, Majjouti M, Alcazar MAA, Mahabir E. Of mice and men: the coronavirus MHV and mouse models as a translational approach to understand SARS-CoV-2. Viruses. 2020;12(8):880.
84. Kraft V, Deeny A, Boot R, Hansen A, Hem A, Vanherck H, Kunstyr I, Maurinblanchet H, Milite G, Needham J. Recommendations for the health monitoring of mouse, rat, hamster, Guinea-pig and rabbit breeding colonies (FELASA). Sci Tech Anim Lab. 1993;18(3):141–63.
85. Larsen SH, Nathans D. Mouse adenovirus: growth of plaque-purified FL virus in cell lines and characterization of viral DNA. Virology. 1977;82(1):182–95.
86. Larsen F, Offersen SM, Li VR, Deng L, Nielsen DS, Rasmussen TS. Choice of ultrafilter affects recovery rate of bacteriophages. Viruses. 2023;15(10): 2051.
87. Lau SK, Woo PC, Li KS, Tsang AK, Fan RY, Luk HK, Cai JP, Chan KH, Zheng BJ, Wang M, Yuen KY. Discovery of a novel coronavirus, China Rattus coronavirus HKU24, from Norway rats supports the murine origin of Betacoronavirus 1 and has implications for the ancestor of Betacoronavirus lineage A. J Virol. 2015;89(6):3076–92.
88. LeDuc JW. Epidemiology of Hantaan and related viruses. Lab Anim Sci. 1987;37(4):413–8.
89. Lee HW, Lee PW, Johnson KM. Isolation of the etiologic agent of Korean hemorrhagic fever. J Infect Dis. 1978;137(3):298–308.
90. Lequin RM. Enzyme immunoassay (EIA)/enzyme-linked immunosorbent assay (ELISA). Clin Chem. 2005;51(12):2415–8.
91. Li Y, Larrimer A, Curtiss T, Kim J, Jones A, Baird-Tomlinson H, Pekosz A, Olivo PD. Influenza virus assays based on virus-inducible reporter cell lines. Influenza Other Respir Viruses. 2009;3(5):241–51.
92. Liang G, Bushman FD. The human virome: assembly, composition and host interactions. Nat Rev Microbiol. 2021;19(8):514–27.
93. Liang CT, Shih A, Chang YH, Liu CW, Lee YT, Hsieh WC, Huang YL, Huang WT, Kuang CH, Lee KH, Zhuo YX, Ho SY, Liao SL, Chiu YY, Hsu CN, Liang SC, Yu CK. Microbial contaminations of laboratory mice and rats in Taiwan from 2004 to 2007. J Am Assoc Lab Anim Sci. 2009;48(4):381–6.
94. Lipman NS, Perkins S, Nguyen H, Pfeffer M, Meyer H. Mousepox resulting from use of ectromelia virus-contaminated, imported mouse serum. Comp Med. 2000;50(4):426–35.
95. Lipton HL, Dal Canto MC. Chronic neurologic disease in Theiler's virus infection of SJL/J mice. J Neurol Sci. 1976;30(1):201–7.
96. Lipton HL, Melvold R. Genetic analysis of susceptibility to Theiler's virus-induced demyelinating disease in mice. J Immunol. 1984;132(4):1821–5.
97. Liu B, Shao N, Wang J, Zhou S, Su H, Dong J, Sun L, Li L, Zhang T, Yang F. An optimized metagenomic approach for Virome detection of clinical pharyngeal samples with respiratory infection. Front Microbiol. 2020;11:1552.
98. Livingston RS, Riley LK. Diagnostic testing of mouse and rat colonies for infectious agents. Lab Anim. 2003;32(5):44–51.
99. Lloyd G, Jones N. Infection of laboratory workers with hantavirus acquired from immunocytomas propagated in laboratory rats. J Infect. 1986;12(2):117–25.
100. Mahler Convenor M, Berard M, Feinstein R, Gallagher A, Illgen-Wilcke B, Pritchett-Corning K, Raspa M. FELASA recommendations for the health monitoring of mouse, rat, hamster, Guinea pig and rabbit colonies in breeding and experimental units. Lab Anim. 2014;48(3):178–92.
101. Manjunath S, Kulkarni PG, Nagavelu K, Samuel RJ, Srinivasan S, Ramasamy N, Hegde NR, Gudde RS. Sero-prevalence of rodent pathogens in India. PLoS One. 2015;10(7):e0131706.
102. Marchal J. Infectious ectromelia. A hitherto undescribed virus disease of mice. J Pathol Bacteriol. 1930;33(3):713–28.
103. Martinez D, Wolanski B, Tytell AA, Devlin RG. Viral etiology of age-dependent polioencephalomyelitis in C58 mice. Infect Immun. 1979;23(1):133–9.
104. Masopust D, Sivula CP, Jameson SC. Of mice, dirty mice, and men: using mice to understand human immunology. J Immunol. 2017;199(2):383–8.
105. Mastriani E, Bienes KM, Wong G, Berthet N. PIMGAVir and Vir-MinION: two viral metagenomic pipelines for complete baseline analysis of 2nd and 3rd generation data. Viruses. 2022;14(6):1260.
106. Mayneris-Perxachs J, Castells-Nobau A, Arnoriaga-Rodríguez M, Garre-Olmo J, Puig J, Ramos

R, Martínez-Hernández F, Burokas A, Coll C, Moreno-Navarrete JM, Zapata-Tona C, Pedraza S, Pérez-Brocal V, Ramió-Torrentà L, Ricart W, Moya A, Martínez-García M, Maldonado R, Fernández-Real JM. Caudovirales bacteriophages are associated with improved executive function and memory in flies, mice, and humans. Cell Host Microbe. 2022;30(3):340–356.e348.
107. McCann A, Ryan FJ, Stockdale SR, Dalmasso M, Blake T, Ryan CA, Stanton C, Mills S, Ross PR, Hill C. Viromes of one year old infants reveal the impact of birth mode on microbiome diversity. Peer J. 2018;6:e4694.
108. McInnes EF, Rasmussen L, Fung P, Auld AM, Alvarez L, Lawrence DA, Quinn ME, del Fierro GM, Vassallo BA, Stevenson R. Prevalence of viral, bacterial and parasitological diseases in rats and mice used in research environments in Australasia over a 5-y period. Lab Anim (NY). 2011;40(11):341–50.
109. McLaughlin MR, Balaa MF, Sims J, King R. Isolation of salmonella bacteriophages from swine effluent lagoons. J Environ Qual. 2006;35(2):522–8.
110. Meyer BJ, Schmaljohn CS. Persistent hantavirus infections: characteristics and mechanisms. Trends Microbiol. 2000;8(2):61–7.
111. Mohammadi H, Sharif S, Rowland RR, Yoo D. The lactate dehydrogenase-elevating virus capsid protein is a nuclear–cytoplasmic protein. Arch Virol. 2009;154(7):1071–80.
112. Moskophidis D, Lechner F, Pircher H, Zinkernagel RM. Virus persistence in acutely infected immunocompetent mice by exhaustion of antiviral cytotoxic effector T-cells. Nature. 1993;362(6422):758–61.
113. Mullis KB. The unusual origin of the polymerase chain reaction. Sci Am. 1990;262(4):56–61, 64–55.
114. Mullis KB, Faloona FA. Specific synthesis of DNA in vitro via a polymerase-catalyzed chain reaction. Methods Enzymol. 1987;155:335–50.
115. Navarro-Marí JM, Sanbonmatsu-Gámez S, Pérez-Ruiz M, Rosa-Fraile MDL. Rapid detection of respiratory viruses by Shell vial assay using simultaneous culture of HEp-2, LLC-MK2, and MDCK cells in a single vial. J Clin Microbiol. 1999;37(7):2346–7.
116. Nicklas W, Kraft V, Meyer B. Contamination of transplantable tumors, cell lines, and monoclonal antibodies with rodent viruses. Lab Anim Sci. 1993;43(4):296–300.
117. Nicklas W, Baneux P, Boot R, Decelle T, Deeny A, Fumanelli M, Illgen-Wilcke B. Recommendations for the health monitoring of rodent and rabbit colonies in breeding and experimental units. Lab Anim. 2002;36(1):20–42.
118. Norman JM, Handley SA, Baldridge MT, Droit L, Liu CY, Keller BC, Kambal A, Monaco CL, Zhao G, Fleshner P, Stappenbeck TS, McGovern DP, Keshavarzian A, Mutlu EA, Sauk J, Gevers D, Xavier RJ, Wang D, Parkes M, Virgin HW. Disease-specific alterations in the enteric virome in inflammatory bowel disease. Cell. 2015;160(3):447–60.
119. Notkins AL. Enzymatic and immunologic alterations in mice infected with lactic dehydrogenase virus. Am J Pathol. 1971;64(3):733–46.
120. Ohsawa K, Yamada A, Takeuchi K, Watanabe Y, Miyata H, Sato H. Genetic characterization of parainfluenza virus 3 derived from Guinea pigs. J Vet Med Sci. 1998;60(8):919–22.
121. Olivo PD. Transgenic cell lines for detection of animal viruses. Clin Microbiol Rev. 1996;9(3):321–34.
122. Pandolfo M, Telatin A, Lazzari G, Adriaenssens EM, Vitulo N. MetaPhage: an automated pipeline for analyzing, annotating, and classifying bacteriophages in metagenomics sequencing data. mSystems. 2022;7(5):e0074122.
123. Parker JC, Igel HJ, Reynolds RK, Lewis AM Jr, Rowe WP. Lymphocytic choriomeningitis virus infection in fetal, newborn, and young adult Syrian hamsters (Mesocricetus auratus). Infect Immun. 1976;13(3):967081.
124. Parker JC, Whiteman MD, Richter CB. Susceptibility of inbred and outbred mouse strains to Sendai virus and prevalence of infection in laboratory rodents. Infect Immun. 1978;19(1):123–30.
125. Percy DH, Barta JR. Spontaneous and experimental infections in scid and scid/beige mice. Lab Anim Sci. 1993;43(2):127–32.
126. Perdue KA, Green KY, Copeland M, Barron E, Mandel M, Faucette LJ, Williams EM, Sosnovtsev SV, Elkins WR, Ward JM. Naturally occurring murine norovirus infection in a large research institution. J Am Assoc Lab Anim Sci. 2007;46(4):39–45.
127. Plyusnin I, Kant R, Jääskeläinen AJ, Sironen T, Holm L, Vapalahti O, Smura T. Novel NGS pipeline for virus discovery from a wide spectrum of hosts and sample types. Virus Evol. 2020;6(2):veaa091.
128. Pritchett-Corning KR, Cosentino J, Clifford CB. Contemporary prevalence of infectious agents in laboratory mice and rats. Lab Anim. 2009;43(2):165–73.
129. Rasmussen TS, Koefoed AK, Jakobsen RR, Deng L, Castro-Mejía JL, Brunse A, Neve H, Vogensen FK, Nielsen DS. Bacteriophage-mediated manipulation of the gut microbiome – promises and presents limitations. FEMS Microbiol Rev. 2020a;44(4):507–21.
130. Rasmussen TS, Mentzel CMJ, Kot W, Castro-Mejía JL, Zuffa S, Swann JR, Hansen LH, Vogensen FK, Hansen AK, Nielsen DS. Faecal virome transplantation decreases symptoms of type 2 diabetes and obesity in a murine model. Gut. 2020b;69:2122–30. https://doi.org/10.1136/gutjnl-2019-320005.
131. Rasmussen TS, Jakobsen RR, Castro-Mejía JL, Kot W, Thomsen AR, Vogensen FK, Nielsen DS, Hansen AK. Inter-vendor variance of enteric eukaryotic DNA viruses in specific pathogen free C57BL/6N mice. Res Vet Sci. 2021;136:1–5.
132. Rasmussen TS, Forster S, Larsen SB, Münchow AV, Tranæs KD, Brunse A, Mejia JLC, Adamberg S, Hansen AK, Adamberg K, Hansen CHF, Nielsen DS. Development of safe and effective bacteriophage-mediated therapies against *C. difficile* infections – a

proof-of-concept preclinical study. bioRxiv. 2023a; https://doi.org/10.1101/2023.03.17.532897.
133. Rasmussen TS, Koefoed AK, Deng L, Muhammed MK, Rousseau GM, Kot W, Sprotte S, Neve H, Franz C, Hansen AK, Vogensen FK, Moineau S, Nielsen DS. CRISPR-Cas provides limited phage immunity to a prevalent gut bacterium in gnotobiotic mice. ISME J. 2023b;17(3):432–42.
134. Rein A. Murine leukemia viruses: objects and organisms. Adv Virol. 2011;2011:403419.
135. Rigatti LH, Toptan T, Newsome JT, Moore PS, Chang Y. Identification and characterization of novel rat polyomavirus 2 in a Colony of X-SCID rats by P-PIT assay. mSphere. 2016;1(6):e00334-16.
136. Rinninella E, Raoul P, Cintoni M, Franceschi F, Miggiano GAD, Gasbarrini A, Mele MC. What is the healthy gut microbiota composition? A changing ecosystem across age, environment, diet, and diseases. Microorganisms. 2019;7(1):14.
137. Roble GS, Gillespie V, Lipman NS. Infectious disease survey of Mus musculus from pet stores in New York City. J Am Assoc Lab Anim Sci. 2012;51(1):37–41.
138. Rose R, Constantinides B, Tapinos A, Robertson DL, Prosperi M. Challenges in the analysis of viral metagenomes. Virus Evol. 2016;2(2):vew022.
139. Rosshart SP, Vassallo BG, Angeletti D, Hutchinson DS, Morgan AP, Takeda K, Hickman HD, McCulloch JA, Badger JH, Ajami NJ, Trinchieri G, Pardo-Manuel de Villena F, Yewdell JW, Rehermann B. Wild mouse gut microbiota promotes host fitness and improves disease resistance. Cell. 2017;171(5):1015–1028 e1013.
140. Rosshart SP, Herz J, Vassallo BG, Hunter A, Wall MK, Badger JH, McCulloch JA, Anastasakis DG, Sarshad AA, Leonardi I, Collins N, Blatter JA, Han SJ, Tamoutounour S, Potapova S, St Claire MBF, Yuan WX, Sen SK, Dreier MS, Hild B, Hafner M, Wang D, Iliev ID, Belkaid Y, Trinchieri G, Rehermann B. Laboratory mice born to wild mice have natural microbiota and model human immune responses. Science. 2019;365(6452):eaaw4361.
141. Rothenburger JL, Himsworth CG, Clifford CB, Ellis J, Treuting PM, Leighton FA. Respiratory pathology and pathogens in wild urban rats (Rattus norvegicus and Rattus rattus). Vet Pathol. 2015;52(6):1210–9.
142. Rousseau MC, Saron MF, Brouqui P, Bourgeade A. Lymphocytic choriomeningitis virus in southern France: four case reports and a review of the literature. Eur J Epidemiol. 1997;13(7):817–23.
143. Roux S, Adriaenssens EM, Dutilh BE, Koonin EV, Kropinski AM, Krupovic M, Kuhn JH, Lavigne R, Brister JR, Varsani A, Amid C, Aziz RK, Bordenstein SR, Bork P, Breitbart M, Cochrane GR, Daly RA, Desnues C, Duhaime MB, Emerson JB, Enault F, Fuhrman JA, Hingamp P, Hugenholtz P, Hurwitz BL, Ivanova NN, Labonté JM, Lee KB, Malmstrom RR, Martinez-Garcia M, Mizrachi IK, Ogata H, Páez-Espino D, Petit MA, Putonti C, Rattei T, Reyes A, Rodriguez-Valera F, Rosario K, Schriml L, Schulz F, Steward GF, Sullivan MB, Sunagawa S, Suttle CA, Temperton B, Tringe SG, Thurber RV, Webster NS, Whiteson KL, Wilhelm SW, Wommack KE, Woyke T, Wrighton KC, Yilmaz P, Yoshida T, Young MJ, Yutin N, Allen LZ, Kyrpides NC, Eloe-Fadrosh EA. Minimum information about an uncultivated virus genome (MIUViG). Nat Biotechnol. 2019;37(1):29–37.
144. Roux S, Camargo AP, Coutinho FH, Dabdoub SM, Dutilh BE, Nayfach S, Tritt A. iPHoP: an integrated machine learning framework to maximize host prediction for metagenome-derived viruses of archaea and bacteria. PLoS Biol. 2023;21(4):e3002083.
145. Rowe WP, Hartley JW, Estes JD, Huebner RJ. Studies of mouse polyoma virus infection: I. procedures for quantitation and detection of virus. J Exp Med. 1959;109(4):379–91.
146. Saiki RK, Scharf S, Faloona F, Mullis KB, Horn GT, Erlich HA, Arnheim N. Enzymatic amplification of beta-globin genomic sequences and restriction site analysis for diagnosis of sickle cell anemia. Science. 1985;230(4732):1350–4.
147. Saiki RK, Gelfand DH, Stoffel S, Scharf SJ, Higuchi R, Horn GT, Mullis KB, Erlich HA. Primer-directed enzymatic amplification of DNA with a thermostable DNA polymerase. Science. 1988;239(4839):487–91.
148. Santos-Medellin C, Zinke LA, Ter Horst AM, Gelardi DL, Parikh SJ, Emerson JB. Viromes outperform total metagenomes in revealing the spatiotemporal patterns of agricultural soil viral communities. ISME J. 2021;15(7):1956–70.
149. Scherneck S, Ulrich R, Feunteun J. The hamster polyomavirus—a brief review of recent knowledge*. Virus Genes. 2001;22(1):93–101.
150. Scholz JAC, Garg R, Compton SR, Allore HG, Zeiss CJ, Uchio EM. Poliomyelitis in MuLV-infected ICR-SCID mice after injection of basement membrane matrix contaminated with lactate dehydrogenase-elevating virus. Comp Med. 2011;61(5):404–11.
151. Schoondermark-van de Ven EM, Philipse-Bergmann IM, van der Logt JT. Prevalence of naturally occurring viral infections, mycoplasma pulmonis and clostridium piliforme in laboratory rodents in Western Europe screened from 2000 to 2003. Lab Anim. 2006;40(2):137–43.
152. Schotmüller H. Zur Ätiologie und Klinik der Bisskrankheit (Ratten-,Katzen-, Eichhörnchen-Bisskrankheit). Dermatoligische Wochenschrift. 1914;58:77.
153. Sebesteny A, Hill AC. Hepatitis and brain lesions due to mouse hepatitis virus accompanied by wasting in nude mice. Lab Anim. 1974;8(3):317–26.
154. Shah SA, Deng L, Thorsen J, Pedersen AG, Dion MB, Castro-Mejía JL, Silins R, Romme FO, Sausset R, Ndela EO, Hjemlsø M, Rasmussen MA, Redgwell TA, Vestergaard G, Zhang Y, Sørensen SJ, Bisgaard H, Enault F, Stokholm J, Moineau S, Petit M-A, Nielsen DS. Hundreds of viral families in the healthy infant gut. bioRxiv. 2021; https://doi.org/10.1101/2021.07.02.450849.

155. Shkoporov AN, Hill C. Bacteriophages of the human gut: the “known unknown” of the microbiome. Cell Host Microbe. 2019;25(2):195–209.
156. Shkoporov AN, Ryan FJ, Draper LA, Forde A, Stockdale SR, Daly KM, McDonnell SA, Nolan JA, Sutton TDS, Dalmasso M, McCann A, Ross RP, Hill C. Reproducible protocols for metagenomic analysis of human faecal phageomes. Microbiome. 2018;6(1):68.
157. Shkoporov AN, Clooney AG, Sutton TDS, Ryan FJ, Daly KM, Nolan JA, McDonnell SA, Khokhlova EV, Draper LA, Forde A, Guerin E, Velayudhan V, Ross RP, Hill C. The human gut Virome is highly diverse, stable, and individual specific. Cell Host Microbe. 2019;26(4):527–541.e525.
158. Shkoporov AN, Stockdale SR, Lavelle A, Kondova I, Heuston C, Upadrasta A, Khokhlova EV, van der Kamp I, Ouwerling B, Draper LA, Langermans JAM, Paul Ross R, Hill C. Viral biogeography of the mammalian gut and parenchymal organs. Nat Microbiol. 2022;7(8):1301–11.
159. Simmons JH, Purdy GA, Franklin CL, Trottier P, Churchill AE, Russell RJ, Besch-Williford CL, Riley LK. Characterization of a novel parainfluenza virus, caviid parainfluenza virus 3, from laboratory Guinea pigs (*Cavia porcellus*). Comp Med. 2002;52(6):548–54.
160. Smith T. Some bacteriological and environmental factors in the pneumonias of lower animals with special reference to the Guinea-pig. J Med Res. 1913;29(141):291–U227.
161. Smith W, Andrewes C, Laidlaw P. A virus obtained from influenza patients. Lancet. 1933;225:66–8.
162. Spackman E, Sitaras I. Hemagglutination inhibition assay. In: Spackman E, editor. Animal influenza virus: methods and protocols. New York: Springer US; 2020. p. 11–28.
163. Stabell EC, Olivo PD. Isolation of a cell line for rapid and sensitive histochemical assay for the detection of herpes simplex virus. J Virol Methods. 1992;38(2):195–204.
164. Stockdale SR, Draper LA, O’Donovan SM, Barton W, O’Sullivan O, Volpicelli-Daley LA, Sullivan AM, O’Neill C, Hill C. Alpha-synuclein alters the faecal viromes of rats in a gut-initiated model of Parkinson’s disease. Commun Biol. 2021;4(1):1140.
165. Taguchi F, Hirano N, Kiuchi Y, Fujiwara K. Difference in response to mouse hepatitis virus among susceptible mouse strains. Jpn J Microbiol. 1976;20(4):293–302.
166. Takeuchi A, Hashimoto K. Electron microscope study of experimental enteric adenovirus infection in mice. Infect Immun. 1976;13(2):569–80.
167. Tanaka M, Kuramochi M, Nakanishi S, Kuwamura M, Kuramoto T. Rat polyomavirus 2 infection in a colony of X-linked severe combined immunodeficiency rats in Japan. J Vet Med Sci. 2018;80(9):1400–6.
168. Tao L, Reese TA. Making mouse models that reflect human immune responses. Trends Immunol. 2017;38(3):181–93.
169. Theiler M. Spontaneous encephalomyelitis of mice, a new virus disease. J Exp Med. 1937;65(5):705–19.
170. Tisza MJ, Buck CB. A catalog of tens of thousands of viruses from human metagenomes reveals hidden associations with chronic diseases. Proc Natl Acad Sci USA. 2021;118(23):e2023202118.
171. Toolan HW, Ledinko N. Inhibition by H-1 virus of the incidence of tumors produced by adenovirus 12 in hamsters. Virology. 1968;35(3):475–8.
172. Toolan HW, Rhode SL, Gierthy JF. Inhibition of 7,12-dimethylbenz(a)anthracene-induced tumors in Syrian hamsters by prior infection with H-1 parvovirus. Cancer Res. 1982;42(7):2552–5.
173. Turnbaugh PJ, Ley RE, Mahowald MA, Magrini V, Mardis ER, Gordon JI. An obesity-associated gut microbiome with increased capacity for energy harvest. Nature. 2006;444(7122):1027–31.
174. Tzeng WP, Zhou Y, Icenogle J, Frey TK. Novel replicon-based reporter gene assay for detection of rubella virus in clinical specimens. J Clin Microbiol. 2005;43(2):879–85.
175. Ueno Y, Sugiyama F, Sugiyama Y, Ohsawa K, Sato H, Yagami K. Epidemiological characterization of newly recognized rat parvovirus, "rat orphan parvovirus". J Vet Med Sci. 1997;59(4):265–9.
176. Vonderfecht SL, Huber AC, Eiden J, Mader LC, Yolken RH. “Infectious diarrhea of infant rats produced by a rotaviruslike agent.” JVirol. 1984 52(1): 94–98.
177. Ward JM, Wobus CE, Thackray LB, Erexson CR, Faucette LJ, Belliot G, Barron EL, Sosnovtsev SV, Green KY. Pathology of immunodeficient mice with naturally occurring murine norovirus infection. Toxicol Pathol. 2006;34(6):708–15.
178. Werner RM, Allen AM, Small JD, New AE. Clinical manifestations of mousepox in an experimental animal holding room. Lab Anim Sci. 1981;31(5 Pt 2):590–4.
179. Williamson JSP, Stohlman SA. Effective clearance of mouse hepatitis-virus from the central-nervous-system requires both CD4+ and CD8+ t-cells. J Virol. 1990;64(9):4589–92.
180. Yalow RS, Berson SA. Immunoassay of endogenous plasma insulin in man. J Clin Invest. 1960;39(7):1157–75.
181. Yan A, Butcher J, Mack D, Stintzi A. Virome sequencing of the human intestinal mucosal-luminal Interface. Front Cell Infect Microbiol. 2020;10:582187.
182. Yilmaz S, Allgaier M, Hugenholtz P. Multiple displacement amplification compromises quantitative analysis of metagenomes. Nat Methods. 2010;7(12):943–4.
183. Yuan W, Wang J, Xu FJ, Huang BH, Lian YX, Rao D, Yin XQ, Wu ML, Zhu YJ, Zhang Y, Huang R, Guo PJ. Development of a duplex real-time RT-PCR for the simultaneous detection and differentiation of Theiler’s murine encephalomyelitis virus and rat theilovirus. J Virol Methods. 2016;236: 139–46.

184. Zaragoza-Solas A, Haro-Moreno JM, Rodriguez-Valera F, López-Pérez M. Long-read metagenomics improves the recovery of viral diversity from complex natural marine samples. mSystems. 2022;7(3):e0019222.
185. Zhai X, Castro-Mejía JL, Gobbi A, Aslampaloglou A, Kot W, Nielsen DS, Deng L. The impact of storage buffer and storage conditions on fecal samples for bacteriophage infectivity and metavirome analyses. Microbiome. 2023;11(1):193.
186. Zhang C, Burch M, Wylie K, Herter B, Franklin CL, Ericsson AC. Characterization of the eukaryotic virome of mice from different sources. Microorganisms. 2021;9(10):2064.
187. Zhou F, Gan R, Zhang F, Ren C, Yu L, Si Y, Huang Z. PHISDetector: a tool to detect diverse in silico phage-host interaction signals for virome studies. Genomics Proteomics Bioinformatics. 2022;20(3):508–23.
188. Zuo T, Lu XJ, Zhang Y, Cheung CP, Lam S, Zhang F, Tang W, Ching JYL, Zhao R, Chan PKS, Sung JJY, Yu J, Chan FKL, Cao Q, Sheng JQ, Ng SC. Gut mucosal virome alterations in ulcerative colitis. Gut. 2019;68(7):1169–79.

10 Bacteria and Bacterial Diagnostics

Henrik Christensen and Werner Nicklas

Abstract

Diagnostics of bacterial agents is important for health surveillance and to investigate diseases in laboratory rodent colonies. A basic principle of laboratory animal health monitoring is that animals are free from all microorganisms which can interfere with experimental results including their reproducibility. Sentinel animals are usually kept in separate cages supplied with bedding from colony cages to be tested in order to pick up infectious agents in the colonies. To monitor the status of gnotobiotic rodents including those with a defined microbiota, metagenomic techniques have been developed. The most prevalent bacteria reported in both mice and rats have been *Helicobacter* spp., *Rodentibacter pneumotropicus*, *Rodentibacter heylii* or other *Rodentibacter* spp., *Staphylococcus aureus*, and beta-hemolytic streptococci of Lancefield type B. For bacteria which can be cultured, MALDI-TOF MS is preferred as a first-line detection method. For *Chlamydia* spp., DNA sequencing requires some kind of preliminary propagation of the agent in cell culture. A major limitation of 16S rRNA gene sequence-based methods for identification including PCR is that some species cannot be separated, for instance, within *Bordetella*. The same problem is relevant for MALDI-TOF MS, and in such cases, whole genome sequencing is needed for species-level identification.

Keywords

Rodents · Bacteria · Specific pathogen-free organisms · Diagnosis · Host microbial interactions

10.1 Introduction

Diagnostics of bacterial agents is important for health surveillance and to investigate diseases in laboratory rodent colonies. First, we briefly describe under which conditions it is necessary to identify bacterial agents in rodent colonies. Further information about health monitoring programs and international standards covering all microorganisms is provided in Chap. 6 of this chapter. Here, we provide a background for the most relevant bacterial agents before we proceed with the diagnostic tests. The aim of bacterial diagnostics is to accurately identify the agents either involved in disease or with a potential to cause disease to genus and species levels.

H. Christensen (✉)
Department of Veterinary and Animal Sciences, University of Copenhagen, Frederiksberg, Denmark
e-mail: Hech@sund.ku.dk

W. Nicklas
Microbiological Diagnostics, German Cancer Research Centre, Heidelberg, Germany

F. Benavides, A. K. Hansen (eds.), *Rodent Quality Control: Genes and Bugs*, Laboratory Animal Science and Medicine 2, https://doi.org/10.1007/978-3-031-59103-7_10

Although diseases rarely occur in well-managed colonies, observations of clinical disease are essential for health monitoring and reporting of colonies. Bacterial infections may compromise certain types of experiments even though the animals are not affected. Routine health monitoring is needed with gnotobiotic (germ-free or associated with a defined microbiota), SPF animals, and imported animals kept in quarantine. A basic principle of laboratory animal health monitoring is that animals are free from all microorganisms which can interfere with experimental results including their reproducibility. It is not a requirement that animals used in biomedical research are free of certain microorganisms under all conditions. However, the health status of the experimental animals including prevalence of opportunistic bacteria should be considered when the experimental results are evaluated including final reporting [52]. Pritchett-Corning et al. [51] provides a systematic overview of infectious agents reported in a diagnostic laboratory based on more than 500,000 mice and 80,000 rats from Europe and North America. Here *Helicobacter* spp., *[Pasteurella] pneumotropica* (now *Rodentibacter pneumotropicus*, *R. heylii* or other *Rodentibacter* spp.), *Staphylococcus aureus,* and beta-hemolytic streptococci of Lancefield type B were the most prevalent in both mice and rats. All other bacterial agents occurred with low prevalence. Hayashimoto et al. [31] reported similar types of data obtained during a 1-year period of observation investigating various sample types and from the study, the dominant bacterial agents were reported as *S. aureus*, *[P.] pneumotropica* (now *R. pneumotropicus*, *R. heylii* or other *Rodentibacter* spp.) and *Helicobacter hepaticus* in mice and *Pseudomonas aeruginosa* in rat colonies. Most traditional "pathogens" such as *Bordetella bronchiseptica*, *Chlamydia* spp., *Clostridium piliforme* (Fig. 10.1), *Citrobacter rodentium*, *Corynebacterium kutscheri*, *Filobacterium rodentium*, *Leptospira* spp., *Listeria monocytogenes*, *Mycobacterium tuberculosis*, *Mycoplasma* spp., *Salmonella enterica*, *Streptobacillus* spp., *Yersinia* spp. do no longer play a role, as they have been eradicated from well managed laboratory rodent populations with the exceptions of *Klebsiella pneumonia* and *Klebsiella oxytoca*.

Anecdotic descriptions exist for a large number of additional bacterial agents that have occasionally been responsible for disease. Also, bacterial agents have been found associated with clinical signs or lesions which can still be found. Among them is *Bordetella pseudohinzii* [33]. It was earlier referred to as *Bordetella hinzii* [30] and has repeatedly been found in high numbers as a causative agent of respiratory disease in mice and has also been shown to confound pulmonary research [20]. Also, *Streptococcus agalactiae* (serogroup B) (Fig. 10.2) is still prevalent in populations of laboratory mice and rats [28, 38, 56] and may cause abscesses, septicemia, and can colonize various organs. Furthermore, it has some zoonotic potential and can cause severe disease especially in human neonates. The natural role of *Streptococcus pneumonia* in rodent colonies has rarely been reported during the last two

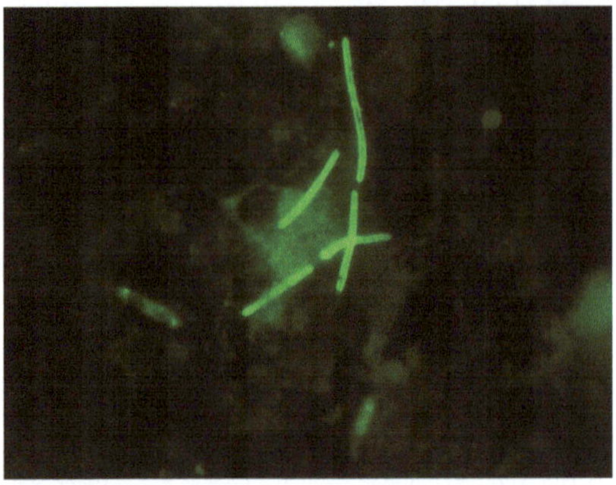

Fig. 10.1 *Clostridium piliforme* demonstrated by indirect immunofluorescence assay in the liver of a naturally infected rat

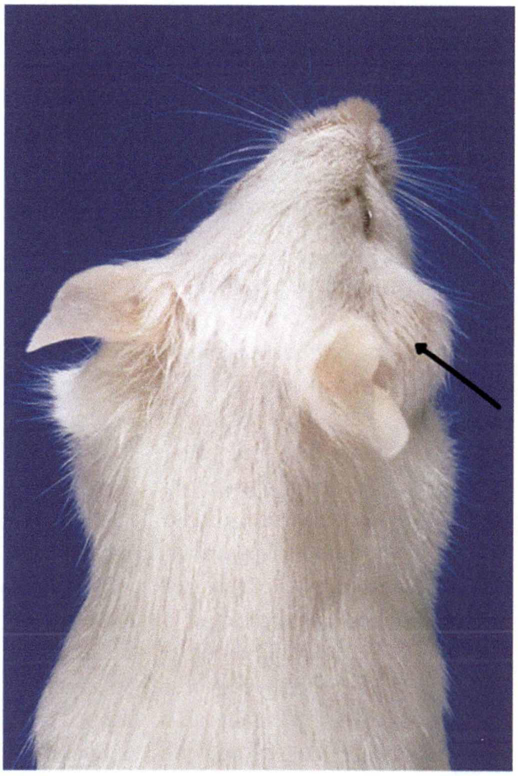

Fig. 10.2 Abscess in a BALB/c mouse upon receipt from a commercial breeder caused by *Streptococcus agalactiae*. (Photo Werner Nicklas)

decades [57] although this bacterium has frequently been used as a model organism in animal experiments.

A number of new genera of *Pasteurellaceae* and also other bacteria isolated from rodents have recently been validly published (*Caviibacterium, Conservatibacter, Cricetibacter, Mesocricetibacter, Necropsobacter, Rodentibacter,* and *Muribacter* (previously named as *Actinobacillus muris*). With the exception of *Rodentibacter* most of these are without reports of associated lesions [19]. However, these agents are of differential diagnostic importance and might also cause lesions or disease as has already been shown for *R. rarus* [43]. Additional hitherto undescribed *Pasteurellaceae* species have been mentioned in the literature such as "*Haemophilus influenzae-murium*" [49] and a species colonizing rats [55], which is still prevalent in contemporary populations of laboratory rodents.

10.2 Sample Types for Diagnostics of Bacterial Agents

Sentinel animals are usually kept in separate cages supplied with bedding from colony cages to be tested in order to pick up infectious agents in the colonies. Environmental sampling of the animal facilities such as soiled surfaces, IVC (individually ventilated cages), rack plenums, and filters of exhaust air dust may supplement sentinel or colony animal testing [41, 52]. Miller et al. [47] preferred exhaust air sampling over sentinel animals and gave several recommendations with respect to the types of filters sampled. The health monitoring of gnotobiotic rodents typically derived from aseptic hysterectomy or embryo transfer requires special focus on agents such as *Helicobacter* spp. and *Mycoplasma* spp. [50]. To monitor the status of such animals including those with a defined microbiota, metagenomic techniques have been developed [13]. The calculation of samples sizes has been described in Chap. 7.

10.3 Reference Strains

Type strains of bacterial species are preferred as controls of the diagnostic tests. Type strains can be obtained from the major culture collections. To locate type strains, the species name is first looked up in LPSN (List of Prokaryotic Names with Standing in Nomenclature) (https://lpsn.dsmz.de/) and here the type strain numbers can be identified in the different culture collections. The type strain of *Bordetella bronchiseptica* is, for instance, CCUG 219 from the culture collection Culture Collection University of Gothenburg. For most species, at least two versions of the type strain can be obtained from different culture collections. An exception is *Clostridium piliforme* and other agents which cannot easily be cultured for which no cultivable type strains exist. Here reference DNA sequences may be accessed from the nucleotide databases such as NCBI (https://www.ncbi.nlm.nih.gov/). Reference strains of *Rodentibacter* including 16S rRNA and whole genomic sequence accession numbers can be found in Table 3 in Benga et al. [8].

10.4 Note on Agents

In Table 10.1 the most relevant bacterial agents have been listed according to disease, host, and relative importance. For some agents additional information has been provided in the following.

10.4.1 *Helicobacter* spp.

Helicobacter spp. are among the bacterial agents most frequently detected contemporarily in laboratory mouse colonies. The genus includes Gram negative spiral—to curve shaped bacteria with a microaerophilic growth requirement and gastro intestinal habitat [60]. The species *Helicobacter bilis* and *Helicobacter hepaticus* have been isolated from hepatitis and hepatic neoplasia and have been associated with hepatic tumors. The latter has also been associated with inflammatory bowel disease. In A/JCr mice, *H. hepaticus* is strongly associated with hepatic cancer [60].

10.4.2 *Chlamydia psittaci* (*Chlamydophila psittaci*) and *Francisella tularensis*

These agents are rarely identified, however, if they are detected in the diagnostic laboratory special precautions need to be taken since they are considered zoonotic organisms of high biosecurity risk in most countries. The consequence may be that all material containing the agent needs to be autoclaved immediately or that special permissions should be obtained from the relevant authorities to continue to work with these organisms or to store them.

10.4.3 *Corynebacterium bovis*

This species was first identified as causative agent of hyperkeratosis in nude mice by [21]. Lesions ("scaly skin disease") are also observed in other immunodeficient (e.g., SCID) and in hairless mice (Fig. 10.3). The bacteria survive well in the environment, and immunocompetent animals can act as asymptomatic carriers. Furthermore, the bacteria can be transmitted through transplantable tumor lines [42]. Due to their high environmental resistance the agents are difficult to eradicate [15, 16].

10.4.4 *Rodentibacter* spp.

The species *Rodentibacter pneumotropicus* and *Rodentibacter heylii* have mainly been reclassified from former [*Pasteurella*] *pneumotropica* biovars Jawetz and Heyl, respectively. In addition, seven new species and two genomospecies exist within the genus *Rodentibacter* mostly

Table 10.1 Bacterial agents of potential relevance to be identified in laboratory animal colonies. (Material from Hayashimoto et al. [31], Mahler et al. [40], Pritchett-Corning et al. [52], Hansen et al. [29])

Species	Former names	Disease	Hosts	Role
Bordetella bronchiseptica		Lesions in airways including pneumoniae	Guinea pigs and pigs, dogs, cats	Low significance in mice and rats
Chlamydia spp.	*Chlamydophila* spp.	Persistent, subclinical infections, conjunctivitis, pneumoniae	Many different animals	Obligate intracellular, importance probably low, may have zoonotic potential
Citrobacter rodentium	*Citrobacter freundii* 4280	Transmissible colonic hyperplasia	Mice	Rare in laboratory mice, but can be found in mice from pet shops and in wild mice
Clostridium piliforme	"*Bacillus piliformis*"	Tyzzer's disease	Several species, gerbils highly susceptible	Obligate intracellular
Corynebacterium bovis		"Scaly skin disease," corynebacterial hyperkeratotic dermatitis	$Foxn1^{nu}$ mice, $Prkdc^{scid}$ mice	May be carried asymptomatic by immunocompetent mice
Corynebacterium kutscheri		"Pseudotuberculosis"	Mice, rats, guinea pigs	May be present in animals symptomless and only cause problems in immunodeficient animals
Escherichia coli		Colibacillosis (diarrhea, dehydration, mortality in young animals)	Rabbits, many different animals	
Filobacterium rodentium	Cilia associated respiratory (CAR) bacillus	Involved in chronic respiratory disease, rats with certain susceptibility	Mice, rats, and other species	May strongly interfere with immune responses
Francisella tularensis		Tularemia	Wild lagomorphs, other rodents and animals including humans	Rare, zoonotic
Helicobacter spp.		Clinical disease especially *H. hepaticus* causing inflammation induced colon carcinogenesis	Many different animals	*H. hepaticus*, *H. bilis* and *H. typhlonius* may impact animal experiments
Klebsiella oxytoca		Generalized infections, genital lesions, pneumonia, suppurative processes. Otitis media by *K. oxytoca* in C3H/HeJ mice	Rodents and rabbits	May interfere with certain animal models and cause disease in immunodeficient animals
Klebsiella pneumonia				
Lawsonia intracellularis		Proliferative enteritis "wet tail"	Hamsters	Intracellular
Leptospira spp.		Leptospirosis	Rats Zoonotic	Extremely rare in laboratory rodents
Mycoplasma pulmonis		Involved in chronic respiratory disease	Mice, rats	May strongly interfere with immune responses. Now rare
Pasteurella multocida		Snuffles	Rabbit	Rabbit may be severely affected

(continued)

Table 10.1 (continued)

Species	Former names	Disease	Hosts	Role
Proteus spp.		Opportunistic infections	Mice	
Pseudomonas aeruginosa		Disease in immunocompromised animals, low significance in immunocompetent	Many different animal species	Only of major concern for immunodeficient and selected groups of immunocompetent animals
Rodentibacter pneumotropicus and *Rodentibacter heylii*	[*Pasteurella*] *pneumotropica* and unnamed taxa	Respiratory -, ocular-, and genital infections, multiple abscesses. Mortality in immunocompetent and immunodeficient animals	Mice and rats	Among most prevalent bacteria in laboratory rodent colonies. Normally opportunistic pathogens, some members reported as primary pathogens
Salmonella enterica		Zoonotic potential	Many different animal species	Infrequently identified
Staphylococcus aureus		Dermatitis, abscesses including facial, zoonotic potential	Many different animal species	Ubiquitous in rodent populations where there is direct contact between humans and animals. Only of major concern for immunodeficient and selected groups of immunocompetent animals
Coagulase negative staphylococci (*S. xylosus*)			Different animal species	
Streptobacillus moniliformis		Rat bite fever, "Haverhill fever"	Rats, occasionally also mice	Rare in laboratory rodents
Streptococcus pneumoniae (alpha-haemolytic)		Important in immunodeficient animals	Most frequently in guinea pigs, also mice and rats	Zoonotic potential
Beta-haemolytic (*Streptococcus agalactiae*)				
Treponema paraluiscuniculi		Rabbit syphilis, "treponematosis"	Rabbit	
Yersinia pseudotuberculosis		Enteritis in acute form, caseous necrotic lesions in lymph nodes and abdominal organs in chronic form	Guinea pigs, rats, rabbits, humans	

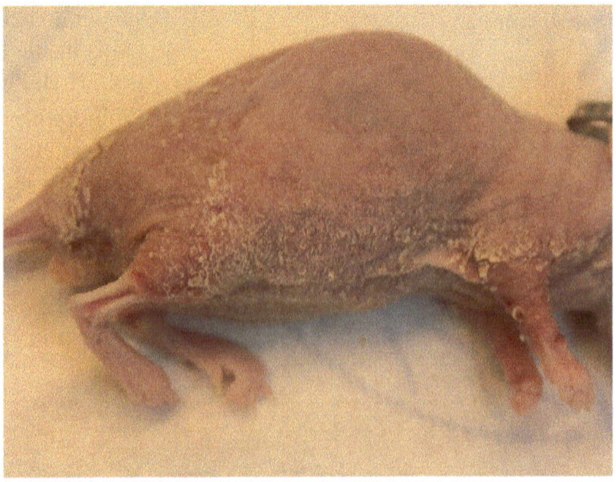

Fig. 10.3 Scaly skin disease in a nude mouse caused by *Corynebacterium bovis*

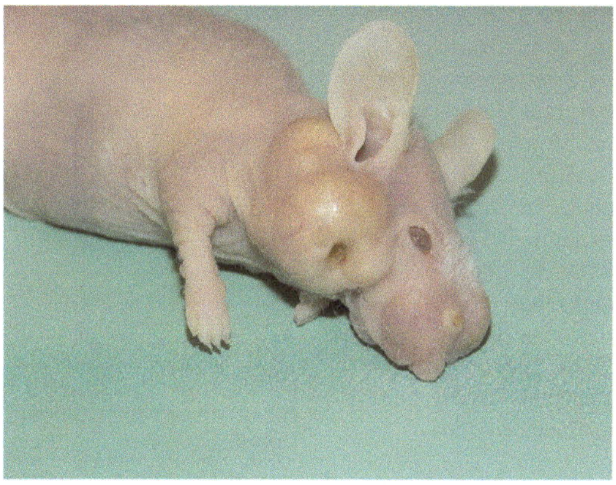

Fig. 10.4 Abscesses in the head region of an immunodeficient nude mouse caused by *Staphylococcus aureus*

without pathogenic potential reported [1, 11]. One of the new species, *R. rarus*, a previously rarely reported taxon, may have a potential to cause disease since it was reported from a street rat [43].

Members of *Rodentibacter* spp. may colonize the upper respiratory tract, conjunctival mucosa and middle ears and the genital mucosa and the lower intestinal tract and can also be isolated from feces [8]. Infections are usually clinically silent. The two species *R. pneumotropicus* and *R. heylii* have been reported as some of the most prevalent infections in laboratory mice and rats [8, 40]. *R. pneumotropicus* and *R. heylii* have mainly been isolated from mice while *R. ratti* and less frequently *R. heylii* have been isolated from rats [8, 9]. Clinical disease is expressed as conjunctivitis, abscesses, metritis, mastitis, and pneumonia. Immunodeficiency contributes to infections with members of *Rodentibacter* and such animals should be kept free of these agents [8]. The morbidity related to members of *Rodentibacter* is normally low. The development of clinical disease will most often be secondary to virus or mycoplasma infections. In addition, *Rodentibacter* spp. have been isolated in mice from pet shops and feral rodents [8]. Under normal conditions the agents are frequently tolerated as part of the normal microbiota, and they are considered as opportunistic pathogens. However, some members of both *R. pneumotropicus* and *R. heylii* have been described as primary pathogens [2, 26].

10.4.5 *Staphylococcus aureus*

S. aureus is regarded as an opportunistic pathogen for immunocompetent animals (Fig. 10.4). It seems to be extremely rare in wild mice and is probably introduced from the human staff. Unexpected high prevalence of *S. aureus* was observed in the study of [31], in which it was the most frequently isolated agent in both mice and rats.

10.5 Isolation

Isolation of the bacterial agent is needed if identification has to be based on cultivation (Fig. 10.5). The sample types used for isolation include fecal pellets and oral-pharyngeal swabs, swabs of the nasal cavities, swabs of the genital mucosa, and adhesive swabs of the fur. In addition, isolation can be carried out from bronchial or nasopharyngeal washes or *postmortem* from samples of lung specimens in addition to specimens of organs affected by lesions. *Postmortem* analysis needs to be performed by a veterinarian skilled with the work of laboratory animals including competences in isolation of bacterial agents and diagnostic methods. Specific detection methods are favored for selected agents (Table 10.2). Some agents cannot be cultured and here molecular-based detection, such as PCR, is the only option for a direct verification. Testing for anaerobic bacteria is usually not performed;

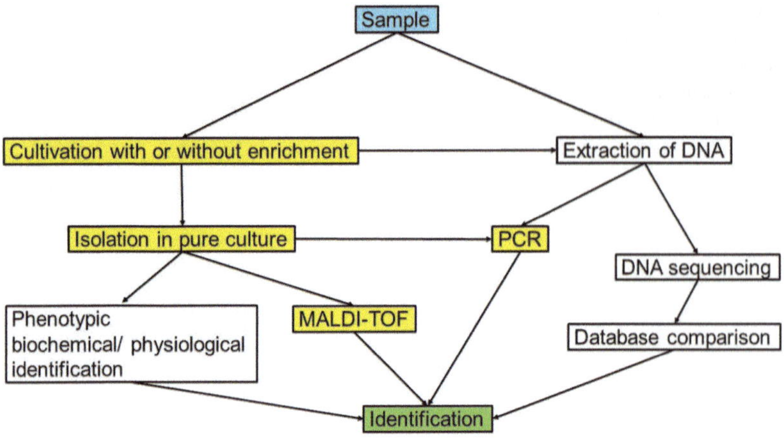

Fig. 10.5 *Flowchart for the identification of bacterial agents of potential importance for health monitoring disease in rodents.* The choice between the major methods is shown with respect to cultivation or direct detection from extracted DNA. The most frequently used methods are labeled yellow

Table 10.2 Diagnostics possibilities of bacterial agents to be identified in laboratory animal colonies. (Material from Hayashimoto et al. [31], Mähler et al. [40], Pritchett-Corning et al. [52], Hansen et al. [29]. Suitable tests are indicated with X with limitations (X))

Species	Phenotypic/ biochemical based on cultivation	Serology	Histology	PCR	MALDI-TOF	16S rRNA gene sequencing	Whole genome sequencing
Bordetella bronchiseptica	X			X	X		X
Chlamydia spp. and *Chlamydophila* spp.		X		X			
Citrobacter rodentium	X			(X)	X		X
Clostridium piliforme		X	X	X		X	
Corynebacterium bovis	X			X	X	X	X
Corynebacterium kutscheri	X			(X)	X	X	X
Escherichia coli	X				X	X	X
Filobacterium rodentium	(X)	X	X	X		X	X
Francisella tularensis	X				X	X	X
Helicobacter spp.	(X)			X		X	X
Klebsiella oxytoca	X			X	X	X	X
Klebsiella pneumonia	X			X	X	X	X
Lawsonia intracellularis		X		X		X	
Leptospira spp.	(X)	X	X	X	X		
Mycoplasma pulmonis	(X)	X		X		X	
Pasteurella multocida	X	X		X	X	X	X
Proteus spp.	X			X	X	X	X
Pseudomonas aeruginosa	X			X	X	X	X
Rodentibacter pneumotropicus and *Rodentibacter heylii*	X	X		X	X	X	X
Salmonella enterica	X	X		X	X	X	X
Staphylococcus aureus	X			X	X	X	X
Coagulase negative staphylococci (*S. xylosus*)				X	X	X	X
Streptobacillus moniliformis	X	X		X	X	X	X
Streptococcus pneumoniae (alpha-hemolytic)	X			X	X	X	X
Streptococcus spp. (beta-hemolytic)	X			X	X	(X)	X
Treponema paraluiscuniculi		X	X			X	
Yersinia pseudotuberculosis	X			X	X	X	X

however, strict anaerobic incubation is needed with suspicion. *Rodentibacter* spp. is best isolated from the nasopharynx and the nasal cavities as well as the genital tract. It is more difficult to obtain isolates from the intestinal site since members of *Rodentibacter* are easily overgrown during the isolation process by other bacteria that grow faster. Growth of *Pasteurellaceae* from the intestinal tract (and also from the oral cavity and the nasopharynx) can be supported by use of commercially available chocolate agar which contains bacitracin or other antibiotics that inhibit growth of Gram-positive bacteria ("Haemophilus agar"). The most common procedure for isolation is to use swabs for collection of material from the nasopharynx or genital tract and plating the material on blood agar plates. Columbia sheep or bovine blood agar is preferred. Some members of the species *R. ratti* and *R. heidelbergensis* as well as *R. trehalosifermentans* require NAD (V-factor) that is supplied by a nurse culture of *Staphylococcus hyicus* or other bacteria releasing the V-factor when cultivated on blood agar. Direct addition of NAD to the growth medium or use of chocolate agar is also a possibility to support the growth of these bacteria. A more selective medium including potassium tellurite and antibiotics has been described by Mikazuki et al. [45] and may be required for isolation of members of *Rodentibacter* from feces. *Haemophilus parainfluenzae* is V-factor requiring and has been isolated from laboratory animals although it is mainly known as a human pathogen [48].

10.5.1 Storage of Isolates

Storage of selected isolates is needed for documentation of the health monitoring protocol and to have references for internal as well as external documentation. Storage can be at −80 °C or in liquid nitrogen in appropriate media/matrices. A special 7.5% glucose serum medium for storage of the more sensitive cultures at −80 °C is recommended [39]. The glucose solution is sterilized by filtration and mixed with sterile calf serum and stored at −20 °C until use. Bacteria to be frozen are subcultured from a single colony which is spread on one or two blood agar plates. Following overnight incubation at 37 °C, the glucose serum medium is used for suspension of the growing culture, which subsequently is stored by freezing at −80 °C. Storage in liquid nitrogen can also be used as a backup to freezing at −80 °C for storage.

10.6 Methods for Identification

10.6.1 Phenotypic Identification

Phenotypic identification was the only choice for diagnostics of bacteria until the molecular and especially DNA-based methods took over three decades ago and until MALDI-TOF (Bruker, Nehren, Germany) came into common use. Phenotypic identification kits are still used a lot related to simplicity and direct way of analysis. Unfortunately, most conventional identification kits are difficult to use with *Rodentibacter* spp. and some other bacteria of veterinary importance. This is both related to the fastidious nature and long incubation times required for some reactions as well as for limited information about the organisms in the databases (e.g., *Citrobacter rodentium*). If phenotypic identification is needed for these organisms, traditional tests are recommended (Fig. 10.6)—and not quick tests [17]. The traditional tests are performed as, for instance, described in [4] and require incubation times up to 2 weeks whereas the quick test typically are of the API (Analytical Profile Index) (Biomérieux, Durham, USA) type where results are obtained immediately or after overnight incubation. Phenotypic identification of *Pasteurellaceae* isolated from rodents performed only with a limited number of tests was shown often to be erroneous [12]. Even with the classical biochemical identification definitive identification of most members of *Pasteurellaceae* to the species level is difficult. For instance, only few phenotypic characteristics separate *Rodentibacter* from *Muribacter*. *Rodentibacter* is negative in acid formation from cellobiose whereas *Muribacter* is positive and in addition the enzymatic tests of beta-glucosidase and alpha-

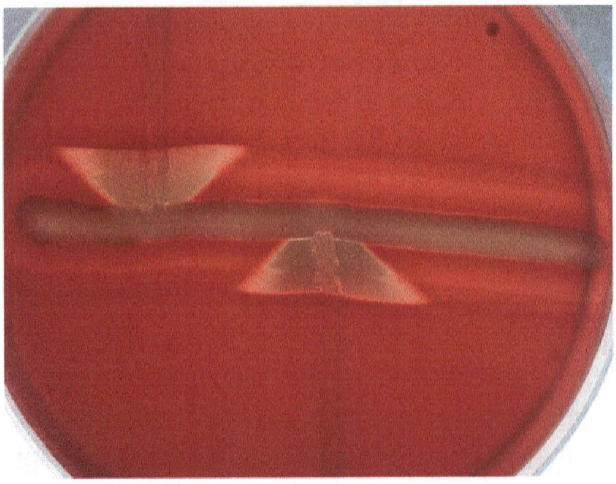

Fig. 10.6 *CAMP (Cristie Atkins Munch-Petersen) test.* Synergistic hemolysis on a blood agar plate as a result from interaction of the beta-hemolysin of *Staphylococcus aureus* and an extracellular protein produced by *Streptococcus agalactiae* (Lancefield group B)

fucosidase differ between the two genera. Characteristics for phenotypic separation of species of *Helicobacter* can be found in Zenner [60].

10.6.2 Antibiotic Resistance

Phenotypic antibiotic resistance of isolates can be measured as minimal inhibitory concentrations (MIC). Detailed protocols are available as Clinical Laboratory Standards Institutes (CLSI) standards. Results can be evaluated in CLSI and EUCAST (https://mic.eucast.org/Eucast2/SearchController/search.jsp?action=init).

The antibiotic resistance can be predicted based on the genes in the whole genomic sequence (WGS) (see section of DNA-based identification). On the webserver ResFinder (http://cge.cbs.dtu.dk/services/ResFinder/) the genomic sequences can be scanned for antibiotic resistance genes. The WGS needs to be uploaded in FASTA format (a standard text format for DNA sequence representation) where the relevant bacterial agent is selected. It is possible both to scan for "Acquired antimicrobial resistance genes" and to evaluate "Chromosomal mutations." The default threshold is 90% identity and 60% length coverage which is appropriate under most conditions [59]. The predicted antibiotic resistances can then be compared to the phenotypic antibiotic resistance and the two measures should correspond for most parameters.

10.6.3 MALDI-TOF

Matrix Assisted Laser Desorption Ionization (MALDI) time-of-flight mass spectrometry (TOF MS) has good specificity, however, the equipment is expensive. Detection systems are offered by the two companies Bruker (Billerica, Massachusetts, USA) and Bioméri ux, Durham, NC, USA). The analysis is often offered as core university of research institution facilities. A limitation is that the databases might be limited with respect to bacterial and fungal agents isolated from rodents. This can be overcome by setting up local databases complementing the commercial ones.

For MALDI-TOF, colony material to be analyzed is dispersed or dissolved in a matrix and thereafter exposed to high-intensity photons allowing the detection of mass to charge ratio when ionized proteins are traveling in an electric field in vacuum. A representation in the form of a characteristic mass spectrum will have peaks specific to genera and species unique to types of microorganisms. The identification can be done comparing the spectra to a database of known microbial isolates. Procedures have been described by Frey and Kuhnert [27]. Identification has been implemented by Adhikary et al. [2] and Kähl et al. [36] for members of *Rodentibacter* and *M. muris*.

10.7 ELISA and Immunofluorescence Tests

These tests allow the indirect testing of specific antibodies in the host against relevant bacteria. Such tests are not without problems and are recommended to be complemented by other tests such as PCR [8, 48]. Shortcomings are, for instance, cross-reactions between different bacteria and that seroconverted animals may remain seropositive after the agent has been eliminated from the individual. Another problem is that antibody formation may be delayed for bacteria living on mucus membranes. Serology has been used for *F. rodentium* (CAR bacillus), *C. piliforme* and *M. pulmonis* [32], and many of the other agents (Table 10.2). A new ELISA test based on a specific antigen from *R. pneumotropicus* was described by Fingas et al. [25].

10.8 Identification Based on PCR and DNA Sequencing

10.8.1 Isolation of DNA for PCR and DNA Sequencing

DNA can be extracted from cultivated bacteria using standard protocols. Boiling lysate methods are usually sufficient for PCR detection of Gram negative bacteria. For Gram positives, DNeasy Blood & Tissue (Qiagen, Germantown, MD, USA) or a similar kit including lysozyme treatment need to be used. For DNA extraction for whole genomic sequencing, a DNA extraction method is normally required which includes steps to avoid protein and RNA in the final DNA preparation. Also here, DNeasy Blood & Tissue, Maxwell RSC Cultured Cells DNA kit (Promega) or similar type of kits can be used.

10.8.2 Isolation of DNA for PCR Detection Without Previous Cultivation and Isolation of Bacteria

Direct extraction of DNA from animals or environmental samples with subsequent detection of relevant bacteria provides additional opportunities for monitoring. For the detection of nonculturable bacteria DNA always has to be extracted without culture. DNA can be extracted without previous cultivation from fecal pellets, tissue or from oral swabs or nasopharyngeal or bronchial washes [32]. For isolation of DNA from samples from animals and their environment species kits are needed such as PowerLyzer PowerSoil DNA isolation kit (Mobio, USA).

10.8.3 PCR Methods for Diagnostics

During the past three decades, PCR tests have been the preferred way to detect most of the bacterial agents of importance for health monitoring of laboratory rodents (Table 10.2). Amplification of the target sequence can be verified by DNA sequencing (Sanger). PCRs were early implemented and preferred for the identification of species of *Helicobacter* spp. related to the fastidious nature of these organisms [54, 60].

The use of the internal transcribed spacer (ITS) as target for identification has mainly been developed for use with rodent members of *Pasteurellaceae* [5, 8]. The internal transcribed spacer region between 16S and 23S rRNA genes varies in length and DNA sequence both within the same bacterial strain related to numerous ribosomal operons and between different species of bacteria. PCRs have been designed for diagnostic purpose utilizing these differences. For *Rodentibacter*, diagnostic PCRs have been described at the species level based on ITS allowing identification of *R. pneumotropicus*, *R. heylii*, *R. rarus*, *R. ratti*, *R. heidelbergensis,* and *R. haemolyticus* [10]. The PCR is simple to perform using the same conserved primers; however, the precise reading of fragment length requires accuracy and inclusion of reference strains. The ITS PCR was presented in a multiplex format by Benga et al. [6].

PCR tests have been further developed as quantitative PCR (qPCR) assays which has improved the sensitivity and shortened analysis time. Such applications will of course require the availability of more specialized testing equipment compared to the standard PCR machine. [23] described a qPCR test for the identification of [*P.*] *pneumotropica* biovars Jawetz and Heyl, respectively (now *R. pneomotropicus*, *R. heylii,* and *Rodentibacter* spp.). qPCR combined with high-resolution melting curve analysis [46, 47] enabled testing for the most relevant members of rodent *Pasteurellaceae* at one time. Henderson et al. [32] described the use of a high-density PCR array including most of the pathogens described in this chapter based on fluorogenic TaqMan PCR reactions.

[14] developed a qPCR test for *R. pneumotropicus* and *R. heylii* based on the virulence gene *ibpA*. The *ibpA* gene is predicted to encode for an inclusion body protein A. The PCR was used for routine hygienic monitoring with an environmental sampling strategy and compared to sentinel-based isolation of agents.

The majority of PCRs have been developed for testing of isolates of bacteria, however, it is also possible to test pharyngeal or genital swabs and other samples directly both originating from clinical samples and for monitoring. Here DNA can be isolated directly from oral swabs, gauze, or prefilters [47]. For estimating the number of organisms in the original sample, DNA standard curves are needed. A drawback of the methods using DNA directly from the sample as template is that DNA from dead organisms might be included in the detection and quantification.

PCRs for the detection of *S. aureus* including various toxins and antibiotic resistance types have been published during the past 3 decades [34]. Many PCRs for identification of *Mycoplasma* spp. have been published [37]. PCR for *C. piliforme* has been published [53]. A PCR test was published for the detection of [*Actinobacillus*] *muris*—now *M. muris* [7].

10.9 DNA Sequence-Based Methods of Detection

10.9.1 16S rRNA Gene Sequencing

The traditional starting point is PCR amplification of the 16S rRNA gene with conserved primers that in principle can be used for all bacteria [18]. The most simple starting point is a colony from an agar plate transferred under sterile conditions to the tube with the PCR mix. For some bacteria including Gram positives, DNA needs to be purified before the PCR. After PCR amplification, the DNA in the PCR tube is then purified to reduce the content of PCR primers and sent to a sequencing laboratory with the sequencing primers for Sanger sequencing. The DNA sequence is typically returned by email and assembled by a relevant computer program. Unfortunately, nearly all programs used for assembly require some sort of paid license. One such program is CLC Genomic Workbench http://www.clcbio.com/products/clc-genomics-workbench/, which is very user friendly including use on conventional Windows PCs. It is important to assemble the consensus sequence from reads made both by forward and reverse sequencing primers. The consensus sequence is generated and stored in text format and used as a query for comparison in a database. Comparison in EzBioCloud (https://www.ezbiocloud.net/) [58] allows comparison to all type strains of bacterial species with a validated published name at genus and species level. A phylogenetic 16S rRNA gene sequence-based analysis can be needed if similarity is obtained to more closely related species. The main limitation with16S rRNA-based identification is the limited resolution to identify some species which is why whole genomic sequencing is used.

10.10 Whole Genomic Sequencing (WGS)

Whole genomic sequencing has now become the primary method for confirmation and identification of bacterial strains. TYGS (type strain genome server) allows identification to species level based on a WGS uploaded as contigs in

FASTA format [44]. The benefit with TYGS is that the query WGS is automatically compared to WGS of all type strains of prokaryotes. WGS-based identification has higher species level specificity within certain genera compared to 16S rRNA gene sequence-based identification.

10.11 Population Genetic Investigations and Epidemiology

10.11.1 Multilocus Sequence Typing

Multilocus sequence typing (MLST) is used to identify and trace clonal groups and complexes. MLST is based on the comparison of DNA sequences of seven conserved genes in a range of strains in a species. For each gene (locus) all different version of sequences are scored as alleles and designated a random number. This number is not reflecting the qualitative difference between the sequences just stating that the sequences are different. An allele profile for instance 1, 3, 4, 5, 2, 3, 1 is defined as a sequence type (ST). MLST typing has mainly been performed on the servers https://pubmlst.org/general.shtml dedicated to a series of species. The databases on these servers have been maintained by curators who have uploaded new sequence types and information of isolates. Users are able to upload sequences to the servers and to obtain a sequence type. With respect to the diagnostics of bacteria associated disease in rodents, MLST is possible for *E. coli*, *C. trachomatis*, *H. cinaedi*, *H. suis*, *K. oxytoca*, *Mycoplasma* spp., *P. multocida*, *P. aeruginosa*, *S. enterica*, *S. aureus* and some species of streptococci such as *S. pneumonia* and *S. agalactiae* as a secondary option in the diagnostic procedure. MLST of *R. pneumotropicus* is possible at ivsmlst.sund.ku.dk based on WGS [3]. For *R. pneumotropicus*, 20 sequence types have been identified.

10.11.2 Single Nucleotide Polymorphisms (SNP)

For the most specific tracing of clones of bacterial agents, single nucleotide polymorphism (SNP) analysis based on whole genomic sequencing is recommended. SNP analysis is also the most sensitive method for strain level identification. In principle all nucleotide differences between two or more genomes can be scored. However, the concept is only operational with closely related genomes and used in regard to minor genetic changes (point mutations). To exclude insertions and deletions, nucleotide (nt) changes are only considered as SNPs if they are spaced at least 10. Sequence reads of strains to be compared are mapped to a reference sequence. The reference is best chosen as a fully closed WGS, however, if not available, the best WGS on contig form with a low number of contigs can substitute. The reference should be closely related to the strains being compared. SNP analysis can be performed at Center for Genomic Evolution (CGE): https://cge.cbs.dtu.dk/services/CSIPhylogeny/ [35]. First the reference WGS is uploaded in FASTA format. Then all isolates (three or more) to be compared to the reference are uploaded. For each strain sequenced by Illumina short read paired end technology, there will be two files (R1 and R2) in fastq format. Read files on "zipped" form are preferred for upload related to smaller file size. All sequences need to be uploaded at one time and a link will be returned with the results by email. The file "Snp-tree-aln.fasta" (alignment file in .fasta format) is downloaded for analysis of number of SNPs between strains. The SNPs can be enumerated from this multiple alignment. If a phylogeny is needed it is recommended to use the aligned SNPs downloaded (Snp-tree-aln.fasta) and use ClustalX or similar multiple alignment program to construct the phylogeny and FigTree or (http://tree.bio.ed.ac.uk/software/figtree/) or other similar tool like MEGAX to illustrate the phylogeny.

10.12 Concluding Remarks

In this chapter, we have tried to outline the most relevant tests to identify bacterial agents in health screenings of laboratory rodents including limitations with the different tests. For most bacterial agents, between three and six different methods for identification are available (see Table 10.2). If the agents are not easily cultivated, the number of methods available for identification are more limited. For *Chlamydia* and *Chlamydophila* spp., DNA sequencing requires some kind of preliminary propagation of the agent in cell culture which is not convenient for routine diagnostics. For bacteria which can be cultured, MALDI-TOF is preferred as a first line detection method if the equipment is available. Control identification for selected samples is recommended using PCRs or DNA sequencing (Fig. 10.1). Most of the PCR tests have been developed based on the 16S rRNA gene sequence as target. Some tests are now considered unspecific, since they were developed based on limited DNA sequence information and uncertain classification of strains used to develop the tests [8]. The problems of misidentification of members of *Pasteurellaceae* were described by [22] based on an unnamed species of *Rodentibacter* (now *R. haemolyticus*, [11]. Isolates including reference strains were submitted to three independent diagnostic laboratories and the outcome was divergent even for the reference strains. It was recommended to identify all species of *Pasteurellaceae* and to report all PCR primers used.

Full characterization of the microbiota of laboratory rodents has been suggested to take the full consequence of the diversity of functions that the microbiota may affect for research [13]. This is of particular importance for animals with a defined microbiota such as Altered Schaedler Flora. Also, specific diagnostics of agents which may be important for experiments even though they are not directly harmful to the laboratory animals (e.g., *Akkermansia muciniphila* [13]. When it comes to specificity, a major limitation of 16S rRNA gene sequence-based methods for identification—either PCR based or based on DNA sequencing—is that some species cannot be separated for instance within *Bordetella* (*B. avium, B. bronchiseptica, B. hinzii, B. parapertussis, B. pertussis, B. pseudoninzii*), *Streptobacillus* (*S. canis, S. felis, S. moniliformis, S. notomytis, S. ratti*) [24] and some species members of streptococci within the Mitis group where *S. pneumonia* is located. The same limitation is to some extent relevant for MALDI-TOF. The sensitivity may be important in certain diagnostic tests. Although some PCR tests are reported with very low limit of detection, they rarely beat direct cultivation where a single bacterium in principle can be detected. Before too much focus is placed on lowering the limit of detection, it needs to be considered if it is of any importance for the given problem.

References

1. Adhikary S, Bisgaard M, Dagnaes-Hansen F, Christensen H. Clonal outbreaks of [*Pasteurella*] pneumotropica biovar Heyl in two mouse colonies. Lab Anim. 2017;51(6):613–21.
2. Adhikary S, Nicklas W, Bisgaard M, Boot R, Kuhnert P, Waberschek T, Aalbaek B, Korczak B, Christensen H. *Rodentibacter* gen. nov. including *Rodentibacter pneumotropicus* comb. nov., *Rodentibacter heylii* sp. nov., *Rodentibacter myodis* sp. nov., *Rodentibacter* ratti sp. nov., *Rodentibacter* heidelbergensis sp. nov., *Rodentibacter trehalosifermentans* sp. nov., *Rodentibacter rarus* sp. nov., *Rodentibacter mrazii* and two genomospecies. Int J Syst Evol Microbiol. 2017;67(6):1793–806.
3. Adhikary S, Bisgaard M, Boot R, Benga L, Nicklas W, Christensen H. Development of multi locus sequence typing (MLST) of Rodentibacter pneumotropicus. Vet Microbiol. 2019;231:11–7. https://doi.org/10.1016/j.vetmic.2019.02.025
4. Barrow GI, Feltham RKA. Cowan and Steel's manual for the identification of medical bacteria. Cambridge: Cambridge University Press; 1993.
5. Benga L, Benten WP, Engelhardt E, Christensen H, Sager M. Analysis of 16S-23S rRNA internal transcribed spacer regions in *Pasteurellaceae* isolated from laboratory rodents. J Microbiol Methods. 2012;90(3):342–9.
6. Benga L, Benten WP, Engelhardt E, Bleich A, Gougoula C, Sager M. Development of a multiplex PCR assay based on the 16S-23S rRNA internal transcribed spacer for the detection and identification of rodent *Pasteurellaceae*. J Microbiol Methods. 2013;95(2):256–61.
7. Benga L, Benten WPM, Engelhardt E, Gougoula C, Sager M. Specific detection and identification of

[*Actinobacillus*] muris by PCR using primers targeting the 16S-23S rRNA internal transcribed spacer regions. J Microbiol Methods. 2013;94(2):88–93.
8. Benga L, Sager M, Christensen H. From the [*Pasteurella*] pneumotropica complex to *Rodentibacter* spp.: an update on [*Pasteurella*] pneumotropica. Vet Microbiol. 2018;217:121–34.
9. Benga L, Knorr JI, Engelhardt E, Gougoula C, Benten PM, Christensen H, Sager M. Current distribution of *Rodentibacter* species among the mice and rats of an experimental facility. J Am Assoc Lab Anim Sci. 2019;58(4):475–8.
10. Benga L, Benten PM, Engelhardt E, Köhrer K, Hueber B, Nicklas W, Christensen H, Sager M. Differentiation among *Rodentibacter* species based on 16S-23S rRNA internal transcribed spacer analysis. Comp Med. 2020;70(6):487–91.
11. Benga L, Nicklas W, Lautwein T, Verbarg S, Gougoula C, Engelhardt E, Benten WPM, Köhrer K, Sager M, Christensen H. *Rodentibacter haemolyticus* sp. nov. isolated from laboratory rodents. Int J Syst Evol Microbiol. 2021;71(8):004947.
12. Boot R, Bisgaard M. Reclassification of 30 *Pasteurellaceae* strains isolated from rodents. Lab Anim. 1995;29(3):314–9.
13. Buchheister S, Bleich A. Health monitoring of laboratory rodent colonies-talking about (r)evolution. Animals (Basel). 2021;11(5):1410.
14. Buchheister S, Roegener F, Zschemisch NH, Talbot SR, Christensen H, Bleich A. One for two: a novel and highly sensitive virulence factor-based quantitative polymerase chain reaction assay for the simultaneous detection of *Rodentibacter pneumotropicus* and *Rodentibacter heylii* in environmental sample material. Lab Anim. 2020;54(3):239–50.
15. Burr HN, Lipman NS, White JR, Zheng J, Wolf FR. Strategies to prevent, treat, and provoke *Corynebacterium*-associated hyperkeratosis in athymic nude mice. J Am Assoc Lab Anim Sci. 2011;50(3):378–88.
16. Burr HN, Wolf FR, Lipman NS. *Corynebacterium bovis*: epizootiologic features and environmental contamination in an enzootically infected rodent room. J Am Assoc Lab Anim Sci. 2012;51(2):189–98.
17. Christensen H, Kuhnert P, Busse HJ, Frederiksen WC, Bisgaard M. Proposed minimal standards for the description of genera, species and subspecies of the *Pasteurellaceae*. Int J Syst Evol Microbiol. 2007;57(Pt 1):166–78.
18. Christensen H, Andersson AJ, Jørgensen SL, Vogt JK. 16S rRNA amplicon sequencing for metagenomics. In: Christensen H, editor. Introduction to bioinformatics in microbiology. Cham: Springer International Publishing; 2018. p. 135–61.
19. Christensen H, Kuhnert P, Busse H-J, Blackall P, Bisgaard M, Nørskov-Lauritsen N. *Pasteurellaceae*. In: Bergey's manual of systematics of archaea and bacteria. New York: Wiley; 2020. p. 1–26.
20. Clark SE, Purcell JE, Sammani S, Steffen EK, Crim MJ, Livingston RS, Besch-Williford C, Fortman JD. *Bordetella pseudohinzii* as a confounding organism in murine models of pulmonary disease. Comp Med. 2016;66(5):361–6.
21. Clifford CB, Walton BJ, Reed TH, Coyle MB, White WJ, Amyx HL. Hyperkeratosis in athymic nude mice caused by a coryneform bacterium: microbiology, transmission, clinical signs, and pathology. Lab Anim Sci. 1995;45(2):131–9.
22. Dafni H, Greenfeld L, Oren R, Harmelin A. The likelihood of misidentifying rodent *Pasteurellaceae* by using results from a single PCR assay. J Am Assoc Lab Anim Sci. 2019;58(2):201–7.
23. Dole VS, Banu LA, Fister RD, Nicklas W, Henderson KS. Assessment of rpoB and 16S rRNA genes as targets for PCR-based identification of *Pasteurella pneumotropica*. Comp Med. 2010;60(6):427–35.
24. Eisenberg T, Heydel C, Prenger-Berninghoff E, Fawzy A, Kling U, Akimkin V, Semmler T, Mühldorfer K, Kämpfer P, Blom J, Ewers C. *Streptobacillus canis* sp. nov. isolated from a dog. Int J Syst Evol Microbiol. 2020;70(4):2648–56.
25. Fingas F, Volke D, Hassert R, Fornefett J, Funk S, Baums CG, Hoffmann R. Sensitive and immunogen-specific serological detection of *Rodentibacter pneumotropicus* infections in mice. BMC Microbiol. 2019;19(1):43.
26. Fornefett J, Krause J, Klose K, Fingas F, Hassert R, Benga L, Grunwald T, Müller U, Schrödl W, Baums CG. Comparative analysis of humoral immune responses and pathologies of BALB/c and C57BL/6 wildtype mice experimentally infected with a highly virulent *Rodentibacter pneumotropicus* (*Pasteurella pneumotropica*) strain. BMC Microbiol. 2018;18(1):45.
27. Frey J, Kuhnert P. Identification of animal *Pasteurellaceae* by MALDI-TOF mass spectrometry. Methods Mol Biol. 2015;1247:235–43.
28. Geistfeld JG, Weisbroth SH, Jansen EA, Kumpfmiller D. Epizootic of group B Streptococcus agalactiae serotype V in DBA/2 mice. Lab Anim Sci. 1998;48(1):29–33.
29. Hansen AK, Nielsen DS, Krych L, Hansen CHF. Bacterial species to be considered in quality assurance of mice and rats. Lab Anim. 2019;53(3):281–91.
30. Hayashimoto N, Yasuda M, Goto K, Takakura A, Itoh T. Study of a *Bordetella hinzii* isolate from a laboratory mouse. Comp Med. 2008;58(5):440–6.
31. Hayashimoto N, Morita H, Ishida T, Yasuda M, Kameda S, Uchida K, Tanaka M, Ozawa M, Sato A, Takakura A, Itoh T, Kagiyama N. Current microbiological status of laboratory mice and rats in experimental facilities in Japan. Exp Anim. 2013;62(1):41–8.
32. Henderson KS, Perkins CL, Havens RB, Kelly M-JE, Francis BC, Dole VS, Shek WR. Efficacy of direct detection of pathogens in naturally infected mice by using a high-density PCR array. J Am Assoc Lab Anim Sci. 2013;52(6):763–72.
33. Ivanov YV, Linz B, Register KB, Newman JD, Taylor DL, Boschert KR, Le Guyon S, Wilson EF, Brinkac

LM, Sanka R, Greco SC, Klender PM, Losada L, Harvill ET. Identification and taxonomic characterization of *Bordetella pseudohinzii* sp. nov. isolated from laboratory-raised mice. Int J Syst Evol Microbiol. 2016;66(12):5452–9.

34. Johnson WM, Tyler SD, Ewan EP, Ashton FE, Pollard DR, Rozee KR. Detection of genes for enterotoxins, exfoliative toxins, and toxic shock syndrome toxin 1 in *Staphylococcus aureus* by the polymerase chain reaction. J Clin Microbiol. 1991;29(3):426–30.
35. Kaas RS, Leekitcharoenphon P, Aarestrup FM, Lund O. Solving the problem of comparing whole bacterial genomes across different sequencing platforms. PLoS One. 2014;9(8):e104984.
36. Kähl S, Maier T, Benga L, Fingas F, Baums CG. Differentiation of *Rodentibacter pneumotropicus*, *Rodentibacter heylii* and *Muribacter muris* by MALDI-TOF MS. J Microbiol Methods. 2020;169:105836.
37. Kim DJ, Park JH, Seok SH, Cho SA, Baek MW, Lee HY, Yang KH, Jang DD, Han BS, Park JH. Differential identification of mycoplasma pulmonis and *M. arthritidis* using PCR-based RFLP. Exp Anim. 2005;54(4):359–62.
38. Koga T, Aoki W, Fujii M, Satou K, Ikeda Y. Spontaneous infection caused by *Streptococcus agalactiae* in KK-A(y) mice. Comp Med. 2017;67(5):416–9.
39. Lapage SP, Redway KF. Preservation of bacteria with notes on other micro-organisms. London: HM Stationery Office; 1974.
40. Mahler M, Berard M, Feinstein R, Gallagher A, Illgen-Wilcke B, Pritchett-Corning K, Raspa M. FELASA recommendations for the health monitoring of mouse, rat, hamster, guinea pig and rabbit colonies in breeding and experimental units. Lab Anim. 2014;48:178–92.
41. Mailhiot D, Ostdiek AM, Luchins KR, Bowers CJ, Theriault BR, Langan GP. Comparing mouse health monitoring between soiled-bedding sentinel and exhaust air dust surveillance programs. J Am Assoc Lab Anim Sci. 2020;59(1):58–66.
42. Manuel CA, Bagby SM, Reisinger JA, Pugazhenthi U, Pitts TM, Keysar SB, Arcaroli JJ, Leszczynski JK. Procedure for horizontal transfer of patient-derived xenograft tumors to eliminate *Corynebacterium bovis*. J Am Assoc Lab Anim Sci. 2017;56(2):166–72.
43. Medkour H, Laidoudi Y, Dahmana H, Salvi B, Lepidi H, Mediannikov O, Davoust B. Severe pneumonia in a street rat (*Rattus norvegicus*) caused by *Rodentibacter rarus* strain RMC2. Open Vet J. 2021;11(1):165–73.
44. Meier-Kolthoff JP, Göker M. TYGS is an automated high-throughput platform for state-of-the-art genome-based taxonomy. Nat Commun. 2019;10(1):2182.
45. Mikazuki K, Takahashi K, Nanba K, Hayashi Y. [Selective media for Pasteurella pneumotropica]. Jikken Dobutsu. 1987;36(3):229–37. Japanese.
46. Miller M, Zorn J, Brielmeier M. High-resolution melting curve analysis for identification of *Pasteurellaceae* species in experimental animal facilities. PLoS One. 2015;10(11):e0142560.
47. Miller M, Ritter B, Zorn J, Brielmeier M. Exhaust air dust monitoring is superior to soiled bedding sentinels for the detection of *Pasteurella pneumotropica* in individually ventilated cage systems. J Am Assoc Lab Anim Sci. 2016;55(6):775–81.
48. Nicklas W. Chapter 19 – *Pasteurellaceae*. In: Fox JG, Davisson MT, Quimby FW, et al., editors. The mouse in biomedical research. 2nd ed. Burlington: Academic; 2007. p. 469–XII.
49. Nicklas W, Bisgaard M, Aalbæk B, Kuhnert P, Christensen H. Reclassification of *Actinobacillus muris* as *Muribacter muris* gen. nov., comb. nov. Int J Syst Evol Microbiol. 2015;65(10):3344–51.
50. Nicklas W, Keubler L, Bleich A. Maintaining and monitoring the defined microbiota status of gnotobiotic rodents. ILAR J. 2015;56(2):241–9.
51. Pritchett-Corning KR, Cosentino J, Clifford CB. Contemporary prevalence of infectious agents in laboratory mice and rats. Lab Anim. 2009;43(2):165–73.
52. Pritchett-Corning KR, Prins J-B, Feinstein R, Goodwin J, Nicklas W, Riley L. AALAS/FELASA working group on health monitoring of rodents for animal transfer. J Am Assoc Lab Anim Sci. 2014;53(6):633–40.
53. Pritt S, Henderson KS, Shek WR. Evaluation of available diagnostic methods for *Clostridium piliforme* in laboratory rabbits (*Oryctolagus cuniculus*). Lab Anim. 2010;44(1):14–9.
54. Riley LK, Franklin CL, Hook RR Jr, Besch-Williford C. Identification of murine helicobacters by PCR and restriction enzyme analyses. J Clin Microbiol. 1996;34(4):942–6.
55. Schulz S, Pohl S, Mannheim W. Mischinfektion von Albinoratten durch *Pasteurella pneumotropica* und eine neue pneumotrope *Pasteurella* species. Zentralbl Veterinarmed B. 1977;24(6):476–85.
56. Shuster KA, Hish GA, Selles LA, Chowdhury MA, Wiggins RC, Dysko RC, Bergin IL. Naturally occurring disseminated group B streptococcus infections in postnatal rats. Comp Med. 2013;63(1):55–61.
57. van der Linden M, Al-Lahham A, Nicklas W, Reinert RR. Molecular characterization of pneumococcal isolates from pets and laboratory animals. PLoS One. 2009;4(12):e8286.
58. Yoon SH, Ha SM, Kwon S, Lim J, Kim Y, Seo H, Chun J. Introducing EzBioCloud: a taxonomically united database of 16S rRNA gene sequences and whole-genome assemblies. Int J Syst Evol Microbiol. 2017;67(5):1613–7.
59. Zankari E, Hasman H, Cosentino S, Vestergaard M, Rasmussen S, Lund O, Aarestrup FM, Larsen MV. Identification of acquired antimicrobial resistance genes. J Antimicrob Chemother. 2012;67(11):2640–4.
60. Zenner L. Pathology, diagnosis and epidemiology of the rodent *Helicobacter* infection. Comp Immunol Microbiol Infect Dis. 1999;22(1):41–61.

11 Parasites and Parasitic Diagnostics

Andrew R. Williams

Abstract

Parasites constitute a diverse highly extremely successful group of organisms, ranging from single-celled protozoa to multicellular helminths. In laboratory rodents, parasitic infections may be commonly found. Often, infections may be asymptomatic, although clinical infections are not uncommon. More importantly, parasites have profound modulatory effects on the host immune system and microbiome, which may drastically confound experimental studies within many fields of biomedicine. For this reason, routine monitoring of rodent facilities for common parasites is warranted. In this chapter, the biology and life cycles of common rodent parasites are reviewed, as well as the most widely used diagnostic techniques. Future perspectives on the monitoring and control of parasites in experimental rodent facilities are discussed.

Keywords

Rodents · Parasites · Specific pathogen-free organisms · Diagnosis · Host microbial interactions

A. R. Williams (✉)
Department of Veterinary and Animal Sciences, University of Copenhagen, Frederiksberg, Denmark
e-mail: arw@sund.ku.dk

11.1 Parasite Infections

Parasitism is one of the most successful forms of life on the plant, with most organisms harbouring one or more parasitic species. Parasitic infestations in animals range from highly pathogenic infections that can cause substantial disease, and even death, through to sub-clinical or asymptomatic infections. Moreover, some parasites have been identified as possible symbionts that may aid in maintaining intestinal health and homeostasis, and indeed controlled infection with parasites has recently been investigated as a tool to prevent harmful inflammation in humans suffering from autoimmune diseases.

Given these profound effects on host gut function and immunity, the importance of parasites goes beyond their potential negative effects on host health and welfare. They also may substantially confound many experimental laboratory investigations in the fields of immunology, endocrinology, nutrition, and microbiome research. For this reason, routine diagnosis and monitoring of parasitic infections in laboratory rodents is warranted. As most parasites are not transmitted directly from animal to animal, but rely on an incubation stage in the external environment, control of transmission is markedly reduced by good hygiene and cleanliness. Thus, inadvertent parasite infection is sometimes not considered a significant issue in SPF lab mice. However, subclinical infections with parasites such as pin-

F. Benavides, A. K. Hansen (eds.), *Rodent Quality Control: Genes and Bugs*, Laboratory Animal Science and Medicine 2, https://doi.org/10.1007/978-3-031-59103-7_11

Table 11.1 Characteristics of common murine parasites

		Pathogenicity	Predilection site	Pre-patent period	Typical diagnosis
Helminths					
Pinworms	*Syphacia obvelata*	Low	Colon	~2 weeks	Tape test
	Aspiculuris tetraptera	Low	Colon	~3 weeks	Faecal egg count
Whipworms	*Trichuris muris*	Medium	Caecum	~4–5 weeks	Faecal egg count
Roundworms	*Heligmosomoides polygyrus*	Low-Medium	Duodenum	~2 weeks	Faecal egg count
Protozoa					
	Giardia muris	Medium	Jejunum	~5 days	Faecal cyst count Intestinal Histology
	Tritrichomonas muris	Low	Colon	~5–7 days	Faecal pseudocyst count Intestinal Histology
	Spironucleus muris	Low-Medium	Duodenum	~3–5 days	Faecal cyst count Intestinal Histology

worms may persist even in highly clean and controlled environments, and parasitological examination of sentinel animals is (during concurrent for screening for viruses, etc.) therefore highly worthwhile.

Here, we will discuss some of the basic biology of parasites of rodents, their transmission and pathological effects on their hosts, and diagnostic methods for common parasites that may be encountered in lab rodents. Features of the most common parasites are summarized in Table 11.1. We focus on enteric parasites as the most common infections that are encountered. Extra-intestinal parasites such as *Toxoplasma gondii*, *Klebsiella* spp., or *Encephalitozoon cuniculi* are rarely present in laboratory rodents, and the reader is referred to other sources [19] for a description of their life cycle and diagnostic procedures.

11.2 Parasitic Disease in Rodents

11.2.1 Helminths

Gastrointestinal worms (helminths) are ubiquitous in wild mice, with studies recording infection prevalence approaching 100% in samples of mice trapped in outdoor areas [4, 21]. Infection can have a large influence on the body condition of the mice, with heavy worm burdens correlating with reduced weight and body condition [29, 31]. Inbred laboratory mice are amenable to experimental infection in order to model the human or livestock response to helminths. These experimental systems have been instrumental in defining many of the core tenets of type-2 immunity. They include the discovery and functional characterization of innate type-2 lymphoid cells, and identification of classical type-2 cytokines, such as interleukin(IL)-4, and IL-13 that underpin our understanding of both helminth immunity and type-2 driven autoimmune disorders such as asthma and food allergies [9]. Moreover, enteric helminths are capable of causing chronic infections that feature a strong regulatory component including the production of T-regulatory cells and the cytokines IL-10 and TGFβ. Indeed, antigens from mouse nematodes are capable of inducing de novo Treg commitment in naïve splenocytes in vitro [17]. Furthermore, alterations in the composition of the commensal gut microbiome are also a feature of enteric helminth infections [7].

As a consequence of these immunomodulatory effects, underlying helminth infections may markedly impact on experimental investigations into metabolic or inflammatory diseases. Indeed, the presence of helminths has been shown to alleviate some indicators of obesity and diabetes in different models, which would have obvious implications for studies assessing novel interventions against these conditions [20]. Underlying helminth infections can also markedly skew the course of infection with other pathogens such as viruses or enteric bacteria, compromising efforts to investigate new vaccines or other immune-based therapeutics for these infections [23].

11.2.2 Protozoa

Unicellular protozoan infections may range from innocuous symbionts that reside in the commensal microbiome through to virulent pathogens which can cause severe disease. Wild mice sampled in the environment offer harbour infections with *Entamoeba muris, Giardia muris*, or *Eimeria* spp. Experimental infections with these pathogens has shown that they can cause pathology including reduced weight gain and soft faeces [24]. In contrast, the presence of commensals such as *Tritrichomonas muris* is not normally associated with pathology.

Protozoan infection with parasites such as *G. muris* has the potential to interfere with research studies on mucosal immunity or gut function as they induce highly polarized Th17 responses, which may potentially interfere with desired immune responses [10]. Furthermore, immunocompromised mice (e.g., SCID or some strains genetically deficient in defined immune molecules) may also spontaneously develop severe disease if inadvertently exposed to parasites such as *G. muris* or *Cryptosporidium* spp. [19, 24].

11.3 Diagnosing Parasitic Infections

11.3.1 Helminths

11.3.1.1 Pinworms

The most common helminths found in laboratory rodent populations are the pinworms or oxyurids. Even rodent facilities with high levels of hygiene and free of other diseases can harbour pinworm transmission. The two main species of pinworms in laboratory mice are *Syphacia obvelata* and *Aspiculuris tetraptera.* Both species share a similar, direct, life cycle. Adult worms reside in the lumen of the caecum or colon where they feed on bacteria. Adult females migrate to the anus in order to lay eggs, which are then transmitted to the environment, where they "embryonate" (become infective) within 1–2 days. Once ingested by a new host, eggs hatch and the larvae establish themselves in the colon, developing to adults within two (*S. obvelata*) or 3 weeks (*A. tetraptera*). Unlike other helminths, *S. obvelata* females die after producing eggs, whereas *A. tetraptera* can have multiple bouts of egg-laying. For both species, the life span of the adult worms is no more than 6 weeks.

Pinworms tend to have little clinical effects on their host. Unlike other enteric helminths, they do not invade host cells or gut tissue, which results in little inflammation or host-driven pathology. However, pinworms antigens are secreted and recognized by the host which can significantly modulate the immune system and has been shown to interfere with experimental models, for example, NOD mice are less prone to develop obesity if pinworms are present [16]. For these reasons, routine monitoring for pinworm infection may be necessary.

Diagnosis of *S. obvelata* is routinely achieved by the "tape test", whereby a small piece of cellophane tape is applied to the perianal area, removed, and then fixated to a glass microscope slides [11]. Eggs can are observed by direct microscopy and can be identified by their oval shape and embryo contents (Figs. 11.1 and 11.2). *S. obvelata* eggs are significantly larger than *A. tetrapera* eggs (length of 140 μm vs. 80 μm, respectively). In contrast, *A. tetrapera* eggs are shed in faeces and therefore the tape test is not particularly useful for assessing egg output; a faecal egg count (FEC) test using saturated salt flotation of eggs from faecal homogenates is preferred [30].

While the tape test or FEC tests are quick to perform and the eggs are easily observed by standard microscopy, they have the drawback of being unable to detect infections in the pre-patent phase. Moreover, for *S. obvelata*, the tape test has rather low sensitivity as eggs may be removed by the grooming of mice, meaning that false nega-

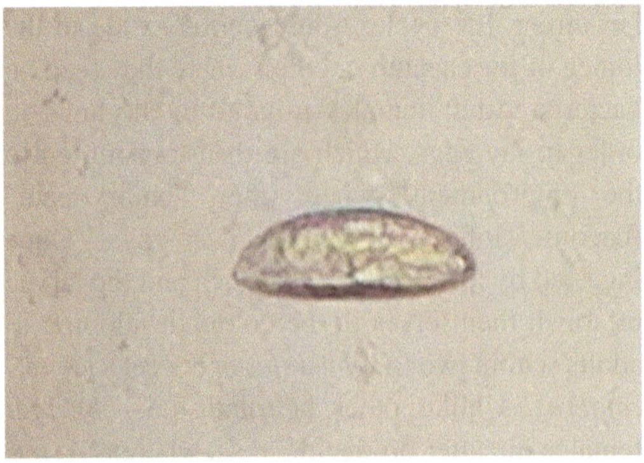

Fig. 11.1 *Syphacia* spp. egg (×10 magnification) [5]

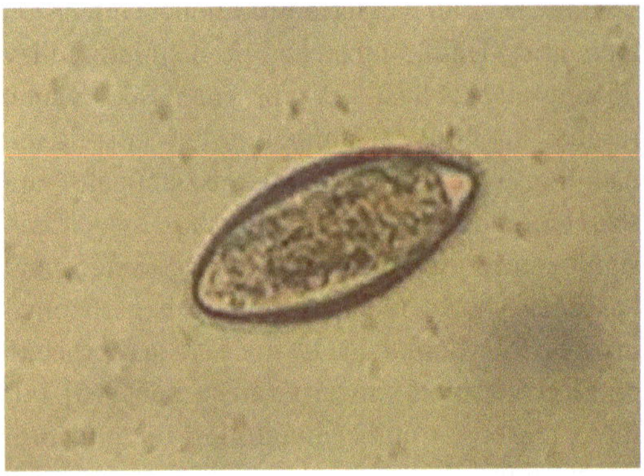

Fig. 11.2 *Aspiculuris* spp. egg (×10 magnification) [5]

tives are common. The gold-standard diagnosis is the observation of adult worms at postmortem examination, where they are readily found in the caecum and colon. However, given that diagnoses may be desired on living animals, molecular methods have been developed that may allow sensitive PCR-based identification of species-specific DNA. *A. tetraptera* infection can be positively identified by PCR assays targeting the 18S rRNA gene in faeces taken from either sentinel animals or mice suspected of infection [8]. Such molecular assays may thus prove to be a valuable adjunct screening tool, in additional to classical parasitological methods.

11.3.1.2 Whipworms

The murine whipworm *Trichuris muris* is a very common parasite of wild mice [4]. However, *T. muris* infestations in SPF mice under laboratory conditions are quite rare, unless the facility is also being used for experimental *T. muris* infections. In these cases, inadvertent transmission of *T. muris* is possible as the embryonated (infective) eggs can remain viable in the environment for many months, or even years when stored under laboratory conditions (unpublished observation). Adult *T. muris* worms reside in the caecum, and eggs are shed in the faeces. The eggs then take approximately 4–6 weeks to develop to the infective stage—for this reason, thorough cleaning and good hygiene practices can effectively disrupt the transmission cycle. Upon ingestion of eggs, adult worms take around 30 days to develop and start laying eggs.

Experimental infections with high doses of *T. muris* eggs normally induce a robust immunity in most mouse strains that is dependent on $CD4^+$ T-helper cells that secrete IL-4 and IL-13 [6]. However, during natural infections, mice are typically exposed to a continual ingestion of a low number of eggs which results in a chronic infection that can last for >6 months [6]. Unlike pinworm infection, whipworms have pronounced effects on the host which can include reduced weight gain, a rough coat, and intestinal inflammation that presents as colitis-like pathology by postmortem examination. However, many infections, particularly with a very low number of worms, can be asymptomatic.

The whipworms are large and readily seen by the naked eye in the caecum at postmortem. Fecal examination is done by the flotation method, and the eggs are lemon-shaped, approx. 55 μm long and 30 μm wide with characteristic "plugs" at either end (Fig. 11.3). To date, reports of molecular detection of *T. muris* eggs in fecal samples are scant, however, loop-mediated isothermal amplification (LAMP)-based assays have been described that may be able to detect pre-patent positivity in urine samples [14]. If *T. muris* eggs are observed in the feces during routine microbiological screening, the facility may need to be deep-cleaned with alcohol-based disinfectants to remove the hardy and resistant eggs.

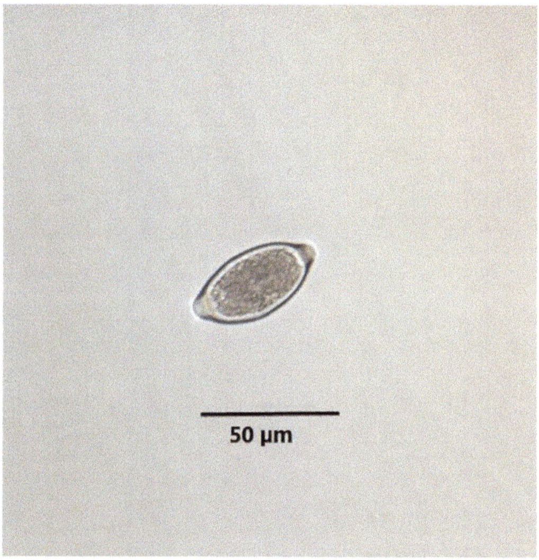

Fig. 11.3 *Trichuris muris* egg. (Source: author)

11.3.1.3 Roundworms

The most common roundworm of mice is *Heligmosomoides polygyrus*, which, similar to whipworms, is extremely common in outdoor wild mice but rarely seen in SPF mice. These worms live in the proximal small intestine and in most strains of laboratory mice they can establish long-term infections that may last several months, in part due to the induction of a T-regulatory cell response which inhibits worm expulsion [17]. The pre-patency phase is around 10 days, and egg excretion may last for many weeks. Unlike pinworms and whipworms, *H. polygyrus* has a free-living larval stage whereby the eggs hatch in the environment (typically within 24 h after excretion) and then undergo two larval moults to reach the third-larval stage which is the infective stage (within 7 days). Whilst the larvae can survive for several weeks if environmental conditions are optimal, unlike the hardy infective eggs of pinworms the larvae are fragile and easily killed by cleaning—thus, standard hygiene practices such as frequent changing of bedding and cleaning of cages should negate the chance of transmission. Diagnosis is by FEC test, whereby the eggs are oval shaped and around 80 µm long (Fig. 11.4). The reddish-coloured worms are readily observable in the duodenum at postmortem examination.

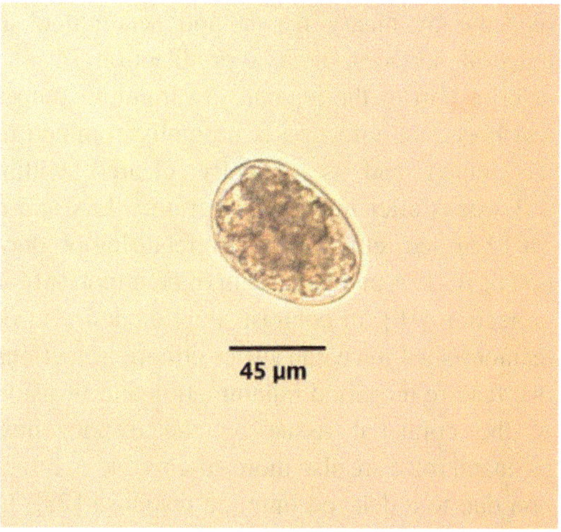

Fig. 11.4 *Heligmosomoides polygyrus* egg. (Source: author)

Moreover, the formation of large granulomas in the duodenum, particularly in Balb/c mice, is also a characteristic of *H. polygyrus* infection.

11.3.1.4 Tapeworms

Cestode parasites (tapeworms) are not commonly observed in laboratory rodents. The rat tapeworm *Hymenolepis diminuta* requires an intermediate arthropod host to complete its life cycle and thus is not effectively transmitted under indoor conditions. *Hymenolepis nana*, which also typically uses an intermediate host, may directly transmit between vertebrate hosts but is extremely rare in SPF rodents.

11.3.2 Protozoans

11.3.2.1 *Giardia muris*

G. muris is not a particularly common parasite, even in outdoor wild mice, but is widely used as a model for human giardiasis. *G. muris* is closely related to *G. duodenalis,* which infects a wide range of mammals including humans, and causes persistent, watery diarrhoea in people when ingested from contaminated food, water, or soil. Infection follows the faecal-oral route. Ingestion of cysts from the environment initiates infection. Excystation in the small intestine is followed by the rapid proliferation of trophozoites, which replicate by binary fission and accumulate at mucosal surfaces, with a predilection for the anterior part of the jejunum. In immunocompetent hosts, the infection is normally confined to the lumen and is typically cleared within 2–3 weeks after infection. Secretory IgA and a Th-17 biased cellular response featuring production of IL-17A and IL-6 seem to be important for protection [10]. In contrast, immunodeficient or malnourished mice can suffer chronic infections that feature increased inflammation and invasion of the epithelial tissue by the trophozoites. Neonatal mice are also more susceptible to infection due to a delayed immune response [25]. In otherwise healthy mice, infections are often asymptomatic although reduced weight gain has been observed. Infection markedly alters the host gut microbiome and metabolome, indicating that the presence of subclinical *Giardia* infections may be a confounding factor in various experimental investigations [15].

Cysts shed in faeces remain infective in the environment for several months if conditions are conducive. Prevention of infection is achieved by regular changing of bedding and high levels of hygiene. Diagnosis of *G. muris* can be either ante-mortem or post-mortem. The presence of cysts in faecal samples can be determined using standard flotation techniques, with the cysts presenting as circular structures with four nuclei, measuring around 16 μm in diameter (Fig. 11.5). Post-mortem, the intestine of animals suspected of giardiasis can be briefly agitated in ice-cold saline, which dislodges the trophozoites from the tissue. The trophozoites are then observable in wet mounts under a light microscope—they are pear-shaped, measure around 5 × 10 μm in size, and have a characteristic rolling movement in solution. DNA-based diagnostic methods (and genotyping) are well-advanced for *G. duodenalis* but are not routinely used for *G. muris* [1].

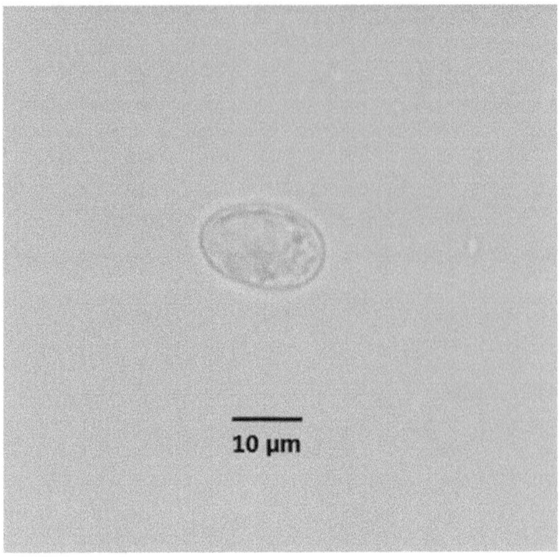

Fig. 11.5 *Giardia muris* cyst. (Source: Charlotte Van Crombrugge, Ghent University, used with permission)

11.3.2.2 *Tritrichomonas muris*

Tritrichomonas muris (Trit. muris) is a common protozoan flagellate found in the caecum of colon of both laboratory and wild mice. The trophozoites lives in the lumen, they are pear shaped and measure around 20 × 15 μm, and divide by binary fission. *Trit. muris* has a simple life cycle—there is no true encystation; however, psuedocysts are formed and both these and trophozoites are passed in faeces and are ingested by other mice by coprophagia, completing the life cycle. Diagnosis can be made by smears of intestinal content, with the aid of Giemsa-Wright staining to identify trophozoites (Fig. 11.6a). Alternatively, pseudocysts can also be identified in faeces (Fig. 11.6b). Molecular methods such as qPCR typing at the 28S rRNA gene can aid in distinguishing *Trit. muris* from other flagellates such as *Giardia* [18].

Trit. muris has not been reported to cause any pathology and is not considered a pathogen. However, the presence of *Trit. muris* has recently been described as a previously unappreciated factor in the phenotypic expression of several disease models. Mice that harbour *Trit. muris* have altered intestinal T-cell profiles in the steady-state and may have exacerbated colitis mediated by T-cell transfer models [12]. The expansion of *Trit. muris* mice in the gut of mice post-weaning has been shown to be responsible for elevated levels of microbial-derived metabolites such as succinate and acetate, which can drive the prolif-

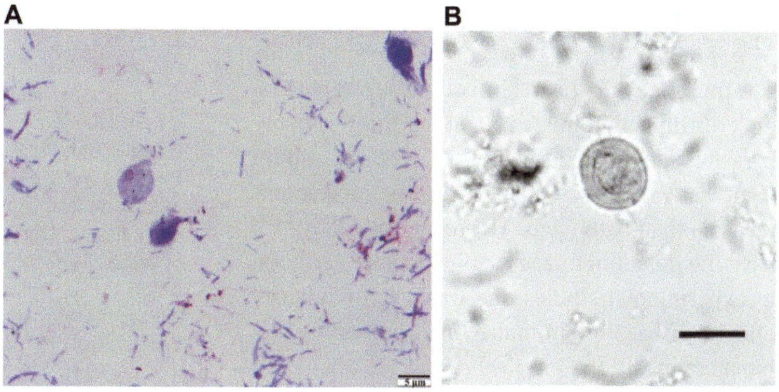

Fig. 11.6 (**a**) *Tritrichomonas muris* trophozoite in gut content stained with Giemsa stain [32] (**b**) *Tritrichomonas muris* pseudocyst in faeces, scale bar = 10 µm [22]

eration of tuft cells in the epithelium [27]. Indeed, the discovery of the role of tuft cells as being able to innately sense intestinal parasites was made, in part, by the serendipitous discovery that tuft cells were higher in mice from one vendor compared to another, which was traced to the presences of *Trit. muris* in one mouse colony but not the other [18]. For this reason, routine evaluation of *Trit. muris* in mouse colonies may be useful to help avoid confounding factors in different disease models.

11.3.2.3 *Entamoeba muris*

E. muris is an amoeba commonly found in both wild and laboratory mice, rats, and hamsters. The parasite has a similar life cycle to *Giardia*. Trophozoites multiply in the lumen of the caecum and colon, followed by encystation and transmission of the cysts in the faeces. The cysts can be observed in the faeces using the flotation technique—they are around 15 µm in diameter and contain around 6–8 nuclei. *E. muris* is considered a commensal and has not been reported to cause pathology or disease. As yet, interactions between the presence of *E. muris* and disease model phenotypes, if they exist, have not been elucidated

11.3.2.4 Cryptosporidium spp.

C. parvum and *C. muris* are uncommon in rodents, and outbreaks in SPF facilities are exceedingly rare. Indeed, experimental infections can be difficult to accomplish, and often require immunosuppression or infection of neonatal mice. Where cryptosporidiosis is suspected, e.g., in SCID mice, the small, spherical oocysts can be determined by flotation and light microscopy, and confirmation achieved by molecular typing of the 18S rRNA genes [2].

11.3.2.5 *Spironucleus muris*

This protozoan flagellate infects both mice and rats and has been reported to cause weight loss, intestinal inflammation, diarrhoea, and enlarged lymph nodes during heavy infections [3]. However, most infections are likely to be subclinical which makes estimates of the prevalence difficult. In some mouse colonies *S. muris* may reside as a commensal member of the microbiota, however, differences in immune response between mice harbouring *S. muris* or not have been reported [28], again indicating that the presence of this parasite has the potential to markedly affect the outcome of experimental immunology studies. The trophozoites reside mainly in the duodenum, and transmission is through cysts passed in the faeces. Diagnosis is based on either wet mounts of intestinal contents, histological examination of intestinal sections (where trophozoites may be located in the crypts of Luberkuhn), or by examination of cysts in faecal samples. The cysts are small in size with a diameter of around 5 µm, being smaller than *G. muris*. Molecular diagnosis has been reported based on amplification of the gene encoding elongation factor 1α gene, and appears to be more sensitive than microscopic examination [13].

11.4 Concluding Remarks

The presence of parasitic infections is likely to be underestimated in many rodent facilities as many parasites do not cause clinical disease and in some cases may be considered to be more commensals than pathogens. However, they may still have the potential to cause considerable variation in experimental studies due to their modulatory effects on baseline immune function and microbiota composition.

Many classical parasitological techniques such as faecal examinations for helminth eggs or protozoan cysts remain the most routine way to diagnose infections. However, molecular diagnosis is likely to overtake these as the most reliable and cost-effective way to detect the presence of common parasites. The advantage of these PCR-based techniques is increased sensitivity, such as the ability to detect pre-patent infections and infections with a very low number of parasites, but also the possibility to multiplex samples and simultaneously assess the abundance of many different parasitic species (as well as concurrent detection of bacterial and viral pathogens). For example, a metagenomics approach using high-throughput sequencing allowed the simultaneous detection of pinworms, several different protozoa, as well as many bacterial pathogens in a single faecal sample [26]. Such technology may still be prohibitively expensive, but the rapid pace of development in this field will likely see such sequencing-based approaches becoming a valuable routine monitoring tool to assess the health status of rodent colonies. Indeed, the term "health status" may be seen to be obsolete and replaced with "microbiota status" as a means of obtaining a snapshot of the underlying composition of the gut environment which can then be used to partition variation in experiments and increase their explanatory power.

References

1. Almeida A, Pozio E, Cacciò SM. Genotyping of *Giardia duodenalis* cysts by new real-time PCR assays for detection of mixed infections in human samples. Appl Environ Microbiol. 2010;76(6):1895–901.
2. Ayinmode AB, Ogbonna NF, Widmer G. Detection and molecular identification of cryptosporidium species in laboratory rats (*Rattus norvegicus*) in Ibadan, Nigeria. Ann Parasitol. 2017;63(2):105–9.
3. Barthold SW. *Spironucleus muris* infection, intestine, mouse, rat, and hamster. In: Jones TC, Popp JA, Mohr U, editors. Digestive system. Berlin/Heidelberg: Springer; 1997. p. 419–22.
4. Behnke JM, Wakelin D. The survival of *Trichuris muris* in wild populations of its natural hosts. Parasitology. 1973;67(2):157–64.
5. Çetinkaya H, Taş T, Vuruşaner C. Determination of the Parasites in the Faeces of Laboratory Animals by Using Flotation Technique in Istanbul, Turkey. J Istanbul Vet Sci. 2017;1(2):35–9.
6. Cliffe LJ, Grencis RK. The *Trichuris muris* system: a paradigm of resistance and susceptibility to intestinal nematode infection. Adv Parasitol. 2004;57:255–307.
7. Cortés A, Peachey L, Scotti R, Jenkins TP, Cantacessi C. Helminth-microbiota cross-talk – a journey through the vertebrate digestive system. Mol Biochem Parasitol. 2019;233:111222.
8. Dole VS, Zaias J, Kyricopoulos-Cleasby DM, Banu LA, Waterman LL, Sanders K, Henderson KS. Comparison of traditional and PCR methods during screening for and confirmation of *Aspiculuris tetraptera* in a mouse facility. J Am Assoc Lab Anim Sci. 2011;50(6):904–9.
9. Douglas B, Oyesola O, Cooper MM, Posey A, Tait Wojno E, Giacomin PR, Herbert DR. Immune system investigation using parasitic helminths. Annu Rev Immunol. 2021;39:639–65.
10. Dreesen L, De Bosscher K, Grit G, Staels B, Lubberts E, Bauge E, Geldhof P. *Giardia muris* infection in mice is associated with a protective interleukin 17A response and induction of peroxisome proliferator-activated receptor alpha. Infect Immun. 2014;82(8):3333–40.
11. Eguíluz C, Viguera E, Pérez J. Modification of the anal tape method for detection of pinworms in rodents. Lab Anim (NY). 2001;30(10):54–5.
12. Escalante NK, Lemire P, Cruz Tleugabulova M, Prescott D, Mortha A, Streutker CJ, Girardin SE, Philpott DJ, Mallevaey T. The common mouse protozoa *Tritrichomonas muris* alters mucosal T cell homeostasis and colitis susceptibility. J Exp Med. 2016;213(13):2841–50.
13. Fain MA, Karjala Z, Perdue KA, Copeland MK, Cheng LI, Elkins WR. Detection of *Spironucleus muris* in unpreserved mouse tissue and fecal samples by using a PCR assay. J Am Assoc Lab Anim Sci. 2008;47(5):39–43.
14. Fernández-Soto P, Fernández-Medina C, Cruz-Fernández S, Crego-Vicente B, Febrer-Sendra B, García-Bernalt Diego J, Gorgojo-Galindo Ó, López-Abán J, Vicente Santiago B, Muro Álvarez A. Whip-LAMP: a novel LAMP assay for the detection of *Trichuris muris*-derived DNA in stool and urine sam-

ples in a murine experimental infection model. Parasit Vectors. 2020;13(1):552.
15. Fink MY, Singer SM. The intersection of immune responses, microbiota, and pathogenesis in giardiasis. Trends Parasitol. 2017;33(11):901–13.
16. Gale EA. A missing link in the hygiene hypothesis? Diabetologia. 2002;45(4):588–94.
17. Grainger JR, Smith KA, Hewitson JP, McSorley HJ, Harcus Y, Filbey KJ, Finney CA, Greenwood EJ, Knox DP, Wilson MS, Belkaid Y, Rudensky AY, Maizels RM. Helminth secretions induce de novo T cell Foxp3 expression and regulatory function through the TGF-β pathway. J Exp Med. 2010;207(11):2331–41.
18. Howitt MR, Lavoie S, Michaud M, Blum AM, Tran SV, Weinstock JV, Gallini CA, Redding K, Margolskee RF, Osborne LC, Artis D, Garrett WS. Tuft cells, taste-chemosensory cells, orchestrate parasite type 2 immunity in the gut. Science. 2016;351(6279):1329–33.
19. Hsu C-K. Chapter 19 – Protozoa. In: Foster HL, Small JD, Fox JG, editors. Diseases. New York: Academic; 1982. p. 359–72.
20. Khudhair Z, Alhallaf R, Eichenberger RM, Field M, Krause L, Sotillo J, Loukas A. Administration of hookworm excretory/secretory proteins improves glucose tolerance in a mouse model of type 2 diabetes. Biomol Ther. 2022;12(5):637.
21. Lewis JW, Morley NJ, Behnke JM. Helminth parasites of the wood mouse *Apodemus sylvaticus* in southern England: levels of infection, species richness and interactions between species. J Helminthol. 2023;97:e18.
22. Lipman NS, Lampen N, Nguyen HT. Identification of pseudocysts of *Tritrichomonas muris* in Armenian hamsters and their transmission to mice. Lab Anim Sci. 1999;49(3):313–5.
23. Loukas A, Maizels RM, Hotez PJ. The yin and yang of human soil-transmitted helminth infections. Int J Parasitol. 2021;51(13–14):1243–53.
24. Owen RL, Nemanic PC, Stevens DP. Ultrastructural observations on giardiasis in a murine model. I. Intestinal distribution, attachment, and relationship to the immune system of *Giardia muris*. Gastroenterology. 1979;76(4):757–69.
25. Paerewijck O, Maertens B, Gagnaire A, De Bosscher K, Geldhof P. Delayed development of the protective IL-17A response following a *Giardia muris* infection in neonatal mice. Sci Rep. 2019;9(1):8959.
26. Scavizzi F, Bassi C, Lupini L, Guerriero P, Raspa M, Sabbioni S. A comprehensive approach for microbiota and health monitoring in mouse colonies using metagenomic shotgun sequencing. Anim Microbiome. 2021;3(1):53.
27. Schneider C, O'Leary CE, von Moltke J, Liang HE, Ang QY, Turnbaugh PJ, Radhakrishnan S, Pellizzon M, Ma A, Locksley RM. A metabolite-triggered tuft cell-ILC2 circuit drives small intestinal remodeling. Cell. 2018;174(2):271–84.e214.
28. Sebesteny A. The transmission of intestinal flagellates between mice and rats. Lab Anim. 1974;8(1):79–81.
29. Taylor CH, Wanelik KM, Friberg IM, Lowe A, Hall AJ, Ralli C, Birtles RJ, Begon M, Paterson S, Jackson JA, Bradley JE. Physiological, but not fitness, effects of two interacting haemoparasitic infections in a wild rodent. Int J Parasitol. 2018;48(6):463–71.
30. Valanparambil RM, Segura M, Tam M, Jardim A, Geary TG, Stevenson MM. Production and analysis of immunomodulatory excretory-secretory products from the mouse gastrointestinal nematode *Heligmosomoides polygyrus* bakeri. Nat Protoc. 2014;9(12):2740–54.
31. Zhang H, Bednář L, Heitlinger E, Hartmann S, Rausch S. Whip- and pinworm infections elicit contrasting effector and distinct regulatory responses in wild house mice. Int J Parasitol. 2022;52(8):519–24.
32. Zhang H, Zhang N, Li J, Zhao P, Li X, Wang X, Zhang X, Yuan B, Gao F, Gong P, Zhang X. Development of nested polymerase chain reaction with novel specific primers for detection of *Tritrichomonas muris* infection in laboratory mice. Animals. 2023;13(20):3177.

Laboratory Animal Pathology in Relation to Spontaneous Infections

12

Henrik Elvang Jensen, Páll Skúli Leifsson, and Louise Kruse Jensen

Abstract

In this chapter, a comprehensive guide for necropsying laboratory animals is given along with sequential illustrations of the different procedures. In this part also important parameters to be determined at the beginning of a necropsy and the necessary equipment for sampling of tissues are stated. The procedures for tissue sampling for histological preparation is followed by sections describing different techniques used for fixation, trimming, embedding, sectioning, and staining of tissues. The second half of this chapter is dedicated to the evaluation of tissue sections with special emphasis on infectious diseases including in situ identification of different infectious agents. In this part of this chapter also different pitfalls with regards to the presence of tissue-artifacts, which may be due to inappropriate handling of the animals ante- and/or postmortem, killing-procedures, tissue handling, fixation, processing and staining are covered. At the end of this chapter, different histochemical and molecular in situ identification techniques are addressed in both text and illustrations, which are followed by the histological demonstration of different examples of spontaneous infections in laboratory animals.

Keywords

Rodents · Pathology · Anatomy · Histology · Diagnosis · Host microbial interactions

12.1 Introduction

To diagnose spontaneous infections in laboratory animals as well as in other animals, it is essential to follow a protocol to carry out a necropsy systematically and correctly. The person carrying out the necropsy should possess knowledge of both anatomy and pathology; therefore, ideally a veterinarian should be responsible for the procedures of necropsies and interpretation of the necropsy results. All necropsies should be carried out as soon as possible after the animal death. The animal should, if not necropsied immediately, be kept in a refrigerator for 1–2 days and not frozen, which causes alterations of tissues [78]. During the necropsy, all changes must be described and

H. E. Jensen (✉) · P. S. Leifsson · L. K. Jensen
Department of Veterinary and Animal Sciences, University of Copenhagen, Frederiksberg, Denmark
e-mail: elvang@sund.ku.dk; ple@sund.ku.dk; louise-k@sund.ku.dk

F. Benavides, A. K. Hansen (eds.), *Rodent Quality Control: Genes and Bugs*, Laboratory Animal Science and Medicine 2, https://doi.org/10.1007/978-3-031-59103-7_12

photo documented following the common terminology applied in pathology with respect to, e.g. location, size, weight, pattern of distribution, and appearance [49]. If a spontaneous infection is suspected during the necropsy and/or due to clinical observations or anamnestic information, specimens should be collected for microbial analysis during the necropsy. Ideally, sterile swabs and instruments should be used for the sampling of material (tissues /whole organs /contents in cavities/body fluid, etc.). However, as several of the laboratory animals are small, a diagnosis or even a suspicion of an infection is often not stated at gross examination of the organs during the necropsy. Therefore, often spontaneous infection suspicion is first evident when histological slides are evaluated, when all materials often have been formalin fixed and therefore are unsuitable for cultivation. The quality of tissue sections together with their staining is of utmost importance for the evaluation. Therefore, committed technicians should carry out these procedures in well-equipped and professional laboratories in order to avoid, e.g., artifacts and staining variations. Due to the occurrence of special reactions within the tissue, the pathologist often will get the suspicion of a spontaneous infection being present. For some viral agents the presence of inclusion bodies and/or necrosis are the hallmarks, in other situations just the presence of an inflammatory reaction may point in the direction of an infection present [2]. In cases of parasites and fungi, the infectious elements are often large enough for detection within standard stained tissues. Bacterial infections should also be suspected in lesions dominated by necrosis and different types of inflammatory exudates and other reactions. Together with fungal elements also bacteria may be visualized by different special stains [56]. Moreover, in order to obtain an etiological diagnosis on formalin fixed and paraffin embedded tissues, several in situ techniques may, combined with e.g. PCR techniques, be applied on tissue sections [66, 71]. This chapter follows the chronology from necropsy of the animal to the evaluation of tissue section to achieve an etiological diagnosis of spontaneous infections in laboratory animals.

12.2 Necropsy Procedures of Laboratory Animals

Live animals are killed by an appropriate method and preferably exsanguinated. The necropsy starts with an external examination of the animal determining the parameters in Box 12.1. At start the animal's postmortem condition must always be noted (fresh, cadaverous to a light, moderate or severe degree), as the state understandably will influence the certainty by which macro- and microscopic conditions are evaluated. The tools in Box 12.2 are usually adequate for rodent and rabbits.

Box 12.1: Parameters to Be Determined at the Beginning of a Necropsy

1. Identity (species, strain/stock, sex (Fig. 12.1), age, body weight, individual identity marking).
2. Body condition (over-/underweight, emaciation, etc).
3. Skin and hair coat condition (reduced/increased haircoat, scales, wounds, etc.).
4. Subcutis condition assessed by palpation (tumors, etc.).
5. Accessible mucous membranes and natural body orifices (wounds, secretion, etc.).
6. In females the mammary glands, and in males, the testicles are examined (tumors, swellings, etc.).

Box 12.2: Tools for Necropsy of Rodents and Rabbits

1. Styrofoam "Flamingo" tray and 4 pins	6. Small bone scissors
2. Scalpel	7. Large and small anatomical forceps
3. Large knife	8. Large and small surgical forceps
4. Large and small straight scissors	9. Movable precision illumination
5. Large and small curved scissors	

Tissues and organs from small rodents should be handled with forceps as far as possible. Organs or tissues should not be directly pinched with the forceps, and harming the delicate organs should be avoided by gripping the surrounding connective tissue. Organs should either be immediately transferred into formalin solution or kept moist by placing them on a tissue moistened with physiological saline, which is also used for rinsing organs, e.g., the cavities of the heart and the opened gastrointestinal tract. The procedure below will note important differences between the rat and the other species (mouse, guinea pig, hamster, and rabbit) [32, 37].

12.3 Removal of Skin and Superficial Glands

12.3.1 Eye and Glands of the Head

The animal is placed on its side, and the skin covering the side of the head, and the outer ear is removed. The tissue surrounding the eye is gripped with the forceps, and the eye is removed with the small curved scissors by cutting along the bony foundation of the eye orbit (orbita) (Fig. 12.1a). The eye is extirpated by transecting the optical nerve. Following removal, the eye is inspected, and the attached glands are identified. The largest, the light yellow Harderian gland, is positioned ventrally to the eye. The somewhat smaller intraorbital light brown lacrimal gland (gl. lacrimalis intraorbitalis) is positioned laterally to the eye. The intraorbital lacrimal gland is not present in the guinea pig and is difficult to see macroscopically in the mouse. The guinea pig, however, possesses a larger salivary gland, gl. zygomaticus (also called gl. lacrimalis zygomatica). The gland occupies the deepest part of the orbit. The flat, circular extraorbital lacrimal gland (gl. lacrimalis extraorbitalis), positioned caudoventrally to the orbit in regio zygomatica, is identified. This gland is not present in the guinea pig. Gl. parotis (the large parotid salivary gland) is identified. The pink-colored gland is situated caudally and caudo-ventrally to the external auditory tube. Ventrally to the auditory tube the smaller ear glands (gll. zymbali) are situated. The glands produce ear wax (cerumen) and are very difficult to localize with the naked eye. In rabbits, there are two glands associated with the third eyelids, which are referred to as the Harderian gland, and it has the following salivary glands: glandulae buccalis caudalis and ventralis, glandula masseterica, glandula mandibularis, and glandula parotis.

12.3.2 Neck and Abdomen

Place the animal on the Styrofoam tray on its back and fixate it by pinning each foot to the tray. Rabbits are placed in a supine position on a table. The skin covering the entire abdomen and the medial surface of the legs is moistened with alcohol, and an incision is made through the skin along the midline from the tip of the chin (regio mentalis) to the anterior edge of the pelvic bone (pecten ossis pubis). In males the incision is placed on one side of the penis, which is loosened from the abdominal wall and moved caudally. In females the skin incision is extended to the genital opening. Subsequently, the skin is dissected to each side, including the skin on the proximal part of the legs, which can be removed after making a medial skin incision. Following removal of the skin the subcutaneous skin is examined, and the lymph nodes of the groin (lnn. inguinales), often embedded in fatty tissue, are identified. In the hamster brown fat is found between the scapulae and ventrally on the neck. Cranially on the neck the connected salivary glands (gl. sublingualis and gl. submaxillaris (s. mandibularis)) and the attached lymph nodes are identified (lnn. mandibularis cranially and lnn. cervicales caudally) (Fig. 12.1b). In females the mammary glands and the pink clitoridal gland (gl. clitoridis) are examined. The latter is positioned subcutaneously and cranio-laterally to the genital opening on both sides (Fig. 12.1c). In males the preputium, penis, and the subcutaneous preputial gland (gl. preputialis), positioned latero-caudally on both sides of the preputium, are identified (Fig. 12.1d). None of these sexual glands are found in the guinea pig. In the male hamster a large sebaceous gland is found on each flank, the "flank organ" (gl. costovertebralis).

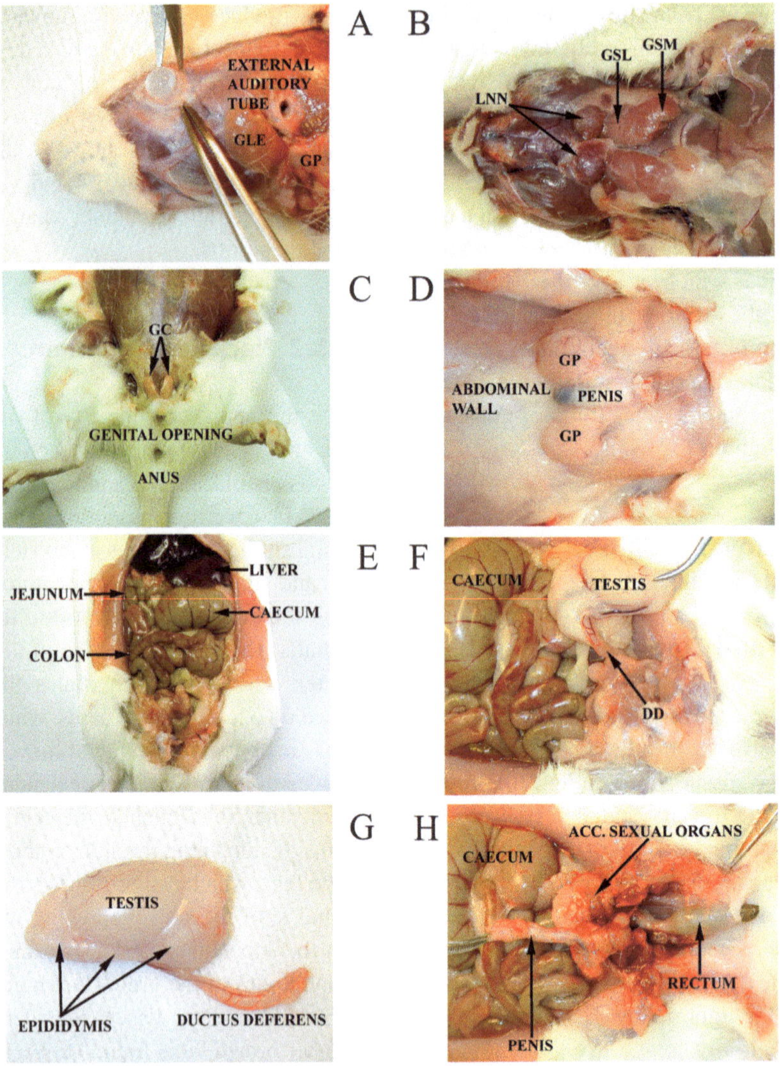

Fig. 12.1 Rodent and rabbit necropsy. (**a**) The eye of the rat is removed by cutting along the bony foundation of the orbita. *GLE* Gl. lacrimalis extraorbitalis, *GP* Gl. parotis. (**b**) On the ventral side of the neck different glands of the rat are identified. *LNN* lymph nodes, *GSL* Gl. sublingualis, *GSM* Gl. submaxilaris. Rat. (**c**) Female rat with clitoridal gland (*GC* Gl. clitoridis). (**d**) Male rat with preputial gland (*GP* Gl. preputialis). (**e**) The opened abdomen of the rat with exposed organs. (**f**) The testicles are returned to the abdominal cavity through the inguinal canal. *DD* Ductus deferens. Male rat. (**g**) Testicle with attached epididymis and ductus deferens. Male rat. (**h**) Removal of the urogenital organs of a male rat. *ACC* accessory. (**i**) Identification of accessory sexual organs in a male rat. *P* prostate gland, *GC* Gl. coagulationis, *GV* Gl. vecicularis, *GB* Gl. bulbourethralis, *U* urinary bladder. (**j**) Organs of the abdominal and pelvic cavity of the female rat. *R* rectum, *OV* ovarium, *SP* symphysis pelvina. (**k**) The opened abdomen of a rat with exposed organs, and lateral displacement of the intestines. (**l**) The opened abdomen of a rat with exposed organs. The esophagus (OE) is cut in order to remove the gastro-intestinal tract. *CX* cartilago xiphoidea, *L* liver. (**m**) Identification of adrenal glands and the kidneys of the rat. (**n**) The ventral part of the neck with exposed organs in the rat. *GT* Gl. thyreoidea, *L* larynx, *T* trachea, *M* musculus masseter. (**o**) Opening the thoracic cavity in the rat. *CX* cartilago xiphoidea, *DI* diaphragma. (**p**) The "plucks" of the rat, i.e., organs removed as a unit from the oral cavity, neck, and the thoracic cavity. (**q**) Opening the cranium of the rat. (**r**) The opened cranium of the rat with the cerebellum and cerebrum exposed. (**s**) The hypophysis is situated in the bottom of the cranium. Rat. (**t**) The nervus ischiadicus (NI) of the rat is found by splitting the muscles on the medial side of the thigh

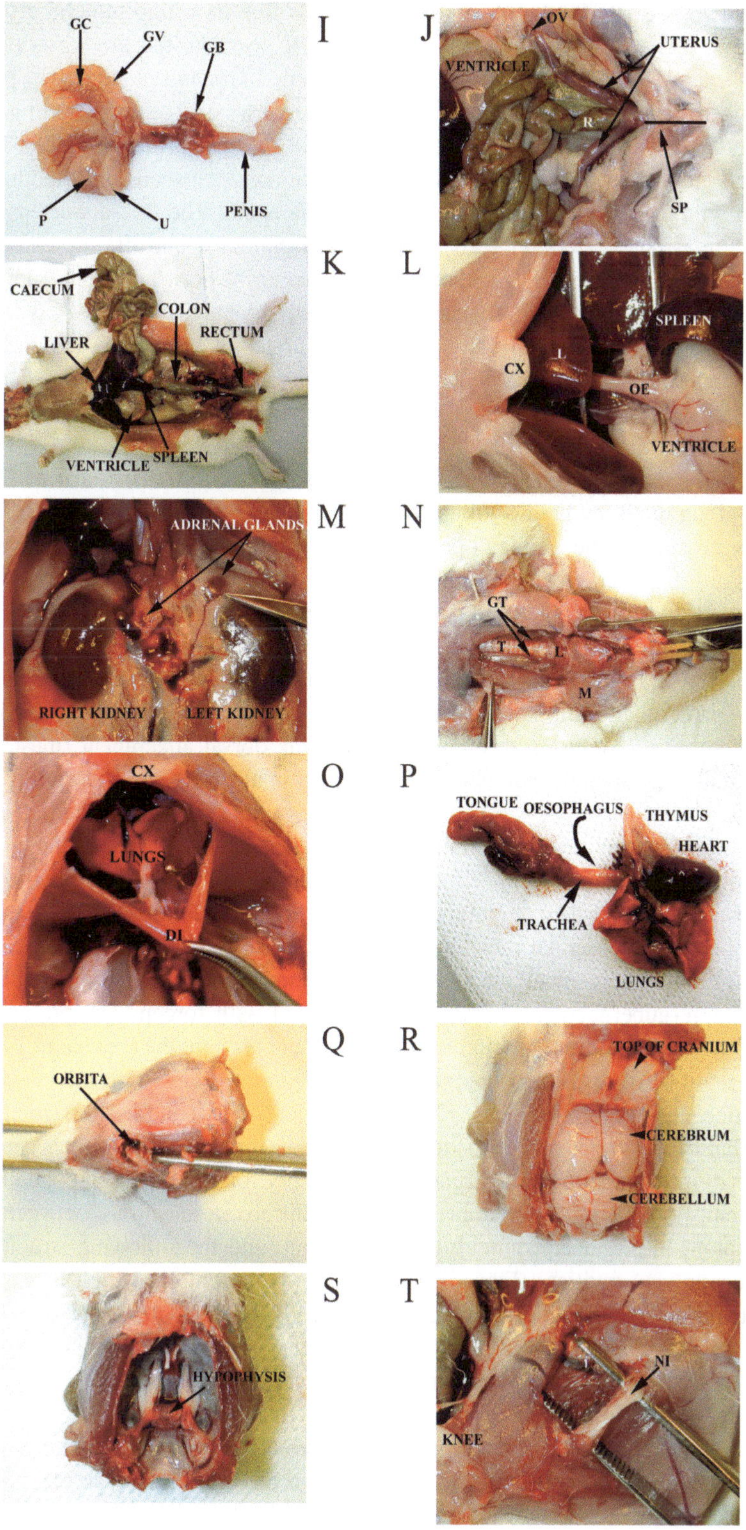

Fig. 12.1 (continued)

12.3.3 Abdominal and Thoracic Organs

12.3.3.1 Opening of the Abdominal Cavity

The abdominal cavity is opened by cutting through the abdominal wall in the linea alba from the tip of the breastbone, sternum (processus xiphoideus) to pecten ossis pubis. The abdominal wall is cut with scissors on both sides, cranially along the rib curvature, and caudally along the groin until the uppermost part of the abdominal cavity (Fig. 12.1e). The ventral abdominal muscles are removed, and the skin surrounding the genital opening/the retroflexed penis and the anus is circumcised along the bony base of the pelvic cavity. By this procedure the genital opening/penis and anus are dissected free as a unit. The pelvic floor is removed following cutting of the pelvis on each side of the midline; the cut is made laterally on each side through the hole in the pubic-ischiatic part of the hipbone, foramen obturatum. Alternatively, cutting through the symphysis sagittally can open the pelvic floor.

12.3.3.2 Sexual Organs and Urinary Bladder

In males, the testicle with attached epididymis on each side is pushed via the inguinal canal into the abdomen (Fig. 12.1f). The ligament attaching the testis to the scrotum (gubernaculum testis) is cut. The ductus deferens (vas deferens) is cut on each side, allowing removal of testes and epididymides for inspection and transection (Fig. 12.1g). Following cutting of the ureter on each side, the remaining parts of the lower urogenital organs on each side are removed as a unit by a dorsal and lateral incision along the organ attachments in the abdominal and pelvic cavities (Fig. 12.1h). Subsequently to removal of the urogenital organs, the accessory sexual glands are localized (gl. vesicularis, s. seminal vesicle (vesicula seminalis), coagulating gland (gl. coagulationis, not present in rabbits), prostate gland, and gl. bulbourethralis), along with the urinary bladder, the urethra, and penis (Fig. 12.1i). Urine may be obtained from the urinary bladder, which is opened. In guinea pigs, the urine is normally cloudy due to a high mineral crystal concentration. If possible, the urethra is opened with the scissors. In rodents, a plug in the proximal part of the urethra is a common finding, the plug may extend into the urinary bladder. In females, the attachment between anus/rectum and vulva/vagina is cut. The free-dissected vulva-vagina is lifted cranially, and the mesenteric attachments of the uterine horns, oviduct (salpinx) and ovaries on both sides are cut, starting caudally (Fig. 12.1j). The ureter in both sides is cut, and the sexual organs and urinary bladder can be removed together as a unit for inspection and opening.

12.3.3.3 Gastro-Intestinal Tract and Spleen

The spleen is dissected from the surrounding tissues by cutting the omentum majus and the attachment (lig. gastrolienalis) to the stomach. Following spleen removal, the intestines are pushed forward or laterally to expose the rectum (Fig. 12.1k). The rectum is fixed with forceps, and the dorsal attachment (mesentery) is cut in the cranial direction. The cutting of the mesentery is continued in the cranial direction to the stomach, thereby freeing the small and large intestine from their attachment to the uppermost part of the abdominal cavity. In the oral end it must be ensured that the pancreas accompanies the duodenum in its mesentery. The ventricle is grabbed with forceps and pulled slightly in the caudal direction; the esophagus is identified at its exit from the diaphragm and cut (Fig. 12.1l). The ligaments between the stomach and liver are cut, and the stomach and intestines can be removed as an entity for inspection and opening (Fig. 12.2a). All fibrous connections between the intestines are cut, and the following structures are identified: (1) mesenteric lymph nodes (lnn. mesenteriales), (2) stomach (ventriculus), which in the mouse, rat, hamster and rabbit is divided into an oral non-glandular part (white) and an aboral grey glandular part (Fig. 12.2b). The gastro-intestinal tract of the guinea pig and rabbit is shown in Fig. 12.2c, d, respectively. The guinea pig stomach is entirely glandular, (3) the small intestine (intestinum tenue) consisting of the duodenum, to which the pancreas is attached, the

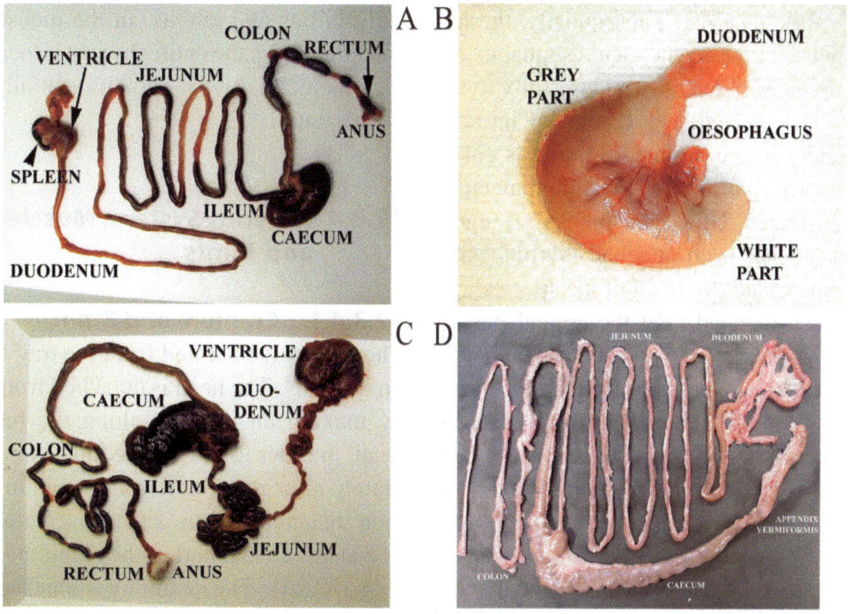

Fig. 12.2 Gastrointestinal anatomy. (**a**) The gastrointestinal tract and spleen of the rat. (**b**) The stomach of the rat. (**c**) The gastrointestinal tract of the guinea pig. (**d**) The intestinal tract of the rabbit

jejunum and ileum with Peyer patches (not macroscopically visible in mice), (4) the cecum, which is very big in the guinea pig (Fig. 12.2c), is located at the transition to the colon, and (5) colon and rectum are found at the aboral end (Fig. 12.1k). Each single section of the gastrointestinal tract is cut open in their entire length opposite to the mesenteric attachment.

12.3.3.4 Liver, Kidneys, and Adrenals

The ligaments extending from the diaphragm (lig. falciforme hepatis) and the dorsal abdominal wall (lig. triangulare dex. et sin.) to the liver are cut. The liver is removed for inspection and transection (Fig. 12.1l). The rat has no gall bladder. From each side the kidney and adrenal gland (positioned craniomedially to the cranial kidney pole on each side) is removed as a unit for inspection and transection (Fig. 12.1m).

12.3.3.5 Oral Cavity, Neck, and Thoracic Cavity

The external masticatory muscles are transected, and the jaw is moved caudo-ventrally (opened) for inspection of the oral cavity. In the hamster the cheek pouches (bursae paraorales) are inspected, located laterally in each cheek. The pouch entrance is placed quite rostrally. Rabbits have collar bones (claviculae) and some breeds, and especially in the female rabbits may have a big crest of skin (sometimes referred to a "dewlap") under the throat. The larynx, trachea, and esophagus are exposed by making an incision from the larynx to the thoracic inlet (apertura thoracis cranialis) (Fig. 12.1n). The glands, which are split by this procedure, and possibly the thymus, are pushed to each side. The diaphragm is cut along its line of attachment to the rib curvature, starting ventrally (i.e. at the sternum) (Fig. 12.1o). The thoracic cavity is opened by removing the breastbone (sternum), by seizing the bone with the forceps and transecting the ribs in the cranio-caudal direction along the sternal attachment. For bone marrow examination, the sternum is suitable. The tongue is seized with forceps and separated from the mandibular bones by cutting the oral mucosal membrane along the tongue and the inner surface of the mandibular bones. The tongue is pulled ventrally between the mandibular bones and then pulled caudally, allowing transaction of the soft palate (palatum molle). The transection is continued caudally on each side of the pharynx. Subsequently, the larynx together with the trachea and esophagus are loosened by incision laterally and dorsally to the structures. The procedure is continued into the thoracic cavity where the mediastinum is cut in the cranio-caudal direction by simultaneously pulling the tongue. Now the "plucks" (tongue, esophagus, trachea with attached thyroid gland (visible to the naked eye in all of the species), gl. parathyreoidea (located in/at the cranial part of gl. thyreoidea—the gland is normally not visible to the naked eye), and possibly the thymus, mediastinum with lymph nodes (lnn. mediastinales), lungs and heart) is removed as a unit for closer examination (Fig. 12.1p). The pericardial sac is opened and inspected, and the heart is then separated from the lungs at the heart base, whereby part of the large vessels (vena cava caudalis and cranialis, v. pulmonis, truncus pulmonalis, and aorta) accompany the heart. The surface of the heart (the epicardium) is inspected, and the cavities of the heart are opened by cutting along the walls: septum interatriale and interventriculare following the direction of the blood flow. Start on the right side following the sequence: vena cava—atrium cordis dex.—ventriculus cordis dex.—truncus pulmonalis. The heart is turned around, and on the left side the sequence is: v. pulmonalis—atrium cordis sin.—ventriculus cordis sin.—aorta. Coagulated blood in the heart cavities is removed, and the endocardium, valves, and myocardium are inspected. If necessary, additional incisions are made in the myocardium. The esophagus is separated from the trachea, leaving only a connection at the larynx, where gl. thyreoidea is identified. Now the esophagus and trachea are opened in their full length by cutting with scissors. The tracheal incision is continued to the main bronchi of the lungs. Finally, single incisions are made in the lobes of the lung, and the lymph nodes surrounding the bronchi and in the mediastinum are identified. The thymus, which is markedly larger in young animals compared to older animals, where only remnants may be found in the mediastinum, is in the rat located along the trachea extra as well as intrathoracically in rats and rabbits. In the mouse, hamster and guinea pig, the entire thymus is located inside the thoracic cavity, cranio-ventrally in the mediastinum.

12.3.4 Nervous System, Muscles, and Joints

12.3.4.1 Cranium and Brain

The animal is liberated from the tray and turned on the belly. The head is detached from the body by making an incision along the back of the head, in order to cut the medulla spinalis immediately after its exit from the cranium via foramen magnum. Skin and musculature of the dorsal and caudal part of the cranium is removed (Fig. 12.1q). The cranium is opened on each side of the head by cutting the cranial bone from the foramen magnum to the middle of the caudal rim of the orbita, which is the extent of the cavity of the cranium. Now the top of the cranium can be lifted rostrally and removed. The brain, on which the cerebellum and the cerebrum are identified (Fig. 12.1r), is removed from the caudal end, cutting the brain nerves on both sides in the rostral direction, holding the head with the dorsal surface downwards towards the palm of the hand. Following removal of the brain, the hypophysis can be observed, lying in a cavity (sella turcica) ventrally in the cavity of the cranium (Fig. 12.1s). The hypophysis is covered by dura mater, which must be incised and removed before the hypophysis can be lifted up. Bulla tympanica (a part of the middle ear) is identified caudo-laterally on the ventral surface of the cranium. The bulla tympanica, which is thin-walled, is opened with the tip of the scissors or knife, and inspected. The nasal cavity is inspected following longitudinal cutting of the cranium in the midline. In mice and young rats, guinea pigs and hamsters the medulla can only be removed without damage if great care is taken. In these animals medulla spinalis should therefore be fixated in situ and removed afterwards.

12.3.4.2 Muscles, Joints, and Peripheral Nervous System

Following removal of skin and fascia, the skeletal muscles are examined by external inspection and comparison with the opposite corresponding muscle. Incisions in muscles should always be made longitudinally as well as transversely with respect to the orientation of the muscle. Before the large joints (shoulder, elbow, hip and knee) are opened, they must be inspected for obvious changes and mobility. On the hind leg, n. ischiadicus is found by splitting the muscles caudally and parallel to the femur on the medial side of the thigh (Fig. 12.1t).

12.4 Removal of Tissue for Histology

Following death, all organs will be subjected to autolysis caused by liberation of endogenous enzymes. Since especially intestines, liver and pancreas are subject to rapid autolysis, a speedy fixation of these organs is particularly important. If an optimal microbiological examination is the most important rather than the registration of pathological conditions it may, depending on the time gap between death and examination, be appropriate to store the animal in a freezer (app. −20 °C). When macroscopically changed tissue, processes, etc. are removed, the transitory zone to the surrounding normal tissue should always be included. From organs intended to be examined histologically, blocks or sections are taken, the thickness of which should not exceed approximately 1 cm, as the penetration of the fixative otherwise will be too poor. Fixation and artifacts will be further discussed below. Equipment necessary for sampling is shown in Box 12.3.

Box 12.3: Necessary Equipment for Sampling

1. Sterile cotton-tipped applicators and Petri dishes, for microbiological examinations
2. Vacutainer tubes, rack and needles for chemical blood analyses etc.
3. Various single-use syringes (1–10 ml) and needles (14G, 19G, and 27G) for collection of fluids
4. Fixatives (10% buffered neutral formalin and Davidson's fixative)
5. Plastic or glass containers (100 ml) with a large opening and tightly fitting screw caps
6. Self-adhesive labels

12.5 Preparation and Staining of Histological Sections

Histopathological examination of organs and tissues is the most informative method when assessing lesions in small animals as the size of the animals often makes an adequate macroscopical evaluation of their organs difficult or impossible, and, therefore, a thorough postmortem examination of a laboratory animal includes histological evaluation to be completed. Tissues to be examined under the microscope have to be prepared using special techniques to render them suitable for histopathological examination. The process starts with sampling, i.e., the removal of tissues from the animal, and is followed by various steps to end with the production of microscope-ready histological slides. The process of tissue slide production can vary somewhat, depending on the type of tissue and the purpose of the examination.

A few of these variations are mentioned below but the emphasis will be on the standard processing of paraffin-embedded, formalin-fixed tissues.

12.5.1 Fixatives

The first step of tissue processing is the fixation of the tissue (Fig. 12.3). The purpose of the fixation is to stop postmortem decay processes (autolysis and putrefaction) and to preserve the integrity and morphology of the tissues. The most commonly used fixative is 10% formalin (4% aqueous formaldehyde solution). It is cheap and gives good preservation of tissues. It is, however, highly toxic and it often interferes with the molecular structures of the tissues to such a degree that the demonstration of certain epitopes, using immunohistochemical techniques and the use of enzyme histochemical methods can be impossible. Alternative fixatives occasionally used include ethanol (alone or in combination with formalin as alcoholic formalin), Bouin's, Carnoy's, Davidson's and Zenker's fixatives to mention a few. Other fixatives than formalin may be preferable for several reasons. Water-soluble substances like glycogen and other carbohydrates are partly removed from the tissue during formalin fixations so if the intention is to study the glycogen contents of tissues other fixatives, e.g., Carnoy's fixative or alcoholic formalin, are recommended [48]. Bouin's fixative is used if the tissues are to be examined immunohistochemically as this fixative is less damaging to the conformation of protein than formalin [25]. If rapid fixation of eyes is required, e.g., to prevent retinal detachment, Davidson's fixative is recommended, but Zenker's and Bouin's fixatives can also be used [48, 80]. Apart from preserving the normal ocular structures these fixatives render the eyes so hard that they can be trimmed without distorting all their structures. It is recommended that the eye should not be fixed in these fixatives for more than 24–48 h (depending on their size), as they can become so hard that they are difficult to trim. They should therefore be transferred to 70% ethanol before further processing.

12.5.2 Fixation Procedures

12.5.2.1 Immersion Fixation

Immersion fixation is usually the method of choice. Here the tissue is simply placed in a container and immersed in the fixative. The volume of the fixative should preferably be 10–20 times the volume of the tissue and the thickness of the

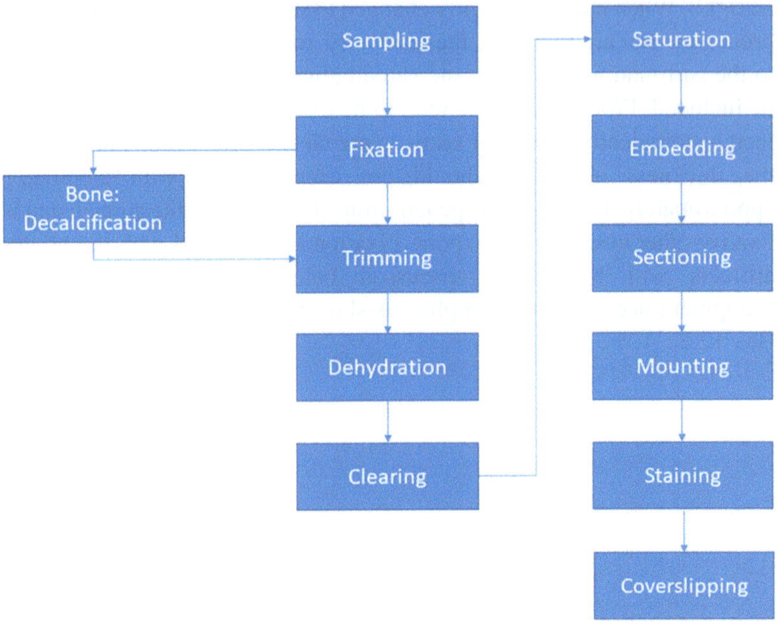

Fig. 12.3 An overview of histopathological processing of tissues from sampling to the final tissue slide

tissue sample should not exceed 6 mm (the brain is an exception as it is normally fixed in toto). Too large liver samples are notoriously slow to become adequately fixed, as the outer, fixed part of the sample tends to form a shell that hinders further diffusion of the fixative into the inner part. It is important to ensure fixation of the tissue from all directions. If the tissue lies on the bottom of the container or it is attached to its walls, fixation will be slower. This can be prevented by putting a paper towel or surgical gauze between the tissue and the bottom of the container and by shaking the container gently to remove the tissues from its walls. The duration of the fixation depends highly on the type of tissue and the size of the sample. Lipid-rich tissues (e.g., brain) are penetrated slowly by the aqueous solution and more time is needed for the fixative to reach the center of a large sample. Under most circumstances 24 h of fixation is sufficient, and often a shorter duration is preferred in order to reduce the likelihood of interference with molecular structures mentioned above.

12.5.2.2 Perfusion Fixation

For rapid fixation of tissues perfusion fixation can be used. The fixative is passed under pressure through the blood vessels of an organ, a part of the body (e.g., the head) or even the whole animal. Instillation fixation or infusion fixation, also known as "pseudo-perfusion" can be used alone or in combination with immersion fixation. Here the fixative is poured directly or instilled in the lumen of the organ. This technique can, e.g., be used when fixing the lungs, stomach or urinary bladder. Similarly, the eyes can be fixed by injecting the fixative directly into the anterior chamber or the vitreous followed by immersion fixation.

12.5.3 Trimming, Embedding and Sectioning of Tissues

After fixation, tissue samples are trimmed, i.e., fresh, flat cut surfaces are made, and they are placed in marked tissue cassettes (Fig. 12.4a) for the next steps, i.e., dehydration and paraffin embedding. In modern laboratories these steps are performed using special equipment (often overnight) where the tissues are passed through series of solvents with the purpose of removing water and replacing it with ascending grades of ethanol (70–99%), followed by xylene and finally warm, liquid paraffin (Fig. 12.4b). When the tissues have been saturated with liquid paraffin, they are taken to the embedding station (Fig. 12.4c) where they are placed in a metal mold, which is filled with liquid paraffin at 60 °C. When the warm paraffin is cooled, it solidifies into tissue blocks, which are the final product before the sectioning or cutting (Fig. 12.4d). Tissue blocks can be stored for decades in a dry, cool, and dark place with only minimal changes in the tissue embedded in the paraffin. They are therefore important as historical reference material. The tissue blocks are usually sectioned at 2–3 μm on a microtome (Fig. 12.4e) and mounted on tissue slides made of glass (Fig. 12.4f). The slides with the tissue sections (Fig. 12.4g) are heated to melt the paraffin in order to ensure a good attachment between the tissue and the slide. The tissue is now ready for the last steps of the process, which are rehydration and staining. The rehydration process is basically the reversed dehydration process described above. The tissue slides are passed through xylene, descending grades of ethanol (99–70%) and finally water. Now the tissue has been rendered penetrable for the aqueous solutions of staining agents used in the staining process described below. After the staining of the tissue section a coverslip, which is a very thin (usually 0.13–0.16 mm), transparent membrane made of glass or plastic, is put on top of the slide using a mounting medium (a glue-like substance) in order to protect the tissue section and to ensure good optical quality of the slide (Fig. 12.4h).

12.5.3.1 Decalcification of Hard Tissues (Bone and Teeth)

As mentioned above, some tissues call for the use of other techniques. Especially bone tissue can be a challenge. Bones are hard and cannot be processed directly as described above. The same applies to teeth, but the histopathological processing of teeth is a separate science that is beyond the scope of this text. In order to render bones soft

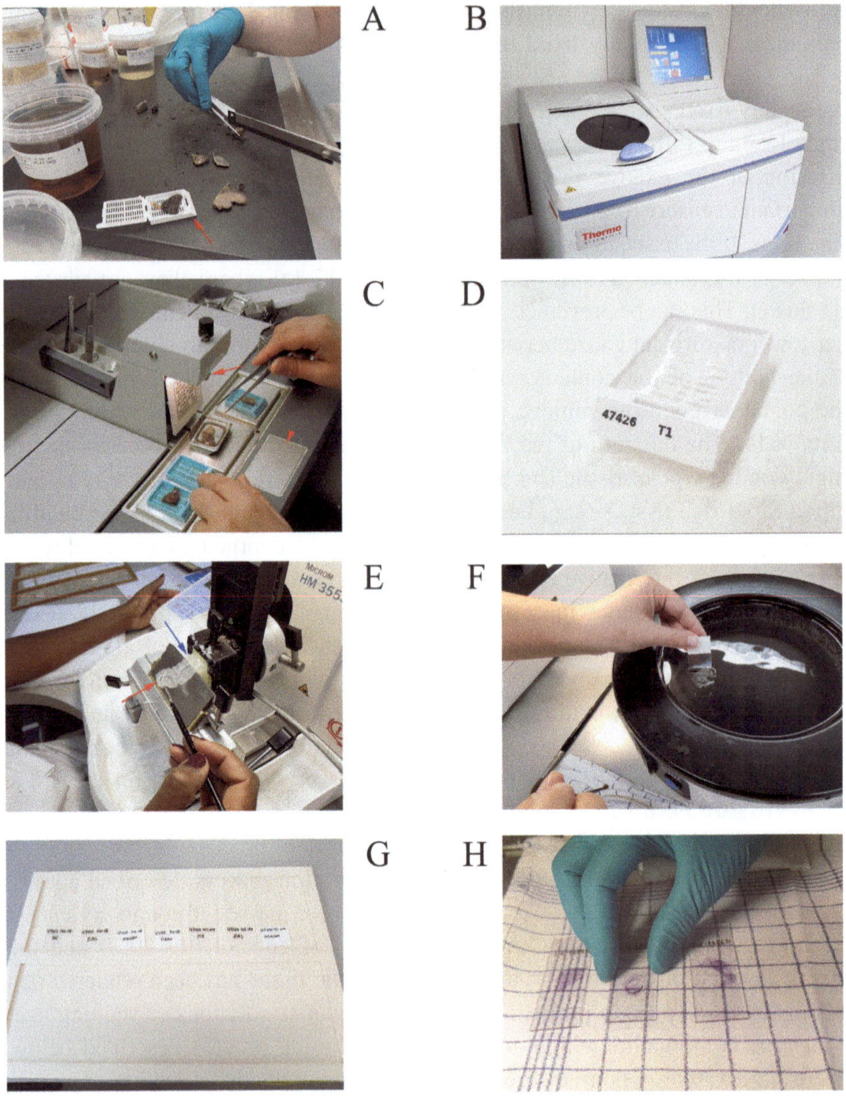

Fig. 12.4 Preparing slides for histopathology. (**a**) After fixation, tissues are trimmed, i.e., they are cut into smaller pieces and placed in tissue cassettes (arrow) for further processing. (**b**) In modern laboratories, tissues are processed in specialized equipment. The final products are tissue samples saturated with paraffin. In the tissue processor, the tissue samples are dehydrated and saturated with liquid paraffin. This is done by removing the aqueous formalin and replacing it with ascending grades of ethanol, followed by xylene and finally warm, liquid paraffin. (**c**) Embedding means covering the tissue samples with warm, liquid paraffin from a paraffin tap (arrow). At room temperature, the paraffin solidifies to a paraffin block. The process is accelerated by placing the mold containing the cassette on a cold plate (arrowhead). One half of the corresponding tissue cassette is attached to the block, thus securing correct identification and traceability of the sample. (**d**) This tissue block with identification numbers on an icy surface is ready for sectioning on a microtome. The cold surface makes the paraffin harder, which facilitates the subsequent sectioning. (**e**) The tissue block (blue arrow) is cut on a microtome with a very sharp blade. The 3–5 μm thin sections (red arrow) are then placed on an approximately 40 °C warm water bath. (**f**) In the mounting process, the floating tissue sections are caught with a tissue slide. (**g**) These tissue slides are ready for staining. (**h**) Cover slipping is the final step in the process. A drop of mounting medium ensures a good adhesion and high optical quality

and suitable for histopathological processing they have to be decalcified using decalcifying media. Most decalcifying agents are acids that remove calcium ions from the bone tissue, thus making it softer. Numerous decalcifying media exist. Some are commercially available, ready-to-use products; others are easily made in the laboratory using common reagents such as formic or citric acid, or ethylenediaminetetraacetic acid (EDTA), which is a chelating agent that, although it is an acid, exerts its effect through its high binding affinity for calcium ions. Bones are usually decalcified immediately after fixation. The duration of the decalcifying process is highly dependent on decalcifying agent used, the size of the specimen and the density of the bone tissue. Some fast-acting decalcifying agents, especially those with a very low pH, can do damage to tissue structures, which may ruin cellular details in the final product. Often more gentle, slowly-acting agents like EDTA are preferred. The density or hardness of the bone depends mainly on the size of the animal and the anatomical localization of the bone. Bones from small animals and bones of the skull or rib cage are softer than from other animals or locations. A notable exception is the petrous part of the temporal bone surrounding the inner ear, which is much harder than other bones of the skull. From small laboratory animals bones can be decalcified within a period of a few days. The decalcifying process can be accelerated by keeping the container with the decalcifying media in constant motion on a platform shaker. Heating the media will also accelerate the process.

12.5.3.2 Cryosectioning of Tissues

As described above, standard processing of tissues for histopathology involves removal of water, in order to replace it with paraffin, and then removal of paraffin in order to make the tissue slide penetrable for aqueous solutions. In the latter process, all soluble lipids that may have been in the tissue will be removed, and if fat cells or soluble fatty substances in the tissue are the scope of the histopathological examination, other techniques have to be used. In most cases the frozen section procedure or cryosectioning is the method of choice. The tissue, either fresh or fixed, is frozen at approximately −20 °C and then cut on a cryostat. The frozen section is then mounted on a tissue slide, and then it can be stained as preferred. Apart from being the method of choice when examining fat cells or soluble fatty substances, cryosectioning is the method of choice when preservation of antigenic structures is vital. As mentioned above, fixation, especially in formalin, often interferes with the molecular structures of the tissues to such a degree that the demonstration of certain epitopes, using immunohistochemical techniques can be impossible. This problem can be avoided using cryosectioning. Furthermore, tissue enzyme activity is preserved in non-fixed tissues and can be measured with enzyme histochemical methods, which is especially important for muscle tissues.

12.5.4 Staining of Tissue Sections

The staining of tissue sections relies on the tissues' histochemical properties, i.e., the ability to bind certain dyes depending on their charge, pH or other chemical properties. The number of available staining methods is very high, and most laboratories only use a very limited number of the methods described in the literature. Normally, laboratories use a standard staining technique, most frequently hematoxylin and eosin (HE), which can be supplemented with special staining methods. The dyes hematoxylin and eosin are alkaline and acidic, respectively. The alkaline and bluish hematoxylin binds to acidic structures in the cell. These consist mainly of DNA in the cell nucleus but also high mRNA abundances in the cytoplasm can be demonstrated. The acidic, pink eosin binds to alkaline structures. These are typically cytoplasmic proteins and extracellular matrix. When combined in the HE method, the result is a good overall view of the tissue structure and components (Fig. 12.5a).

12.5.4.1 Special Tissue Stains

As staining with HE does not always provide enough information about the tissue examined, a large number of special tissue staining techniques have been developed. Often the staining methods are aimed at demonstrating certain tissue structures like carbohydrates, basement membranes,

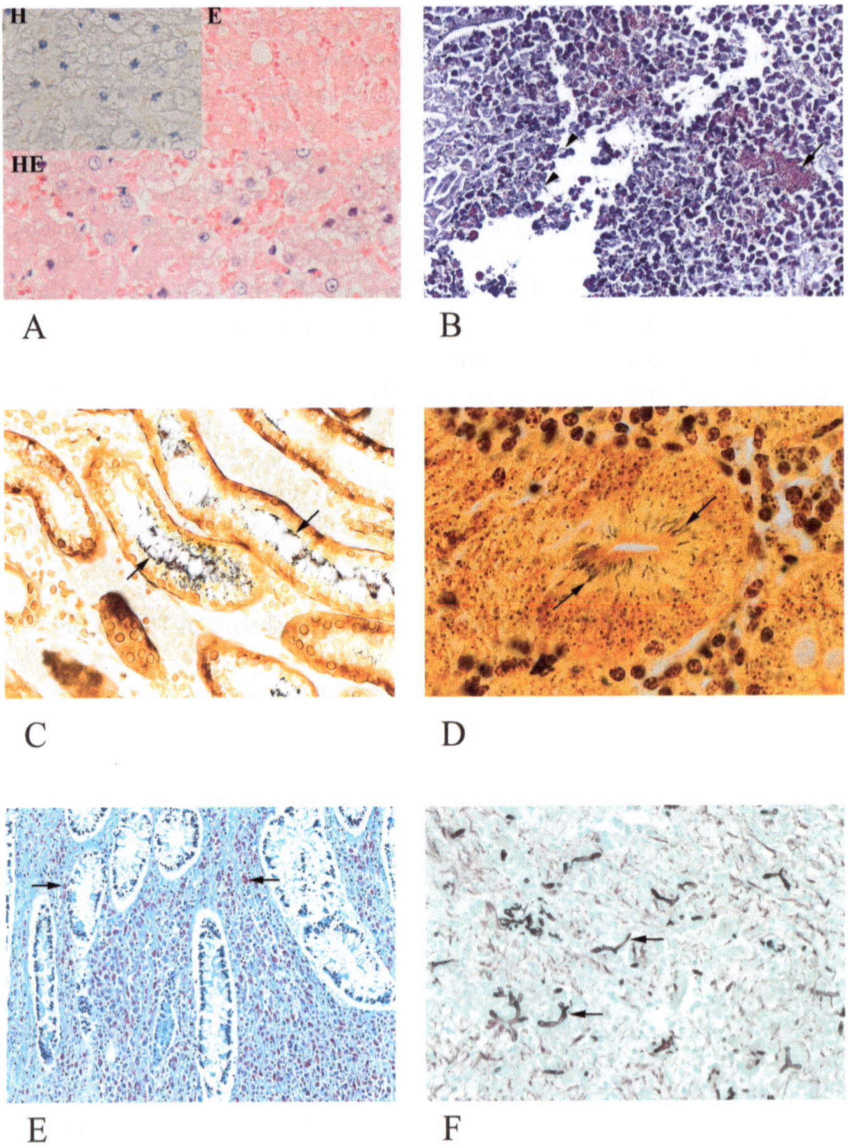

Fig. 12.5 Different staining techniques. (**a**) When combined, hematoxylin (top left (H)) and eosin (top right (E)) provide a good overall view (bottom) of this liver section. HE. (**b**) The Gram staining method can differentiate between Gram positive (blue) and Gram negative (red) bacteria. In this tissue sections most bacteria are Gram negative cocci (arrow). A few Gram positive rods can also be seen (arrowheads). Gram. (**c**) In this kidney slide stained with the Levaditi method, spirochetes (*Leptospira* sp.) can be seen in the kidney tubules (arrows). Levaditi's silver stain. (**d**) In this gut section stained with the Warthin-Starry method, spirochetes can be seen in the epithelial cell cytoplasm (arrows). Warthin-Starry. Ziehl–Neelsen. (**e**) The Ziehl–Neelsen stain is the method of choice when acid fast bacteria like mycobacteria are to be demonstrated. Here, macrophages laden with red *Mycobacterium avium* ssp. *paratuberculosis* can be seen (arrows). (**f**) Filamentous fungi (arrows) are often stained with the Grocott method. Other organisms with carbohydrate-rich walls can also be demonstrated. GMS

elastic fibers, mucins or neuronal axons. Lipids can be demonstrated using lipophilic stains like oil red O or Sudan Black on frozen sections as described above. Other methods stain specific cell populations such as eosinophils, mast cells or plasma cells. Some stains demonstrate specific pathological changes such as extracellular fibrin or amyloid depositions, mineralization or fibrosis. Bacteria, fungi, and protozoa can be demonstrated using various staining methods. The Gram stain enables the differentiation between Gram positive and Gram negative bacteria. The Ziehl–Neelsen stain is used to demonstrate acid fast bacteria such as mycobacteria. Fungi are most often demonstrated using the Periodic acid-Schiff (PAS) stain or various silver stains such as Grocott's methenamine silver reaction (GMS). In Fig. 12.5b–f a few examples of staining methods used to demonstrate microorganisms in tissue sections can be seen. Lately, molecular in situ methods (see below) to a large degree have replaced the use of special stains to demonstrate microorganisms in tissue section. However, they are still widely used to demonstrate the presence of microorganisms in tissues. As a token of absence of microorganisms a negative staining is of no value. Table 12.1 gives an overview of various frequently used staining methods.

Table 12.1 Frequently used tissue staining methods

Staining	Structures/tissue/cells stained	Reference
General overview		
Hematoxylin and eosin	All tissue structures	Bancroft and Layton [6]
Microorganisms		
Giemsa	Microorganisms	Morris et al. [56]
Gram	Bacteria	Morris et al. [56]
Grocott's methenamine silver reaction	Fungi	Morris et al. [56]
Levaditi's	Bacteria	Bancroft and Cook [5]
Periodic acid-Schiff	Fungi	Layton and Bancroft [47]
Phloxine-tartrazine	Viral inclusion bodies	Morris et al. [56]
Southgate's mucicarmine	Cryptococci	Bancroft and Cook [5]
Warthin-Starry	Bacteria esp. spirochetes	Morris et al. [56]
Wright and Craighead	Protozoa	Lillie [53]
Ziehl–Neelsen	Acid fast bacteria	Morris et al. [56]
Amyloid		
Congo red	Amyloid	Gilbertson [26]
Carbohydrates		
Alcian blue	Carbohydrates	Layton and Bancroft [47]
Best carmine	Carbohydrates	Luna [55]
Periodic acid-methenamine silver	Basement membranes	Bancroft and Layton [7]
Cell populations		
Luna	Eosinophils	Luna [55]
Methyl green pyronin	Plasma cells	Bancroft and Cook [5]
Toluidine blue	Mast cells; cartilage	Pearse [61]
Connective and other mesenchymal tissues		
Gomori	Reticular fibers (type III collagen)	Bancroft and Layton [7]
Gordon and Sweets'	Reticular fibers	Bancroft and Layton [7]
Masson trichrome	Collagen	Bancroft and Layton [7]
Movat pentachrome	Collagen	Bancroft and Layton [7]
Phosphotungstic acid-hematoxylin (PTAH)	Fibrin	Layton and Bancroft [47]
Picrosirius red	Collagen	Pearse [61]
Safranin O	Cartilage	Lillie [53]
Van Gieson	Collagen	Bancroft and Layton [7]
Verhoeff-van Gieson	Elastic fibers	Bancroft and Layton [7]
Lipids		
Oil red O	Lipids	Highley and Sullivan [31]
Sudan black	Lipids	Highley and Sullivan [31]
Metals, minerals, and pigments		
Perls' Prussian blue	Ferric iron	Orchard [59]
Rubeanic acid	Copper	Orchard [59]
Schmorl's	Lipofuscin	Bancroft and Cook [5]
Von Kossa	Calcium ions	Lillie [53]

12.6 Evaluation of Histological Sections in Relation to Infections

Histological examination of tissue samples is important for the diagnosis of spontaneous infections. Histology can in particular be useful, when conventional culture methods fail to identify the infectious agent [27]. Most infective microorganisms cause patho-morphological changes, and histology can be used to identify and characterize these changes. Commonly, the patho-morphological changes of infections will be related to the inflammatory response initiated by the pathogen. Only a correct histological evaluation will give reliable information about a feasible pathogenesis, etiology and diagnosis.

12.6.1 Evaluation of Tissue Sections

Histological sections should be evaluated as an "inverted pyramid" leaving the specific characterization of lesions, i.e., content, location, and distribution at the end. First, the microscopic anatomy of the tissue should be recalled. Based on this, the entire slide must be evaluated consistently at low magnification (×10/20), i.e., by breaking the tissue into its major components to evaluate each component in detail. For instance, when evaluating a section of a lung one should follow the air direction from the bronchus to the alveolar space including, alveolar septa, vasculature and pleura. Furthermore, all hollow organs have both a lumen and an inside-outside wall with multiple different layers/components, which must be evaluated consistently from the inside and out ensuring that lesions are not overlooked. Once identified changes or lesions must be examined at higher magnification (×40). Lesions in routine HE stained sections from infected laboratory animals will consist of the infecting microorganisms and patho-morphological changes. Of note, the lesion may only consist of an inflammatory reaction and/or changed host tissue morphology, as the microorganism might not always be visible. Both cellular and non-cellular inflammatory components of a lesion will give information about duration and severity of the infection. Following identification of a lesion, the localization in relation to microscopic anatomical structures and the distribution (focal, multifocal, disseminated, and diffuse) within the organ must be evaluated. Both the content, localization and distribution of a lesion provide valuable information about the possible pathogenesis, etiology, and diagnosis.

12.6.2 How to Get a Suspicion of Infection

Clearly, visible microorganisms in HE stained sections will indicate an infectious diagnosis. In HE stained sections most microorganisms will be colored by hematoxylin and thus appear purple (Fig. 12.6a, b). If microorganisms are observed, the size and shape (bacilli, cocci, fungal hyphae, parasite, etc.) must be evaluated [15]. Afterwards, it should be decided if additional histochemical staining's should be performed in order to identify the microorganisms better (Table 12.1) [15]. Moreover, molecular techniques may be applied in order to identify the specific infective agent (Fig. 12.6c, d) (see Tables 12.2, 12.3, and 12.4). Of note, molecular techniques can also be useful for visualization and identification of microorganisms which are non-visible in routine HE stained sections (see Sect. 12.6.4). Inflammation is a hallmark of most infectious diseases. However, it is important to remember that inflammation also can be associated with non-infectious conditions like neoplasia, allergy, autoimmune disorders, and idiopathic disorders [15]. Most infectious microbes initiate an acute inflammatory response just after they are encountered by the host animal. Briefly, the acute inflammatory response consists of vascular and cellular changes which result in accumulation of fluid and plasma proteins, as well as leucocytes, in the extracellular tissue [2]. The accumulation of fluid, plasma proteins, and cells occurs due to increased permeability of the local capillary endothelium. Fibrinogen and other plasma proteins will leak into the local area and fibrinogen will be formed into fibrin [2]. Fibrin can be recognized in HE

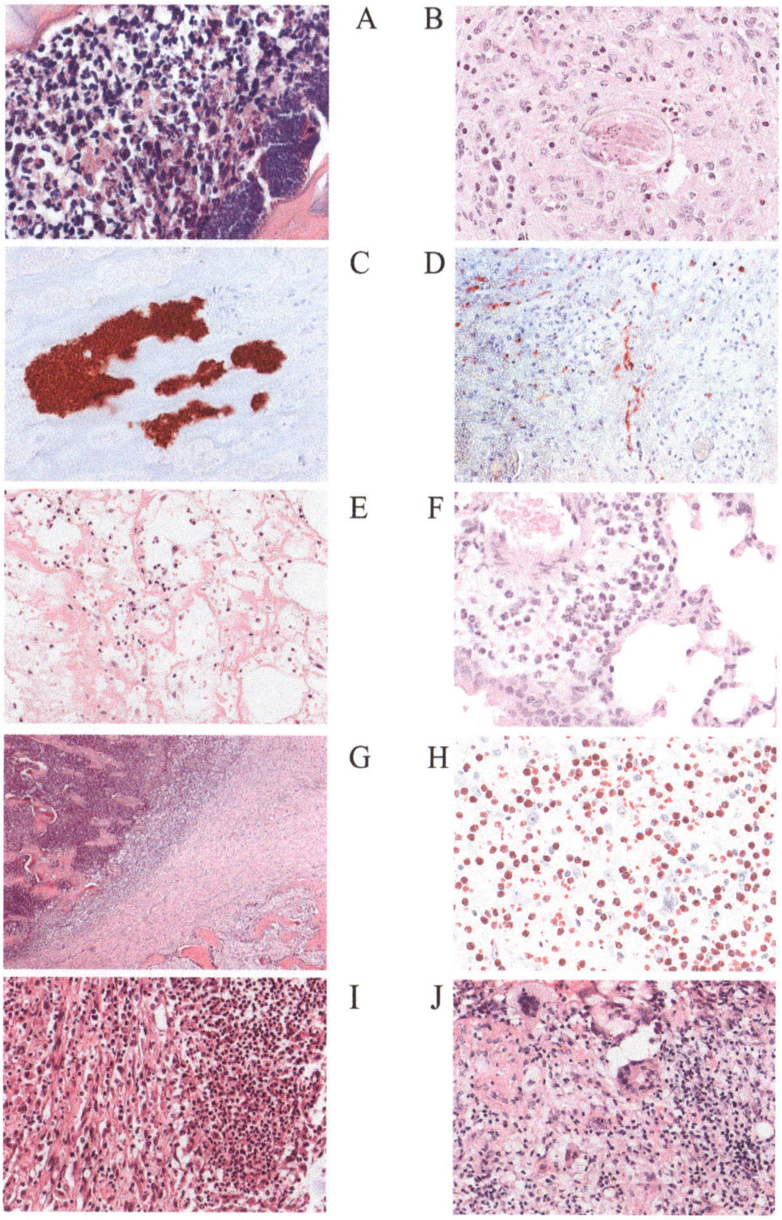

Fig. 12.6 Examples of histopathology relevant in relation to infections in laboratory animals. (**a**) Acute, suppurative osteomyelitis, pig. Infective bacteria can be seen as purple colonies in HE stained sections. HE. (**b**) Chronic, hepatic schistosomiasis, pig. In tissues, parasites and parasitic elements are often of a size making them easily detectable in HE stained sections. HE. (**c**) Acute, suppurative osteomyelitis, pig. Identification of *Staphylococcus aureus* by immunohistochemistry using a specific antibody towards the bacterium. (**d**) Chronic, mycotic dermatitis. Rabbit. Identification by *Aspergillus fumigatus* hyphae by in situ hybridization using a DNA probe. (**e**) Acute, subcutaneous cellulitis. Rabbit. The presence of a diffuse fibrino-suppurative inflammation should raise the suspicion of an infection. HE. (**f**) Accumulation of neutrophils in the lung of a mouse infected with *Pasteurella multocida*. Neutrophils should always be suspected for an infectious cause. HE. (**g**) Chronic, osteomyelitis, pig. Encapsulation of a lesion by connective tissue demonstrate a chronic inflammatory response. HE. (**h**) Chronic, eosinophilic hepatitis. Pig. An eosinophilic inflammation should be suspected for a parasitic cause or a result of a hypersensitivity reaction. Luna. (**i**) Chronic, pyo-granulomatous dermatitis. Rat. A pyo-granulomatous inflammation with presence of macrophages and epithelioid cells and/or giant cells should be suspected for an infectious cause. HE. (**j**) Chronic, granulomatous dermatitis. Rat. The presence of a granulomatous inflammation with macrophages, epitheloid cells and giant cells should be suspected for an infectious cause. HE

Table 12.2 Molecular in situ demonstration (immunohistochemistry and in situ hybridization) of pathogenic bacteria in laboratory mice, rats, and rabbits and their primary target organs

Pathogen	Species	Primary target organ (secondary targets)	In situ identification
Filobacterium rodentium (prev. *Cilia-associated respiratory bacillus*)	Mice, rats	Respiratory system	No description of IHC or ISH. Detectable as a filamentous argyrophilic bacillus with Warthin-Starry silver stain (non-specific) [60]
Klebsiella pneumoniae	Mice, rats	Respiratory system	Immunohistochemistry and in situ hybridization [75]
Mycoplasma pulmonis	Mice, rats	Respiratory system (joints and genitalia)	Immunofluorescence (in vitro) [68] and immunohistochemistry [52]
Streptococcus pneumoniae	Mice, rats	Respiratory system (middle ear)	Immunohistochemistry [21] and in situ hybridization [82]
***Helicobacter* spp.** (*H. hepaticus, H. bilis, H. muridarum, H. trogontum, H. rodentium,* and *Flexispira rappini*)	Mice, rats	Hepatic system (gastrointestinal tract)	In situ hybridization of *H. hepaticus* [28], Immunohistochemistry only available for *H. pylori* [33]
Citrobacter rodentium (prev. *Citrobacter freundii*)	Mice, rats	Intestinal tract, primarily colon	Indirect immunofluorescence [18] and Immunohistochemistry [11]
Clostridium piliforme (prev. *Bacillus piliformis*)	Mice, rats, rabbits	Intestinal tract (hepatic system and heart)	No description of IHC or ISH for this pathogen. Detectable with Warthin-Starry silver stain as long rod-shaped bacteria that tend to stack on top of each other (non-specific) [76]
Clostridium spiroforme		Gastrointestinal tract	No description of tests for this pathogen
Pseudomonas aeruginosa	Mice, rats, rabbits	Respiratory system	Immunohistochemistry [36]
Salmonella enteritidis* serotype *typhimurium	Mice, rats	Gastrointestinal tract incl. liver and spleen	Immunohistochemistry [64]
***Rodentibacter* spp.** (*prev. Pasteurella pneumotropica*)	Mice, rats	Skin and adnexa	No specific detection method for this pathogen
Staphylococcus aureus	Mice, rats, rabbits	Skin (Respiratory system, etc.)	Immunohistochemistry [41] and In situ hybridization (clinical samples) [67]
***Corynebacterium* spp.**	Mice	Skin—causes hyperkeratosis	No specific detection method for this pathogen, but can be visualized using Gram-stain [43]
Corynebacterium kutscheri	Mice, rats	Liver and kidneys	No specific detection method for this pathogen, but stimulates Galectin-3 secretion from macrophages, which can be visualized by immunohistochemistry [81]
Bordetella bronchiseptica	Rabbits	Respiratory system	Immunohistochemistry [70]
Pasteurella multocida	Rabbits	Respiratory system (dissemination to ears and genitals)	Immunohistochemistry [8] and fluorescence in situ hybridization [62]
Treponema paraluiscuniculi (rabbit syphilis)	Rabbits	Skin and genitalia	No description of IHC or ISH for this pathogen. Detectable with Warthin-Starry silver stain [35]
Listeria monocytogenes	Rabbits	Hepatic system (Spleen and adrenal glands)	Immunohistochemistry [14]
Francisella tularensis	Rabbits	Liver, spleen and bone marrow	Indirect immunofluorescence and immunohistochemistry [30]

Table 12.3 Molecular in situ demonstration (immunohistochemistry and in situ hybridization) of spontaneous viruses in laboratory mice, rats, and rabbits and their primary target organ

Pneumonia virus of mice *(PVM)*	Mice	Respiratory system	Immunohistochemistry [10], In situ hybridization [20]
Murine respirovirus *(prev. Sendai virus)*	Mice, rats	Respiratory system (Reproduction system)	Immunohistochemistry [22, 74]
Mouse cytomegalovirus	Mice	Salivary- and adrenal glands (hematopoietic system)	Immunohistochemistry and immunofluorescence [45]
Rat cytomegalovirus	Rats		
Mouse parvovirus type 1	Mice	Hematopoietic system—cell proliferation	In situ hybridization [34]
Mouse rotavirus	Mice	Gastrointestinal tract	Immunohistochemistry [44, 63] and in situ hybridization [63]
Rat rotavirus-like agent	Rats	Gastrointestinal tract	No description of tests for this pathogen
***Rotavirus* spp.**	Rabbits	Gastrointestinal tract	Immunohistochemistry and in situ hybridization [72]
Mouse thymic virus	Mice	Thymus (lymphoid system)	No specific detection method for this pathogen
Reovirus type 3	Mice, rats	Hepatic system	Immunohistochemistry [79]
Rabbit oral papillomavirus	Rabbits	Oral cavity	No description of IHC or ISH for this pathogen
Cottontail rabbit papillomavirus	Rabbits	Eyelids and ears (lymphoid system and lungs)	Immunohistochemistry and in situ hybridization [17]
Mouse mammary tumor virus *(MMTV)*	Mice	Mammary glands	Immunohistochemistry [84]
Lymphocytic choriomeningitis virus *(LCMV)*	Mice	Hematopoietic system (vascular structures and liver)	In situ hybridization [83] immunofluorescence [46] and immunohistochemistry [23]
Gammaarterivirus lacdeh *(prev. Lactate dehydrogenase-elevating virus)*	Mice	Hematopoietic system (lymphoid and neurologic system)	Immunohistochemistry [13]
Theiler's murine encephalomyelitis virus *(TMEV)*	Mice	Neurological system	Immunohistochemistry [16] and in situ hybridization [58]
Rat Theilovirus	Rats		
Murine mastadenovirus	Mice	Neurological system (Kidneys, heart, spleen, etc.)	Immunohistochemistry [50] and in situ hybridization [69]
***Adenovirus* spp.**	Rabbits	Gastrointestinal tract (potentially liver and lungs)	
Ectromelia virus (*ECTV*, mousepox)	Mice	Skin and lymphoid tissue (Liver and spleen)	Immunohistochemistry [3, 77]
Myxoma virus	Rabbits	Multi-systemic edema, hemorrhages and necrosis	No description of IHC or ISH for this pathogen
Mouse minute virus *(prev. Minute virus of mice)*	Mice	Fetal development and immune system	Immunohistochemistry [54]
Toolan's H-1 virus	Rats	Neurological and skeletal system	No description of tests for this pathogen
Rat virus (prev. Kilham rat virus *(KRV)*	Rats	CNS, GI and reproductive system	Immunohistochemistry [4]
Lapine parvovirus	Rabbits	Gastrointestinal tract	No description of IHC or ISH for this pathogen
Mouse hepatitis virus *(MHV)*	Mice	Polytropic strains: Respiratory system and liver Enterotropic strains: GI tract	In situ hybridization [19] and Immunohistochemistry [52, 65]
Rat coronavirus/Sialodacryoadenitis virus *(SDAV)*	Rats	Eyes, salivary glands	Immunohistochemistry [9]
Coronavirus *(pleural effusion disease/ infectious cardiomyopathy virus)*	Rabbits	Cardiopulmonary system (Liver, lymphoid system, kidneys)	No description of IHC or ISH for this pathogen
Rabbit enteric coronavirus	Rabbits	Gastrointestinal tract	No description of IHC or ISH for this pathogen
Rabbit hemorrhagic disease virus	Rabbits	Vascular system (DIC and multi-organ failure)	Immunohistochemistry and in situ hybridization [1]

Table 12.4 Molecular in situ demonstration (immunohistochemistry and in situ hybridization) of spontaneous parasites and fungi in laboratory mice, rats, and rabbits and their primary target organ

Parasites			
Giardia muris	Mice, rats	Intestinal tract	–
Spironucleus muris (prev. *Hexamita muris)*			–
Syphacia obvelata and ***Aspicularis tetraptera*** (pinworms)	Mice	Intestinal tract	–
Syphacia muris (pinworms)	Rats		–
Passalurus ambiguus (pinworm)	Rabbits		–
Myobia musculi and ***Myocoptes musculinus*** (mites)	Mice	Skin	–
Radfordia affinis (mites)	Rats		–
Cheyletiella parasitivorax	Rabbits		–
Sarcoptes scabiei	Rabbits		–
Psoroptes cuniculi	Rabbits	Ears	–
Encephalitozoon cuniculi	Mice, rats	Gastrointestinal tract (dissemination to CNS, kidneys, liver and lungs)	–
Cryptosporidium parvum	Rabbits	Gastrointestinal tract	–
Eimeria stiedae	Rabbits	Liver	–
***Eimeria* spp.** *(E. intestinalis, E. flavescens, E. magna, E. irresidua, E. piriformis, E. perforans, E. neoleporis, and E. media)*	Rabbits	Gastrointestinal tract	–
Fungi			
Dermatophytes *(Trichophyton mentagrophytes, Microsporum gypseum,* and *Microsporum canis)*	Rabbits	Skin and adnexa	–
Pneumocystis carinii	Mice, rats	Respiratory system	Immunohistochemistry [42]

stained section as eosinophil threads and should cause suspicion of infection (Fig. 12.6e). Furthermore, fibrin threads can also be clearly visualized in blue by PTAH staining (Table 12.1). Vascular changes like vasodilatation and stasis in the acute inflammation phase can result in edema and hyperemia, which histologically are seen as enlargement of interstitial space and erythrocyte filled small vessels and capillaries, respectively. Neutrophils are the first leucocytes to migrate from the microvasculature into a local inflammatory focus. Neutrophils can be recognized in HE stained sections due to their fragmented nucleus and they should always raise the suspicion of an infection (Fig. 12.6e, f) [2]. Infections caused by bacteria will often (but not always!) give an extensive extravasation of neutrophils, i.e., a suppurative inflammatory reaction (Fig. 12.6f). Chronic inflammation can be a sequela to acute inflammation if the microorganisms are not eliminated [2]. Alternatively, some microorganism can evoke chronic inflammation directly, e.g., *Mycobacterium* spp. [2]. In general, chronic inflammation is characterized by lymphocytes and macrophages accompanied by tissue repair seen as granulation tissue formation and fibrosis (Fig. 12.6g) [2]. Often the dominating type of infiltrating immune cells will hint the type of microorganism and thus the diagnosis, e.g. lymphocytes are often seen in viral infections and eosinophil granulocytes in parasitic infections (Fig. 12.6h) [15]. Although, the neutrophil is associated with the initial inflammatory response it can often also be found in chronic inflammation in combination with signs of granulation tissue and fibrosis or in cases of granulomatous inflammation, i.e., pyogranulomatous inflammation (Fig. 12.6i, j). Granulomatous inflammation is a distinct type of chronic inflammation in which cells of the monocyte-macrophage system dominate and take the form of macrophages, epithelioid cells, and giant cells (Fig. 12.6j). Granulomatous inflammation is often associated with specific bacteria, e.g., *Mycobacterium* spp. and some fungal infections.

12.6.3 Tissue Artifacts

One of the most important challenges met by the histopathologist is the formation of tissue artifacts; a term used for all findings that are due to postmortem phenomena or improper processing of the tissue [73]. A few artifacts occur antemortem but many arise in the animal at the time of killing or postmortem before or during the sampling and subsequent handling and processing of the tissue being examined. In every step on the way to the final tissue slide, the possibility of artifact formation is present. Some tissues are more prone to artifact formation and some steps in the processing of tissues tend to cause more artifacts than others. Furthermore, some artifacts are more challenging to the histopathologist than others, and artifacts are a greater challenge to the untrained histopathologist than the experienced one. However, even the experienced histopathologist can sometimes be taken by surprise. The list of possible artifacts in histopathological sections is inexhaustible, and only the most common and challenging will be described below.

12.6.3.1 Antemortem and Killing Artifacts

Antemortem artifacts are not as common in laboratory animals as in animal patients undergone clinical examinations and treatment. Sometimes, however, medical and surgical procedures are performed on laboratory animals giving rise to artifacts. These include the presence of suture material, talcum powder, contrast media or other foreign material in the tissues. It is more common to find artifacts that arise during the killing of the animal. It is, therefore, essential to know how an animal was euthanized to keep related changes in mind. Tissue injections with barbiturates may cause changes histologically resembling necrosis (Fig. 12.7a). They are often found in the heart muscle or liver after intracardiac or intrahepatic injections. Killing of small laboratory animals through cervical dislocation causes severe trauma and hemorrhage in muscles and spinal cord that have to be differentiated from genuine lesions.

12.6.3.2 Postmortem Artifacts

After the animal dies, many physiological processes remain active for some time or at least until the level of available adenosine triphosphate (ATP) in the cells reaches a level where energy demanding processes come to a halt. Consequently, muscle cells retain their ability to contract, even after death. This can be seen as widespread contraction of smooth and striated muscle cells. This causes formation of contraction bands in especially striated muscles (skeletal and cardiac) and reduces the lumen of many hollow organs with a muscular wall so that arteries and arterioles seem to be without a lumen (Fig. 12.7b). The formation of contraction bands can be reduced by postponing the fixation until after the cessation of rigor mortis but this will of course increase the likelihood of other artifacts, namely autolysis and putrefaction. Autolysis is the postmortem decay of the tissues due to the organism's own degradation processes. Autolysis rarely comes alone but is normally accompanied by varying degrees of putrefaction, i.e., invasion of tissue degrading bacteria and other microorganisms from the gut and the environment. Postmortem degradation starts at the time of death so immediate and adequate fixation of the tissues is most important to reduce its occurrence. Some organs are more prone to show signs of postmortem degradation, and especially the intestines, kidneys, and prostate are sensitive, while bones, skin, and skeletal muscle are quite resilient. In tissue sections, postmortem degradation is seen as effacement of cellular structures, loss of intercellular attachments, condensation of nuclei, desquamation of epithelia from underlying basement membranes, and invasion of bacteria, often with accompanying gas production.

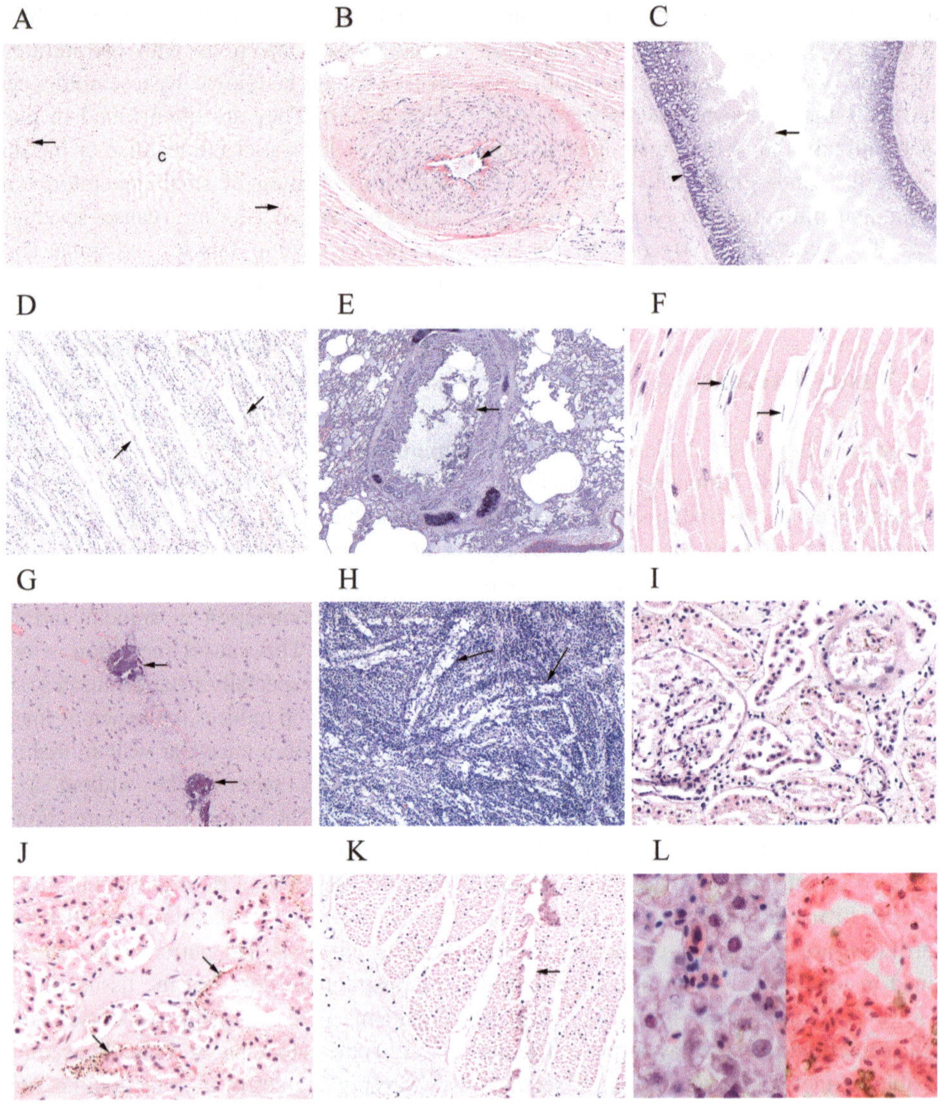

Fig. 12.7 Examples of artifacts in histological preparations. (**a**) Barbiturate damage caused by intrahepatic injection. The liver cells surrounding the central vein (C) appear necrotic and all red blood cells have been erased except those far from the vein (arrows). HE. (**b**) The lumen (arrow) of this artery looks very narrow due to postmortem contraction of the smooth muscle cells in the artery wall. HE. (**c**) Autolysis in a small intestine. The epithelium of the intestinal villi (arrow) has been lost and the remaining part of the luminal mucosa is without cellular details. Only the basal part of the intestinal crypts has been preserved (arrowhead). HE. (**d**) The epithelium of the tubules and collecting ducts in this kidney has become detached from the underlying basement membrane due to postmortem autolysis (arrows). HE. (**e**) This pulmonary bronchus contains sloughed epithelial cells (arrow) due to of autolysis. HE. (**f**) Elongated bacteria (arrows), probably clostridia, have invaded this heart muscle in the process of postmortem putrefaction. Postmortem bacterial growth can usually be differentiated from bacterial infections by the absence of an inflammatory response. HE. (**g**) During the opening of the skull, sawdust in the form of bone fragments (arrows) has been pressed into the brain parenchyma by the saw blade. HE. (**h**) freezing artifacts are seen as irregular lines of cavities in the tissue (arrows). They arise when water forms crystals within the tissue, displacing the cells. Also, note the absence of red blood cells that have undergone lysis. HE. (**i**) Inadequate fixation of tissues causes many of the same artifactual changes as in autolysis. In this kidney section there is detachment of epithelia and condensation of nuclei. HE. (**j**) Acid hematin (arrows) is formed in a chemical reaction between formalin and hemoglobin. It is birefringent so it can easily be differentiated from other pigments. HE. (**k**) Apart from the mechanical damage to the section (arrow) there is severe shrinkage of the smooth muscle cells in this intestinal wall so they seem to be surrounded by empty spaces. HE. (**l**) Sections from this avian liver have been stained with hematoxylin and eosin at two different professional laboratories. Note the difference in color tone and the staining intensity of the various tissue structures. HE

Figures 12.7c–f show some examples of postmortem degradation.

12.6.3.3 Tissue Handling Artifacts

Mechanical damage to fresh tissues is a common source of postmortem artifacts. Tissues should always be treated with utmost care to avoid this, which is especially important when handling the brain, which has very little connective tissue, making it very sensitive to mechanical artifact formation. It is also important not to squeeze, pull, or tear the tissue. When examining brain sections under the microscope it is common to find pieces of bone adjacent to or even embedded within the brain (Fig. 12.7g). This artifact arises during the opening of the skull and represents saw dust from the skull that has been pushed into the brain by the saw blade. Often plant material, hair or other foreign material can be found in tissue sections because of contamination from the environment. This happens especially if the tissue has been in contact with the cut contents or the fur. Often cross contamination of tissues take place so that one type of tissue can be found embedded within another type. For instance, liver cells can be found in the lungs or gut epithelium in the brain. This cross contamination can take place at several stages of the processing but it happens often when the tissues are trimmed before embedding. When the cross contamination takes place during later stages, i.e., mounting and staining, the foreign material will normally be found in a different optical plane than the tissue itself when examined under the microscope, making its interpretation easier. Freezing of tissues prior to histopathological examination should be avoided unless it is done under controlled condition in the production of frozen sections. Freezing can be accidental because of low ambient temperatures or it can be deliberate and necessary because the tissue has to be stored and transported for a long time before it can be examined. Freezing causes damage to the tissue through the production of ice crystals that distort the tissue (Fig. 12.7h). Furthermore, red blood cells undergo hemolysis and become very difficult to see in the normal HE section. Other cellular structures are lost as well, creating a blurred view of the tissue. Drying out of tissues is especially seen in cases of delayed fixation of tissue samples. Drying is seen as condensation and increased staining of the surface part of the tissue and is usually not difficult to interpret. Heat damage is especially seen in biopsies removed by cautery, but sometimes heat is used to sterilize tissue surfaces before sampling for microbiology, and in these cases care should be taken that sampling for histopathology is as far from the sterilized area as possible.

12.6.3.4 Fixation Artifacts

Inadequate fixation often cause tissue artifacts, e.g., due to factors such as too thick tissue samples, too short duration of the fixation or too high tissue/fixative ratio. As the duration of the fixation is highly dependent on the size and type of tissue, the samples, with the notable exception of the brain, should not be thicker than 6 mm and the volume of the fixative should be 10–20 times the volume of the tissue. Inadequately fixed tissues show artifacts like cellular shrinkage, condensation of nuclei and reduced chromatin staining. Intercellular spaces are exaggerated and especially the inner parts of the sample are autolyzed, and cellular details are lost as described above (Fig. 12.7i). Acid hematin is a pigment that forms in formalin fixed tissues due to a chemical formalin-hemoglobin reaction. It is seen as brownish to black, granular or crystalline structures especially in areas with abundance of hemoglobin such as blood vessels and areas of hemorrhage. Although acid hematin formation is a well-known phenomenon, which should not confuse the trained histopathologist, it may sometimes disturb the evaluation of tissue sections. As acid hematin shows birefringence, it is easily be differentiated from other pigments by polarized light (Fig. 12.7j).

12.6.3.5 Tissue Processing and Staining Artifacts

The standard treatment of the tissue with dehydrating agents followed by the removal of soluble lipids can cause considerable changes, usually reduction, of the volume of the tissue. The brain is especially susceptible to shrinkage due to its high lipid content but all soft tissues may shrink, while bone is resistant. In tissue sections, shrinkage can be seen as apparent cavities around blood vessels or formation of spaces where no spaces should be (Fig. 12.7k). Often, however, the phenomenon will go unnoticed, and it is of no consequence unless the volume or size of the tissue is important, such as for stereology and other morphometric disciplines. Various forms of mechanical damage such as folds, wrinkles, streaks, scratches, and cracks often arise during the sectioning of the tissue block on the microtome and the mounting on the tissue slide, but they can also arise later in the process if the tissue section is touched or if it gets partially detached from the slide. Usually these phenomena are not difficult to interpret. The staining process is subject to great variations in staining intensity (Fig. 12.7l). Often solutions vary in concentration, especially if old. It is therefore of utmost importance that all protocols are followed strictly and that a rigorous quality control is applied. During mounting and cover slipping bubbles of air can be caught between the tissue section and the slide or between the tissue section and the cover slip. In the former case, increased staining and rupture of the overlying tissue can be seen. In the latter case, small air bubbles can be seen on the top of the tissue section.

12.6.4 Molecular In Situ Demonstration of Infectious Agents in Laboratory Animals

Several immunohistochemical and in situ hybridization techniques have proven effective for detecting specific species of different infective agents causing spontaneous infections in laboratory animals (Tables 12.2, 12.3, and 12.4). By immunohistochemistry antigens of infective agents, such as bacteria, virus, and fungi, are specifically detected and stained by the application of appropriately labeled antibodies, which visualizes and documents the high-resolution distribution and localization of specific infective agents within their proper histological in situ context. A wide variety of different direct and indirect immunohistochemical staining techniques are used, and special text books should be consulted [40, 66]. In situ hybridization techniques detect specific sequences of nucleic acids by hybridization of enzyme or fluorescently labeled oligonucleotide probes to complementary target sequences of DNA or RNA [51, 57]. For the bacterial identification, ribosomal RNA (rRNA) targeting probes are most commonly applied because of the natural amplification of rRNA molecules [12]. Bacterial rRNA is present in all cells and highly conserved between species, but still contains variable sequences which are target for identification at species level. The 16S rRNA is the preferred target for specific bacterial identification due to its combination of conserved sites and variable regions [24]. When the probes have been designed and evaluated, the in situ hybridization techniques can be completed in a few hours [40]. When enzymes are used for visualization, the technique—as it is for immunohistochemistry—allows direct localization of the reacting agent in the tissue and thereby a histological correlation between the infective agent and the tissue morphology and/or the pathological changes present [71]. As the identification of infective agents by in situ hybridization and immunohistochemical techniques is based upon probe sequence and antigen specificity unknown species are not identified [24, 38]. Although a number of genus and species specific probes and primary antibodies for immunohistochemistry have successfully been applied for the detection and identification of infective agents of importance in laboratory animals (Tables 12.2, 12.3, and 12.4), appropriate positive and negative controls must be applied in parallel [38, 66, 71].

12.7 Histopathological Examples of Spontaneous Infections in Laboratory Animals

Organisms are normally not disclosed in HE-stained sections exemplified by infection with *Pneumocystis carinii*. However, suspicions of the infection should be made when accumulation of foam macrophages within the alveolar lumen is present (Fig. 12.8a). In order to demonstrate an organism such as *P. carinii* especially the application of the GMS-stain is useful (Fig. 12.8b) [39]. Bacteria may be the cause of a

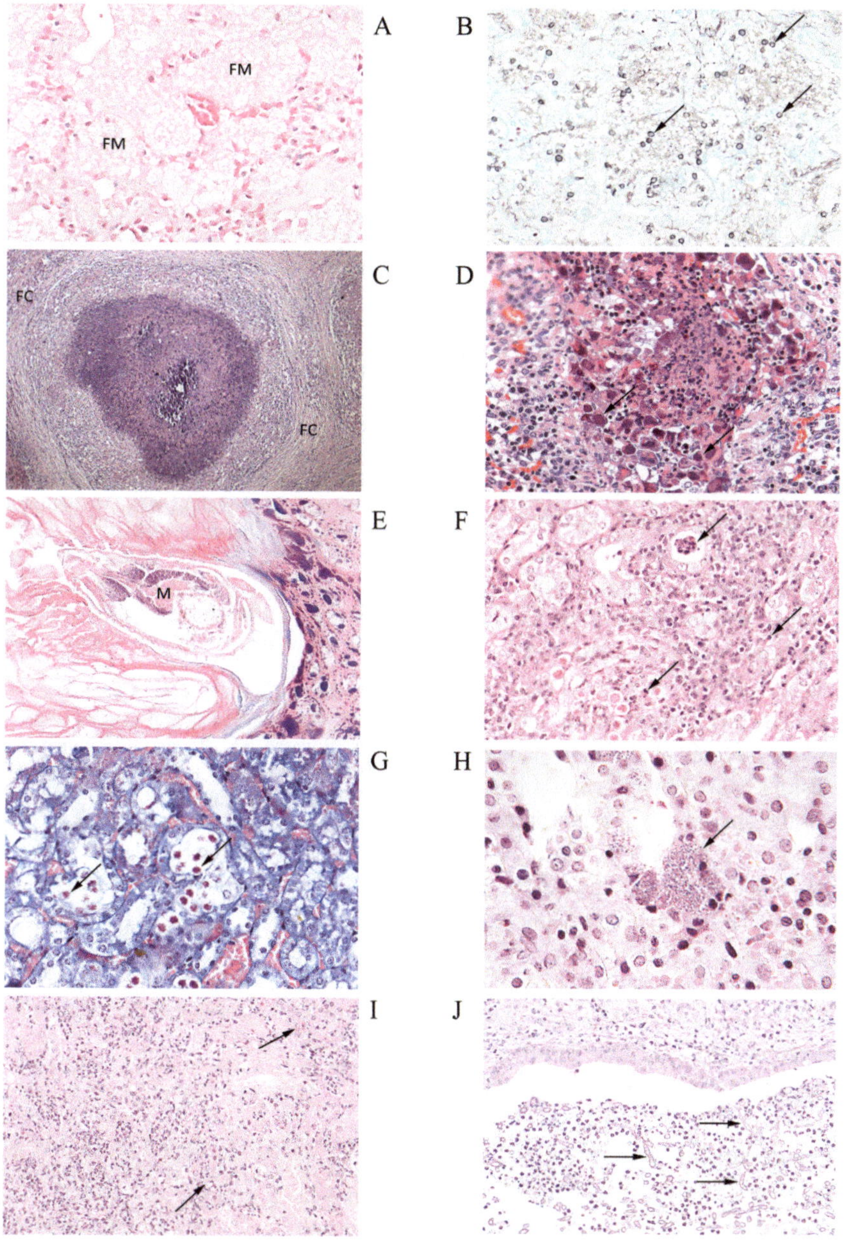

Fig. 12.8 Histopathological examples of spontaneous infections in laboratory animals. (**a**) Rat lung. The alveoli are expanded due to the accumulation of foam macrophages (FM). HE. (**b**) Rat lung. The same as A but stained with the Grocott's methenamine silver stain. Multiple circular elements of *Pneumocystis carinii* are visualized (arrows) within the foamy material of the macrophages. GMS. (**c**) Rabbit, subcutis. A chronic abscess (apostema) encircled by a fibrous capsule (FC) has formed due to infection with *Staphylococcus aureus*. HE. (**d**) Guinea pig lung. Along with the presence of necrotizing alveolitis and bronchopneumonia, large, basophilic intranuclear inclusions bodies (arrows) are seen within the epithelial cells. HE. (**e**) Guinea pig skin. Dermatitis with hyperkeratosis caused by mange due to mite (*Trixacarus caviae*) infestation (M). HE. (**f**) Murine kidney. Chronic, peritubular interstitial and tubular nephritis with infiltration of different inflammatory cells (arrows). HE. (**g**) Murine kidney. Same as F. Within the lumen of the tubular structures multiple *Klossiella muris* parasites (arrows) are seen. PAS. (**h**) Rabbit kidney. Within the epithelial cells lining the renal tubuli, the elements of *Encephalitozoon* (formerly *Nosema*) *cuniculi* (arrow) are seen. HE. (**i**) Acute, diffuse suppurative nephritis is present together with fungal elements (arrows) of *Aspergillus fumigatus*. HE. (**j**) Murine kidney. Within the renal pelvic cavity, *Aspergillus fumigatus* hyphae (arrows) are seen to be intermingled with neutrophils, i.e., mycotic suppurative pyelonephritis is present. HE

variety of inflammatory reactions including the formation of suppurative lesions, when the infection is caused by a pyogenic bacterium like *Staphylococcus aureus*. Suppurative infections in solid tissues are often the cause of abscess formation encapsulated by fibrotic tissue, e.g., following subcutaneous introduction due to skin ulcerations following biting and/or scratches (Fig. 12.8c). A number of viruses are the cause of lesions in laboratory animals [29], and they are often spreading within housing facilities. When scrutinizing tissue sections for virus infections several inflammatory reactions often combined by necrosis may be present. In few cases, the specific viruses may be the reason for inclusion body formation in the cytoplasm or nucleus of the infected cells. In, e.g., adenovirus infection of guinea pigs, the formation of basophilic nuclear inclusions bodies within the cells of the lining epithelium of bronchioles is significant for the infection (Fig. 12.8d). Parasites are also often the cause of spontaneous infections in laboratory animals. As with several of the other causes of infections, lesions caused by parasites are totally directed from the animal species and parasite in question. Parasites may be the cause of skin lesions in the shape of mange caused by mites (Fig. 12.8e), or several of internal organs like the kidneys, e.g. in cases of *Klossiella muris* infection in mice (Fig. 12.8f, g) and, e.g., *Encephalitozoon* (formerly *Nosema*) *cuniculi* infection in rabbits (Fig. 12.8h). The development and application of experimental animal models in mycology are used extensively in the study of a wide range of aspects, including pathogenesis, pathology, virulence, transformative properties, defense mechanisms (immunity), therapy, and prophylaxis [29]. Fungi are generally not host specific, in contrast to many other invasive microorganisms. Therefore, a variety of different animal species are suitable for infection with most of the fungi important in both human and veterinary medicine. Therefore, also disseminated infections with a number of fungi, e.g., *Aspergillus fumigatus*, may be the cause of spontaneous infections in laboratory animals (Fig. 12.8i, j).

References

1. Abrantes J, Lopes AM. A review on the methods used for the detection and diagnosis of rabbit hemorrhagic disease virus (RHDV). Microorganisms. 2021;9(5):972.
2. Ackermann MR. In: Zachary JT, editors. Pathological basis of veterinary disease. Elsevier. 2017. p. 73–131.
3. Alejo A, Saraiva M, Ruiz-Argüello MB, Viejo-Borbolla A, de Marco MF, Salguero FJ, Alcami A. A method for the generation of ectromelia virus (ECTV) recombinants: in vivo analysis of ECTV vCD30 deletion mutants. PLoS One. 2009;4(4):e5175.
4. Alkanani AK, Hara N, Gianani R, Zipris D. Kilham rat virus-induced type 1 diabetes involves beta cell infection and intra-islet JAK-STAT activation prior to insulitis. Virology. 2014;468–470:19–27.
5. Bancroft JD, Cook HC. Manual of histological techniques. Edinburgh/London/Melbourne/New York: Churchill Livingstone; 1984.
6. Bancroft JD, Layton C. Chapter 10 – The hematoxylins and eosin. In: Suvarna SK, Layton C, Bancroft JD, editors. Bancroft's theory and practice of histological techniques. 8th ed. London: Elsevier; 2019. p. 126–38.
7. Bancroft JD, Layton C. Chapter 12 – Connective and other mesenchymal tissues with their stains. In: Suvarna SK, Layton C, Bancroft JD, editors. Bancroft's theory and practice of histological techniques. 8th ed. London: Elsevier; 2019. p. 153–75.
8. Bhat P, Singh ND, Leishangthem GD, Kaur A, Mahajan V, Banga HS, Brar RS. Histopathological and immunohistochemical approaches for the diagnosis of Pasteurellosis in swine population of Punjab. Vet World. 2016;9(9):989–95.
9. Bihun CG, Percy DH. Morphologic changes in the nasal cavity associated with sialodacryoadenitis virus infection in the Wistar rat. Vet Pathol. 1995;32(1):1–10.
10. Bonville CA, Bennett NJ, Koehnlein M, Haines DM, Ellis JA, DelVecchio AM, Rosenberg HF, Domachowske JB. Respiratory dysfunction and proinflammatory chemokines in the pneumonia virus of mice (PVM) model of viral bronchiolitis. Virology. 2006;349(1):87–95.
11. Borenshtein D, Nambiar PR, Groff EB, Fox JG, Schauer DB. Development of fatal colitis in FVB mice infected with *Citrobacter rodentium*. Infect Immun. 2007;75(7):3271–81.
12. Bottari B, Ercolini D, Gatti M, Neviani E. Application of FISH technology for microbiological analysis: current state and prospects. Appl Microbiol Biotechnol. 2006;73(3):485–94.
13. Carlson Scholz JA, Garg R, Compton SR, Allore HG, Zeiss CJ, Uchio EM. Poliomyelitis in MuLV-infected ICR-SCID mice after injection of basement membrane matrix contaminated with

lactate dehydrogenase-elevating virus. Comp Med. 2011;61(5):404–11.
14. Casalone C, Iulini B, Pintore MD, Maurella C, Bozzetta E, Cantile C, Gandini G, Capucchio MT, Calistri A, D'Angelo A, Caramelli M. Zoonoses surveillance in Italy (2000–2009): investigation on animals with neurological symptoms. IntechOpen; 2011. https://www.intechopen.com/chapters/23057
15. Chandler FW. Approaches to the pathologic diagnosis of infectious diseases. In: Connor DH, Chandler FW, editors. Pathology of infectious diseases, vol. 1. Stamford: Appleton and Lange; 1997. p. 3–7.
16. Ciurkiewicz M, Herder V, Khan MA, Uhde AK, Teich R, Floess S, Baumgärtner W, Huehn J, Beineke A. Cytotoxic CD8(+) T cell ablation enhances the capacity of regulatory T cells to delay viral elimination in Theiler's murine encephalomyelitis. Brain Pathol. 2018;28(3):349–68.
17. Cladel NM, Peng X, Christensen N, Hu J. The rabbit papillomavirus model: a valuable tool to study viral-host interactions. Philos Trans R Soc Lond Ser B Biol Sci. 2019;374(1773):20180294.
18. Collins JW, Chervaux C, Raymond B, Derrien M, Brazeilles R, Kosta A, Chambaud I, Crepin VF, Frankel G. Fermented dairy products modulate *Citrobacter rodentium*-induced colonic hyperplasia. J Infect Dis. 2014;210(7):1029–41.
19. Compton SR, Ball-Goodrich LJ, Johnson LK, Johnson EA, Paturzo FX, Macy JD. Pathogenesis of enterotropic mouse hepatitis virus in immunocompetent and immunodeficient mice. Comp Med. 2004;54(6):681–9.
20. Cook PM, Eglin RP, Easton AJ. Pathogenesis of pneumovirus infections in mice: detection of pneumonia virus of mice and human respiratory syncytial virus mRNA in lungs of infected mice by in situ hybridization. J Gen Virol. 1998;79(Pt 10):2411–7.
21. Dietert K, Gutbier B, Wienhold SM, Reppe K, Jiang X, Yao L, Chaput C, Naujoks J, Brack M, Kupke A, Peteranderl C, Becker S, von Lachner C, Baal N, Slevogt H, Hocke AC, Witzenrath M, Opitz B, Herold S, Hackstein H, Sander LE, Suttorp N, Gruber AD. Spectrum of pathogen- and model-specific histopathologies in mouse models of acute pneumonia. PLoS One. 2017;12(11):e0188251.
22. Faísca P, Desmecht D. Sendai virus, the mouse parainfluenza type 1: a longstanding pathogen that remains up-to-date. Res Vet Sci. 2007;82(1):115–25.
23. Fischer SA, Graham MB, Kuehnert MJ, Kotton CN, Srinivasan A, Marty FM, Comer JA, Guarner J, Paddock CD, DeMeo DL, Shieh WJ, Erickson BR, Bandy U, DeMaria A Jr, Davis JP, Delmonico FL, Pavlin B, Likos A, Vincent MJ, Sealy TK, Goldsmith CS, Jernigan DB, Rollin PE, Packard MM, Patel M, Rowland C, Helfand RF, Nichol ST, Fishman JA, Ksiazek T, Zaki SR. Transmission of lymphocytic choriomeningitis virus by organ transplantation. N Engl J Med. 2006;354(21):2235–49.
24. Fraher MH, O'Toole PW, Quigley EM. Techniques used to characterize the gut microbiota: a guide for the clinician. Nat Rev Gastroenterol Hepatol. 2012;9(6):312–22.
25. Gersh I. Histochemistry: theoretical and applied. A. G. Everson Pearse. Boston: Little, Brown; London: Churchill, 1953. 530 pp. Illus. + plates. $12.00. Science. 1953;118(3078):785.
26. Gilbertson JA. Chapter 15 – Amyloid. In: Suvarna SK, Layton C, Bancroft JD, editors. Bancroft's theory and practice of histological techniques. 8th ed. London: Elsevier; 2019. p. 231–53.
27. Gupta E, Bhalla P, Khurana N, Singh T. Histopathology for the diagnosis of infectious diseases. Indian J Med Microbiol. 2009;27(2):100–6.
28. Hamada T, Yokota K, Ayada K, Hirai K, Kamada T, Haruma K, Chayama K, Oguma K. Detection of helicobacter hepaticus in human bile samples of patients with biliary disease. Helicobacter. 2009;14(6):545–51.
29. Hansen AK. Microbiological status. In: Hau J, Schapiro SJ, editors. Handbook of laboratory animal science: essential principles and practices. Boca Raton: Taylor and Francis; 2021. p. 237–68.
30. Hestvik G, Uhlhorn H, Mattsson R, Westergren E, Södersten F, Åkerström S, Gavier-Widén D. Pathology of natural *Francisella tularensis* subsp. holarctica infection in two yellow-necked mice (*Apodemus flavicollis*). Acta Vet Scand. 2018;60(1):26.
31. Highley JR, Sullivan N. Chapter 18 – Neuropathology and muscle biopsy techniques. In: Suvarna SK, Layton C, Bancroft JD, editors. Bancroft's theory and practice of histological techniques. 8th ed. London: Elsevier; 2019. p. 306–36.
32. Iburg TM. Necropsy of the rabbit. In: Jensen HE, editor. Necropsy: a handbook and atlas. Frederiksberg: Biofolia; 2011. p. 207–30.
33. Itthitaetrakool U, Pinlaor P, Pinlaor S, Chomvarin C, Dangtakot R, Chaidee A, Wilailuckana C, Sangka A, Lulitanond A, Yongvanit P. Chronic *Opisthorchis viverrini* infection changes the liver microbiome and promotes helicobacter growth. PLoS One. 2016;11(11):e0165798.
34. Jacoby RO, Johnson EA, Ball-Goodrich L, Smith AL, McKisic MD. Characterization of mouse parvovirus infection by in situ hybridization. J Virol. 1995;69(6):3915–9.
35. Jekl V, Nováková M, Jeklová E, Pospíšilová P, Křenová J, Faldyna M, Škorič M, Šmajs D. Penicillin treatment failure in rabbit syphilis due to the persistence of Treponemes (*Treponema paraluisleporidarum* Ecovar Cuniculus) in the focus of infection. Front Vet Sci. 2021;8:675631.
36. Jennings LK, Dreifus JE, Reichhardt C, Storek KM, Secor PR, Wozniak DJ, Hisert KB, Parsek MR. *Pseudomonas aeruginosa* aggregates in cystic fibrosis sputum produce exopolysaccharides that likely impede current therapies. Cell Rep. 2021;34(8):108782.
37. Jensen HE. Necropsy of rodents (rat, mouse, Guinea pig and hamster). In: Jensen HE, editor. Necropsy – a handbook and atlas. Frederiksberg: Biofolia; 2011. p. 187–205.

38. Jensen HE. Histopathology in the diagnosis of invasive fungal diseases. Curr Fungal Infect Rep. 2021;15(1):23–31.
39. Jensen HE, Barington K. Animal models in mycology. In: Hau J, Schapiro SJ, editors. Handbook of laboratory animal science. Boca Raton: CRC Press; 2021.
40. Jensen HE, Jensen LK, Barington K, Pors SE, Bjarnsholt T, Boye M. Fluorescence in situ hybridization for the tissue detection of bacterial pathogens associated with porcine infections. In: Cunha MV, Inácio J, editors. Veterinary infection biology: molecular diagnostics and high-throughput strategies. New York: Springer; 2015. p. 219–34.
41. Jensen LK, Henriksen NL, Bjarnsholt T, Kragh KN, Jensen HE. Combined staining techniques for demonstration of *Staphylococcus aureus* biofilm in routine histopathology. J Bone Jt Infect. 2018;3(1):27–36.
42. Kim KS, Jung JY, Kim JH, Kang SC, Hwang EK, Park BK, Kim DY, Kim JH. Epidemiological characteristics of pulmonary pneumocystosis and concurrent infections in pigs in Jeju Island, Korea. J Vet Sci. 2011;12(1):15–9.
43. Kim TH, Kim DS, Han JH, Chang SN, Kim KS, Seok SH, Kim DJ, Park JH, Park JH. Detection of *Corynebacterium bovis* infection in athymic nude mice from a research animal facility in Korea. J Vet Sci. 2014;15(4):583–6.
44. Kordasti S, Istrate C, Banasaz M, Rottenberg M, Sjövall H, Lundgren O, Svensson L. Rotavirus infection is not associated with small intestinal fluid secretion in the adult mouse. J Virol. 2006;80(22):11355–61.
45. Kosugi I, Arai Y, Baba S, Kawasaki H, Iwashita T, Tsutsui Y. Prolonged activation of cytomegalovirus early gene e1-promoter exclusively in neurons during infection of the developing cerebrum. Acta Neuropathol Commun. 2021;9(1):39.
46. Kunz S, Rojek JM, Roberts AJ, McGavern DB, Oldstone MB, de la Torre JC. Altered central nervous system gene expression caused by congenitally acquired persistent infection with lymphocytic choriomeningitis virus. J Virol. 2006;80(18):9082–92.
47. Layton C, Bancroft JD. Carbohydrates. In: Bancroft's theory and practice of histological techniques. London: Elsevier; 2019.
48. Layton C, Bancroft JD, Suvarna SK. Chapter 4 – Fixation of tissues. In: Suvarna SK, Layton C, Bancroft JD, editors. Bancroft's theory and practice of histological techniques. 8th ed. London: Elsevier; 2019. p. 40–63.
49. Leifsson P, Jensen HE. Introduction. In: Jensen HE, editor. Necropsy – a handbook and atlas. Frederiksberg: Biofolia; 2011. p. 17–31.
50. Lenaerts L, Verbeken E, De Clercq E, Naesens L. Mouse adenovirus type 1 infection in SCID mice: an experimental model for antiviral therapy of systemic adenovirus infections. Antimicrob Agents Chemother. 2005;49(11):4689–99.
51. Levsky JM, Singer RH. Fluorescence in situ hybridization: past, present and future. J Cell Sci. 2003;116(Pt 14):2833–8.
52. Liang CT, Wu SC, Huang YT, Lin YC, Chang WJ, Chou JY, Liang SC, Liu CH. Immunohistochemical diagnosis of mouse hepatitis virus and mycoplasma pulmonis infection with murine antiserum. J Comp Pathol. 2004;131(2–3):214–20.
53. Lillie RD. Histopathologic technic and practical histochemistry. New York: Blakiston; 1954.
54. López-Bueno A, Segovia JC, Bueren JA, O'Sullivan MG, Wang F, Tattersall P, Almendral JM. Evolution to pathogenicity of the parvovirus minute virus of mice in immunodeficient mice involves genetic heterogeneity at the capsid domain that determines tropism. J Virol. 2008;82(3):1195–203.
55. Luna LG. Manual of histologic staining methods of the Armed forces institute of pathology. New York: McGraw-Hill; 1968.
56. Morris GB, Ridgway EJ, Suvarna SK. Traditional stains and modern techniques for demonstrating microorganisms in histology. In: Bancroft's theory and practice of histological techniques. London: Elsevier; 2018. p. 254–79.
57. Moter A, Göbel UB. Fluorescence in situ hybridization (FISH) for direct visualization of microorganisms. J Microbiol Methods. 2000;41(2):85–112.
58. Murray PD, McGavern DB, Lin X, Njenga MK, Leibowitz J, Pease LR, Rodriguez M. Perforin-dependent neurologic injury in a viral model of multiple sclerosis. J Neurosci. 1998;18(18):7306–14.
59. Orchard GE. Chapter 14 – Pigments and minerals. In: Suvarna SK, Layton C, Bancroft JD, editors. Bancroft's theory and practice of histological techniques. 8th ed. London: Elsevier; 2019. p. 198–230.
60. Orós J, Fernández A, Rodríguez JL, Franklin CL, Matsushita S, Poveda JB. Association of cilia-associated respiratory (CAR) bacillus with natural chronic tracheitis in goats. J Comp Pathol. 1997;117(3):289–94.
61. Pearse AGE. Histochemistry. In: Theoretical and applied. Vol. 2: Analytical technology. Chruchill-Livingstone: Edinburgh; 1985.
62. Pors SE, Hansen MS, Bisgaard M, Jensen HE. Occurrence and associated lesions of *Pasteurella multocida* in porcine bronchopneumonia. Vet Microbiol. 2011;150(1–2):160–6.
63. Reimerink JHJ, Boshuizen JA, Einerhand AWC, Duizer E, van Amerongen G, Schmidt N, Koopmans MPG. Systemic immune response after rotavirus inoculation of neonatal mice depends on source and level of purification of the virus: implications for the use of heterologous vaccine candidates. J Gen Virol. 2007;88(Pt 2):604–12.
64. Rieger J, Janczyk P, Hünigen H, Plendl J. Enhancement of immunohistochemical detection of salmonella in tissues of experimentally infected pigs. Eur J Histochem. 2015;59(3):2516.

65. Rout SS, Singh M, Shindler KS, Das Sarma J. One proline deletion in the fusion peptide of neurotropic mouse hepatitis virus (MHV) restricts retrograde axonal transport and neurodegeneration. J Biol Chem. 2020;295(20):6926–35.
66. Sanderson T, Wild G, Cull AM, Marston J, Zardin GJ. Immunohistochemical and immunofluorescent techniques. In: Bancroft's theory and practice of histological techniques. London: Elsevier; 2019.
67. Sayyadi N, Connally RE, Lawson TS, Yuan J, Packer NH, Piper JA. Time-gated luminescent in situ hybridization (LISH): highly sensitive detection of pathogenic *Staphylococcus aureus*. Molecules. 2019;24(11):1–16.
68. Simmons WL, Bolland JR, Daubenspeck JM, Dybvig K. A stochastic mechanism for biofilm formation by *Mycoplasma pulmonis*. J Bacteriol. 2007;189(5):1905–13.
69. Smith K, Brown CC, Spindler KR. The role of mouse adenovirus type 1 early region 1A in acute and persistent infections in mice. J Virol. 1998;72(7):5699–706.
70. Taha-Abdelaziz K, Bassel LL, Harness ML, Clark ME, Register KB, Caswell JL. Cilia-associated bacteria in fatal *Bordetella bronchiseptica* pneumonia of dogs and cats. J Vet Diagn Invest. 2016;28(4):369–76.
71. Taniere P, O'Sullivan B, Evans M, Hughes F. Chapter 20 – Molecular pathology. In: Suvarna SK, Layton C, Bancroft JD, editors. Bancroft's theory and practice of histological techniques. 8th ed. London: Elsevier; 2019. p. 395–433.
72. Tatti KM, Gentsch J, Shieh WJ, Ferebee-Harris T, Lynch M, Bresee J, Jiang B, Zaki SR, Glass R. Molecular and immunological methods to detect rotavirus in formalin-fixed tissue. J Virol Methods. 2002;105(2):305–19.
73. Thompson SW, Luna LG. An atlas of artifacts encountered in the preparation of microscopic tissue sections. Springfield: Charles C. Thomas; 1978.
74. Tian J, Pinto JM, Cui X, Zhang H, Li L, Liu Y, Wu C, Wei Y. Sendai virus induces persistent olfactory dysfunction in a murine model of PVOD via effects on apoptosis, cell proliferation, and response to odorants. PLoS One. 2016;11(7):e0159033.
75. Tu Y-C, Lu M-C, Chiang M-K, Huang S-P, Peng H-L, Chang H-Y, Jan M-S, Lai Y-C. Genetic requirements for *Klebsiella pneumoniae*-induced liver abscess in an oral infection model. Infect Immun. 2009;77(7):2657–71.
76. Veterinary-Pathology-Forum. Tyzzer's disease in a Foal – *Clostridium piliforme*. 2010. Retrieved 22 July 2021,fromhttps://vetpath.wordpress.com/2010/08/14/tyzzers-disease-in-a-foal-clostridium-piliforme
77. Volz A, Langenmayer M, Jany S, Kalinke U, Sutter G. Rapid expansion of CD8+ T cells in wild-type and type I interferon receptor-deficient mice correlates with protection after low-dose emergency immunization with modified vaccinia virus Ankara. J Virol. 2014;88(18):10946–57.
78. Waggie KS. Postmortem procedures. In: Hau J, Schapiro SJ, editors. Handbook of laboratory animal science. Boca Raton: CRC Press; 2021.
79. Wetzel JD, Barton ES, Chappell JD, Baer GS, Mochow-Grundy M, Rodgers SE, Shyr Y, Powers AC, Thomas JW, Dermody TS. Reovirus delays diabetes onset but does not prevent insulitis in nonobese diabetic mice. J Virol. 2006;80(6):3078–82.
80. Wilcock BP, Njaa BL. Chapter 5 – Special senses. In: Maxie MG, editor. Jubb, Kennedy & Palmer's pathology of domestic animals: volume 1. 6th ed. Edinburgh: W.B. Saunders; 2016. p. 407–508.e402.
81. Won YS, Jeong ES, Park HJ, Lee CH, Nam KH, Kim HC, Park JI, Choi YK. Upregulation of galectin-3 by *Corynebacterium kutscheri* infection in the rat lung. Exp Anim. 2007;56(2):85–91.
82. Xu F, Droemann D, Rupp J, Shen H, Wu X, Goldmann T, Hippenstiel S, Zabel P, Dalhoff K. Modulation of the inflammatory response to *Streptococcus pneumoniae* in a model of acute lung tissue infection. Am J Respir Cell Mol Biol. 2008;39(5):522–9.
83. Yin C, Djavani M, Schenkel AR, Schmidt DS, Pauza CD, Salvato MS. Dissemination of lymphocytic choriomeningitis virus from the gastric mucosa requires G protein-coupled signaling. J Virol. 1998;72(11):8613–9.
84. Zhang G, Chen M, Graham D, Subsin B, McDougall C, Gilady S, Kneteman M, Law L, Swain M, Trauner M, Wrzesinski S, Flavell R, Wasilenko S, Mason A. Mouse mammary tumor virus in anti-mitochondrial antibody producing mouse models. J Hepatol. 2011;55(4):876–84.

MIX
Papier aus verantwortungsvollen Quellen
Paper from responsible sources
FSC® C105338

FSC
www.fsc.org

If you have any concerns about our products,
you can contact us on
ProductSafety@springernature.com

In case Publisher is established outside the EU,
the EU authorized representative is:
Springer Nature Customer Service Center GmbH
Europaplatz 3, 69115 Heidelberg, Germany

Printed by Libri Plureos GmbH
in Hamburg, Germany